GLOBAL ISSUES

NATURAL DISASTERS

GLOBAL ISSUES

NATURAL DISASTERS

Lesli J. Favor

Foreword by Katherine Ellins
University of Texas Institute for Geophysics

Facts On File
An Infobase Learning Company

Facts On File, Inc.
An imprint of Infobase Learning
132 West 31st Street
New York NY 10001

Library of Congress Cataloging-in-Publication Data

Favor, Lesli J.
　Natural disasters / Lesli J. Favor ; foreword by Katherine Ellins.
　　p. cm.—(Global issues)
　Includes bibliographical references and index.
　ISBN 978-0-8160-8260-5
　1. Natural disasters. I. Title.
　GB5014.F38 2011
　904'.5—dc22　　　　　　　　　　2010028996

Facts On File books are available at special discounts when purchased in bulk quantities for businesses, associations, institutions, or sales promotions. Please call our Special Sales Department in New York at (212) 967-8800 or (800) 322-8755.

You can find Facts On File on the World Wide Web at http://www.infobaselearning.com

Text design by Erika K. Arroyo
Illustrations by Dale Williams
Composition by Hermitage Publishing Services
Cover printed by Yurchak Printing, Inc., Landisville, Pa.
Book printed and bound by Yurchak Printing, Inc., Landisville, Pa.
Date printed: May 2011

Printed in the United States of America

10 9 8 7 6 5 4 3 2 1

This book is printed on acid-free paper.

For Steve. Thank you.

CONTENTS

PART II: PRIMARY SOURCES

PART III: RESEARCH TOOLS

List of Maps and Tables

List of Acronyms

BBC	British Broadcasting Corporation
Central Register	Central Register of Disaster Management Capacities
CERF	Central Emergency Response Fund
CHEC	Central Hydroelectric Company of Caldas (Colombia)
CRED	Centre for Research on the Epidemiology of Disasters
CVO	Cascades Volcano Observatory (USGS)
DART	Deep-ocean Assessment and Reporting of Tsunamis
DoD	Department of Defense
DRI	Disaster Risk Index
DVI Guide	Disaster Victim Identification Guide
EF Scale	Enhanced Fujita Scale
ERC	Emergency Relief Coordinator
FAO	Food and Agriculture Organization (UN)
FEMA	Federal Emergency Management Agency
FEWS	Famine Early Warning System
GDACS	Global Disaster Alert and Coordination System
GDP	gross domestic product
HEWS	Humanitarian Early Warning Service
HUD	Housing and Urban Development
IASC	Inter-Agency Standing Committee
IBC	International Building Code
INSARAG	International Search and Rescue Advisory Group
INTERPOL	International Criminal Police Organization

IOC	Intergovernmental Oceanographic Commission
MCEER	Multidisciplinary Center for Earthquake Engineering
MDG	Millennium Development Goal
MPH	miles per hour
MWR	Ministry of Water Resources (China)
NHC	National Hurricane Center
NOAA	National Oceanic and Atmospheric Administration
NRCS	Natural Resource Conservation Service
NSSL	National Severe Storms Laboratory
NWS	National Weather Service
OAS	Organization of American States
OCHA	Office for the Coordination of Humanitarian Affairs
OFDA	Office of Foreign Disaster Assistance (USAID)
ORT	oral rehydration therapy
OSOCC	On-Site Operations Coordination Centre
PA	public assistance
RIS	reservoir-induced seismicity
SBA	Small Business Administration
SCS	Soil Conservation Service
SPC	Storm Prediction Center
TEC	Tsunami Evaluation Committee
TRIAMS	Tsunami Recovery Impact Assessment and Monitoring System
UN-CMCoord	United Nations Humanitarian Civil-Military Coordination
UNDAC	United Nations Disaster Assessment and Coordination
UNDP	United Nations Development Programme
UNESCO	United Nations Educational, Scientific and Cultural Organization
UNHCR	United Nations High Commissioner for Refugees
USAID	United States Agency for International Development
USGS	United States Geological Survey
VDAP	Volcano Disaster Assistance Program
VEI	Volcanic Explosivity Index

List of Acronyms

WFP	World Food Programme
WGCEP	Working Group on California Earthquake Probabilities
WHO	World Health Organization
WMO	World Meteorological Organization

Foreword

A major earthquake and tsunami struck Japan in March 2011, taking the lives of more than 17,000 people and causing nuclear accidents at two power plants. As ordinary folk from around the world moved by the magnitude of the disaster rallied to help survivors, most were oblivious to the risk that natural hazards pose to their own lives. Lesli Favor has distilled this complex topic into a readable book that is an excellent starting point to help students understand the causes of natural disasters and appreciate the importance of planning and preparation in minimizing the loss of life and economic costs when a disaster does strike.

Earth is a complicated system of interrelated processes. Humans are part of this system. We depend on Earth for water to sustain life, support agriculture, and fuel manufacturing; a climate modulated by a vast ocean and a thin protective atmosphere that keeps us comfortable; and an abundance of natural resources that spur technological and economic development. Among our neighbors in the solar system, Earth is unique as a habitable planet. Living on Earth, however, links us inextricably to the dynamic interplay between the processes that operate on our planet and makes us vulnerable to natural hazards, including earthquakes, tsunamis, volcanic eruptions, extreme weather, floods, and droughts. Our position in the solar system also exposes us to the risk of a collision with a comet or meteorite.

In the face of this reality, humans are not powerless. Within the last several decades, a revolution in our understanding of the workings of the Earth has taken place. This revolution began with the emergence of the theory of plate tectonics, which provided a model to explain the occurrence of many of Earth's features and destructive events, such as earthquakes, tsunamis, and volcanic eruptions. Technological advances made possible the retrieval of high-resolution geologic records from lake, estuary, and ocean sediments; ice cores; and soils that document past environmental change and catastrophic natural events. More recently, space exploration has led to the deployment of modern Earth-observing, navigation, and communication satellites, which provide information about the global distribution of natural hazards. They also monitor weather and ocean systems, the extent of forest fires, the changing sizes of ice sheets and glaciers, floods, and droughts. Armed with a new perspective of planet Earth as a system, the latest scientific knowledge, and real-time data, 21st-century scientists are equipped with a better understanding of the physical processes that generate natural hazards and models to develop forecasts about when and where natural hazards are likely to occur. At the same time, preparedness planners and emergency responders have the ability to communicate rapidly between widely dispersed places and to pinpoint the location of disasters even in the most remote parts of the world. Today, we are better prepared than at any time in human history to prevent natural hazards from becoming natural disasters.

The first chapter of *Natural Disasters* clearly explains the fundamental science behind natural hazards and discusses the complex circumstances that lead to natural disasters. One of the key factors that determines whether a natural hazard will evolve into a disaster is human preparedness and response. The author observes that people may have cultural biases or be locked in ways of thinking that blind them to the risks posed by natural hazards or that prevent them from taking actions that would limit their exposure and vulnerability. She uses the fate of Pompeii, a Roman city in the southern part of present-day Italy at the base of Mt. Vesuvius, a stratovolcano, to illustrate this point. Volcanically active for millennia, the area is endowed with rich volcanic soils and agricultural fertility. Pompeii, established initially as a settlement in the Bronze Age, was affected by recurrent volcanic activity prior to the terrible eruption of Vesuvius in 79 C.E. But human memories are short in comparison to geologic time. The eruption was preceded by an extended period of inactivity so that the memory of previous disasters had been erased. Despite our knowledge of the hideous fate suffered by residents of Pompeii and persistent volcanic activity—Vesuvius has erupted more than 40 times since that fateful day in 79 C.E.—3 million people continue to live under the volcano. When Vesuvius erupts again, as it surely will, the potential exists for the loss of life and property to be greater than in 79 C.E., prompting the Italian government to take disaster planning very seriously. Chapter 1 provides a helpful overview of both the history of natural disasters and the human response.

The second chapter underscores the vulnerability of the United States, a technologically advanced and prosperous country, to natural hazards. The author chronicles some of the best-known natural disasters in U.S. history, from the Galveston Hurricane of 1900 to Hurricane Katrina in 2005, recounting both the events leading up to, and the aftermath of, each disaster. As sobering as it is to contemplate past catastrophes, these events serve as cautionary tales. By studying them, we can learn their causes and determine

Foreword

what best efforts to make to safeguard life and property from potential disasters. The second part of the chapter examines the current risks associated with earthquakes, tornadoes, hurricanes, and tsunamis. Although it is impossible to eliminate these hazards, the author explains how scientists, local, state, and national governments, and people who live in hazardous areas can engage cooperatively in efforts to reduce their impacts.

In chapter 3, the author offers case studies from four parts of the world to highlight different types of natural disasters and the effect that levels of economic development and political realities have on disaster prevention and relief efforts. In each case, she considers the accompanying local and global response and reconstruction or engineering mitigation efforts. The case studies are the 2004 Sumatra-Andaman earthquake and tsunami in the Indian Ocean; water resource management in China, a country historically prone to devastating floods; famine and its tragic impact on the Sahel zone; and the 1985 eruption of the Colombian volcano Nevado del Ruiz in the Andes range. As these case studies illustrate, careful monitoring, hazard assessment studies, and the implementation and rehearsal of emergency management procedures are crucial tools for reducing the impact of natural hazards.

Part II of *Natural Disasters* collects relevant primary source documents, such as preparedness resources, firsthand accounts of disasters, governmental reports, and publications of professional organizations. Part III provides useful research tools, including advice on how to get started, statistics and maps, an annotated bibliography, a chronology, an annotated list of organizations, short biographies of key players, and a glossary.

We live on a dynamic, constantly changing planet where natural hazards will continue to affect human life. The world's population currently stands at 6.8 billion people. According to United Nations projections, it is expected to rise to 9 billion people by 2040. Increasingly, the world's population will live in areas that are at high risk of natural disasters. Efforts to promote Earth science and disaster awareness are taking hold in schools throughout the United States, as well as in other countries around the world. There is no doubt that an Earth-literate workforce able to function in interdisciplinary and international settings is vital to address the challenges that natural hazards pose for the 21st century. Not everyone needs to be an expert on the topic, but individuals and communities should have a basic understanding of natural hazards and know how to prepare for and respond to natural disasters. By promoting this awareness, *Natural Disasters* will inspire students to embark on careers related to the topic or to become informed citizens capable of building disaster-resilient communities.

—Katherine Ellins
Program Manager, University of Texas Institute for Geophysics

PART I

At Issue

1

Introduction

At 5:46 A.M. on January 17, 1995, a powerful earthquake shook the city of Kobe, Japan, for 20 seconds. The second-largest metropolitan area in Japan, Kobe had an estimated population of 14 million. Japan had spent decades implementing hazard-reduction programs and policies that had been praised as highly successful. But under the stress of the Kobe earthquake, reinforced structures collapsed and emergency response procedures bogged down. Fires burned rapidly out of control, gas and water mains broke, power outages spread across the city, and more than 200,000 buildings were damaged. During the disaster, more than 5,000 people lost their lives and at least 35,000 were injured. The earthquake caused an estimated $100 billion in damages.[1]

The factors that came together to cause the Kobe disaster highlight the fact that this is a world full of natural hazards. A natural hazard is a dangerous geophysical event (such as the Kobe earthquake) or condition (such as a drought) that could cause a natural disaster. The event or condition becomes a natural disaster when it causes human, economic, or environmental losses that exceed the affected area's ability to cope without outside help. In the years between 1980 and 2000, natural disasters caused the deaths of 1.2 million people worldwide.[2] Each year more than 600 disasters[3] increase the toll by tens of thousands of deaths, hundreds of thousands of injured people, and billions of dollars in economic damages.

People around the world live at risk of being struck by a natural disaster. In 2005, 19 percent of the world's land area and "3.4 billion people (more than half of the world's population) [were] relatively highly exposed to at least one hazard."[4] In some countries, half or more of the population is exposed to three or more natural hazards. For instance, in Taiwan 73 percent of the population is exposed to multiple natural hazards. In Chile, where a magnitude 8.8 earthquake struck in February 2010, 54 percent of the population is exposed to multiple natural hazards.[5]

3

DEFINING THE ISSUE
Natural Hazards

Throughout the history of our planet, forces of nature have influenced the course of life on Earth. Ice ages have displaced humans and wildlife. Changes in water levels in rivers and lakes have destroyed (submerged) cities or nourished fertile farming settlements. Changing sea level has drowned coastal communities or stranded former port cities miles inland (e.g., the ancient city of Miletus). Mudflows and ash from volcanoes or massive tsunami waves have laid waste to cities and land, sometimes hundreds of miles away from the source of the disaster.

In ancient times, people often explained the forces of nature as the intervention of divine beings. The religious beliefs of a culture, such as the Aztecs, may have demanded human or animal sacrifice to calm the wrath of a god, which could be unleashed at any moment as a volcanic eruption, a plague, or a drought. Preserved in the modern legal term *act of God,* these beliefs have faded as extensive scientific research has uncovered physical causes of these events. Scientists now meticulously study these natural hazards, events that could cause harm to people, property, or the environment. Examples of potential threats include avalanches, earthquakes, floods, hurricanes, landslides, tsunamis, volcanoes, wildfires, epidemics, and plant diseases. In many cases, natural hazards occur with no effect on people. For instance, massive avalanches take place frequently around the world, but since they often happen in remote mountain regions, little damage to people or structures results. In contrast, when a natural hazard occurs near a city or close to an area often used by people, the effects can be disastrous.

These natural disasters, with their origins in the natural world, comprise one of the two major categories of disaster, with the other category being human-caused (or man-made) disasters. This second category includes disasters whose origins lie in the actions, errors, or negligence of human beings, with examples including nuclear disasters, acts of terrorism, and toxic chemical spills. Natural and human-caused disasters have some areas of overlap and interconnectivity, which can sometimes blur the lines between these categories.

Natural v. Human-Caused Disasters

Natural disasters are distinguishable from human-caused disasters by their origins. Natural disasters originate in the natural world—in nature—whereas human-caused disasters result from the actions of people. For instance, the natural internal processes of Earth produce geological hazards such as volcanic eruptions and earthquakes that lead to disasters such as the January 2010 earthquake in Haiti that killed nearly a quarter of a million people and destroyed or severely damaged a majority of the structures in the nation's capital. Biological hazards, another type of natural hazard, can result in pandemics such as the Black Death, which swept across Asia and Europe during the late 1340s and killed more than 25 million people. Geological and biological hazards, with their origins in nature, are types of natural hazards.

In contrast, human-caused disasters trace back to the actions, whether intentional or accidental, of one or more human beings. For instance, acts of terrorism, including explosions, biological threats, chemical threats, nuclear blasts, and radiological dispersion devices, frequently result in disaster. The terrorist acts against the United States on September 11, 2001, created a disaster that killed 2,995 people, destroyed the Twin Towers at the World Trade Center, damaged adjacent buildings, and damaged the Pentagon. The Federal Emergency Management Agency (FEMA) declared the attacks a major disaster. Another example is the Cedar Fire of 2003, the second-largest wildfire in California's history at the time. Set by an arsonist, the fire burned more than 280,000 acres, destroyed 2,232 homes, and killed 14 people.[6]

The distinction between natural and human-caused disasters can be as obvious as the difference between a volcanic eruption and the explosion of a nuclear power plant. With other incidents, however, the distinction is more subtle. Famines are a case in point. They can result from drought. But human actions can turn the threat of famine into a disastrous reality. Consider the Bengal Famine of 1770, which resulted in the deaths of 10 million people from starvation and disease.[7] During 1768 and 1769, there was a shortage in crop yields. However, human action led to the catastrophe that followed. Bengal was ruled by the British East India Company, which held taxation rights in the form of land taxes and trade tariffs. By tripling or quadrupling the tax rates that had been in place before its control, the company financially drained landowners. Even during the years of harvest shortfalls, the company continued to raise tax rates. In addition, the company's policy of forbidding hoarding, the storing up of surplus grain, meant that when famine conditions struck, there were no reserves to fall back upon. These two human factors exacerbated the natural famine conditions, tipping the balance toward the destruction of the population of Bengal.

Just as economic or political policies can contribute to a natural hazard, human interactions with the natural world can directly or indirectly cause natural hazards. Geothermal drilling, a type of drilling that extracts heat energy from deep beneath the Earth's crust, has been linked to "induced seismicity," or earthquakes triggered by human actions. As explained by James

Glanz, an energy and environment journalist, "The technique to tap geothermal energy creates earthquakes because it requires injecting water at great pressure down drilled holes to fracture the deep bedrock. The opening of each fracture is, literally, a tiny earthquake in which subterranean stresses rip apart a weak vein, crack, or fault in the rock."[8] In December 2009, a geothermal drilling project in Basel, Switzerland, was shut down because a government study determined that earthquakes caused by the project would cause millions of dollars in damages *each year*. According to a January 2010 report in the *New York Times*, 123 geothermal projects in 38 U.S. states had received funding from the U.S. government.[9]

Environmental disasters are characterized by their impact on the natural world—by the contamination of bodies of water, for example, or the destruction of habitats. While natural disasters often include some degree of environmental damage, environmental disasters are typically classed as a subcategory of human-caused disasters. Environmental disasters may unfold over a period of days, weeks, years, or even decades. A wildfire, for example, may sweep through a forest in a few days. Nuclear contaminants linger for years, as in the case of the Hanford Site in eastern Washington State.

Established in 1943 as part of the Manhattan Project, the Hanford Site produced plutonium-239 and uranium-235 used to produce nuclear bombs, with production of nuclear materials continuing until 1989. Over the decades, nuclear contaminants caused environmental harm that has yet to be reversed. As of 2010, approximately 80 square miles of groundwater in the aquifer that feeds the Columbia River is contaminated beyond state and federal drinking water standards. Aboveground, biological vectors including tumbleweeds, wasps, rabbits, and winds have spread the contamination. Each year, the U.S. Department of Energy prepares a Hanford Site Environmental Report to provide information and data on groundwater-monitoring programs, potential radiation doses to on-site staff and nearby residents, site cleanup and remediation activities, and other vital information.

More recently, on April 20, 2010, an explosion on an oil-drilling rig in the Gulf of Mexico resulted in a massive oil spill, covering a surface area of at least 2,500 square miles. Following the explosion, the oil well, located 5,000 feet (1,500 m) below sea level, continued spilling thousands of barrels of crude oil into the gulf each day. According to the Flow Rate Technical Group, scientists and engineers from the U.S. government, universities, and research institutions put together to estimate the flow of oil, the leak ended up being approximately 205.8 million gallons of oil. Hundreds of species of birds and sea life faced damage to and destruction of their habitats, and commercial fishers faced huge economic losses. The National Oceanic and Atmospheric

Administration (NOAA) placed a temporary restriction on fishing in federal waters affected by the spill, off the coasts of Louisiana, Mississippi, Alabama, and Florida between the mouth of the Mississippi River and Florida's Pensacola Bay. Experts struggled to assess the full scale of environmental damage that the oil spill caused.

Exposure

Complex circumstances from a variety of sources lead to natural disasters. Some circumstances concern people's ties to the land as farmers and fishers. For example, fertile land along rivers draws farmers close to flood-prone areas. Similarly, fishing villages grow up in coastal areas likely to be struck by typhoons, hurricanes, or tsunamis. Other circumstances that can lead to disaster concern the link between people and food. For instance, drought is especially disastrous to populations dependent on annual farming, for they have little or no food stores.

In addition to these kinds of circumstances, society plays a significant role in the severity of a disaster by creating exposure to natural hazards. Exposure refers to the people and property that are at risk of damage from a natural hazard. For example, when a town or city grows up in the level valley formed by a long-forgotten volcanic mudflow, there exists exposure to a mudflow in the future. Exposure can result in significant tolls on human lives and property. For natural disasters occurring around the world between 1991 and 2005, almost 1 million people lost their lives, more than 1 billion people's lives were affected, and the total economic damages topped $1 trillion.[10]

Why do people take the risk of living in an area that is exposed to natural hazards? In many cases, the people who live close to a potential disaster are unaware of the danger. Several hundred years may have passed since the last disaster in the area, allowing people time to forget the dangers and to build cities closer and closer to hazards. This explanation is especially true in the least-developed countries of the world, where 11 percent of the world's population accounts for 53 percent of the casualties from natural disasters.[11]

A classic example of exposure occurred with the eruption of Mt. Vesuvius in 79 c.e. Pompeii, a Roman city in the southern part of present-day Italy, had been flourishing for more than a thousand years at the base of the volcano. An eruption in 1360 b.c.e. apparently wiped out one of the earliest settlements of the city, but after a few hundred years people forgot about the disaster and again, attracted to the area by the rich volcanic soils, settled in the extensive flatlands at the base of Mt. Vesuvius. By 59 c.e., the population of Pompeii had reached 20,000—twice the size of most Roman cities at the time—and it boasted a 20,000-seat amphitheater. Within a decade, the city would be completely destroyed.

The first disaster struck in 62 C.E. An earthquake caused extensive damage, making a large portion of the city too dangerous to enter. Relief efforts began immediately. Workers cleared rubble, and city leaders created a master plan to rebuild the city. After 17 years, most of the reconstruction efforts were complete, and life in the city had returned to normal. Then, the ground began to shake again. A series of small earthquakes, or tremors, lasted for around four days. Then Vesuvius blew open, sending a cloud of ash 17 miles into the sky.

Pumice, small pieces of volcanic rock, began to rain on the city at a rate of five to six inches per hour, and a cloud of ash darkened the sun. By the next morning, an eight-foot layer of pumice covered the city. People who had stayed in their homes were trapped. The remainder of the city's population had attempted to flee in the darkness, but two more eruptions extended the destruction almost 25 miles outward from the volcano. As a result, most of the people attempting to escape also died under the rain of pumice and ash or from breathing in the poisonous gases released from deep within the volcano.

Vulnerability

Pompeii is a reminder of the tragedy that can occur when cities create exposure by growing up too close to an overwhelming threat, such as a volcano. However, not all natural hazards strike with such devastating force. In the same way that society creates exposure to a natural hazard, it may also limit the vulnerability of the people and structures, keeping the destructive forces of nature at bay. Floods can be controlled by dams and levees. Hurricanes, tornadoes, and windstorms can be weathered. Modern construction techniques minimize the damage from earthquakes. Extensive forest management and firefighting efforts can tame the dangers from wildfires. Irrigation channels and rainwater harvesting can alleviate the effects of drought.

Mozambique, an African country with particularly high exposure to natural hazards, experienced major flooding in 2007. Heavy rainfall and a Category 4 cyclone disrupted the lives of between 300,000 and 500,000 people and destroyed countless acres of crops.[12] Despite that, extensive preparation and quick response by aid organizations reduced the population's exposure to danger, and as a result few people perished.

However, hazard-reduction planning also allows people to choose to live near natural hazards, relying on warning systems and preventative measures to protect them. Hazard-reduction systems often perform as expected,

8

reducing a population's vulnerability. A case in point is the evacuation of the area around Mount St. Helens before its eruption in 1980. Often, however, preparations are inadequate, leading to tragedies such as the levee failures and flooding in New Orleans when Hurricane Katrina struck in 2005.

The toll in human life often takes the spotlight in discussions of natural disasters, but economic tolls can have even more of an impact on survivors. Disruptions in health care, transportation, communication links, or energy supplies result when critical aspects of infrastructure are damaged or destroyed. People may lose their homes or become unemployed for long periods. The Haiti earthquake of January 12, 2010, took the lives of at least 222,570 people, but it also destroyed 97,294 homes and displaced 1.3 million people.[13] The 2010 earthquake in Chile displaced a staggering 2 million people.[14]

In the aftermath of a disaster, economic losses can cause a severe decrease in the standard of living in the affected area. Economic losses can delay programs meant to reduce poverty and hunger or to improve sanitation. As well, economic losses can delay efforts to provide education, health services, housing, and safe drinking water.

Despite the best efforts of governments and international organizations, extensive threats to people and economies remain in some areas of the world. Other dangers lurk out of sight, waiting to test the planning of the developed world. Some level of vulnerability to natural disasters will always exist.

TYPES OF NATURAL HAZARDS
Hurricanes

A hurricane is a powerful storm at sea. It begins as a tropical depression, a storm with winds that reach up to 38 miles per hour. The storm's winds rotate in a distinctive counterclockwise direction in the Northern Hemisphere and clockwise in the Southern Hemisphere. As the storm gains intensity, it becomes a tropical storm, with wind speeds of 39 to 73 miles per hour (MPH). At the center of the wind's spiral, an area of calm called the eye develops. Despite the violent winds and torrential rains outside the eye, within the eye the wind is relatively light and skies are fair. The eye is surrounded by the eye wall, an area of convection in which air spirals upward, creating the highest surface winds of the storm. Finally, when the winds exceed 74 MPH, the storm is officially called a hurricane and receives a name for tracking purposes.

The practice of assigning women's names to hurricanes began during World War II. Aside from a period in 1951–53 when the United States named storms using a phonetic alphabet (e.g., Able, Baker, Charlie), this practice continued until 1979, when the procedure was changed to alternate between male and female names. Today, the World Meteorological Organization (WMO) maintains lists of names for storms in various regions of the world, and the names rotate each year.

Hurricanes form in warm waters near the equator—in the southern Atlantic Ocean, the Caribbean Sea, the Gulf of Mexico, and the eastern Pacific Ocean. Similar storms occur around the world, but the location of the storm determines what it is called. Similar storms that form in the western Pacific Ocean are called typhoons, and storms in the Indian Ocean are called cyclones.

These storm systems can grow up to 400 miles across, with winds exceeding 155 MPH. Accompanied by heavy rains and tornadoes, the high winds and pressure from the storm can produce a coastal storm surge, a wall of water driven ashore at high speed that can devastate everything in its path. In 1970, a storm surge in Bangladesh killed an estimated 300,000 people.[15] The 1900 Galveston hurricane was the deadliest in U.S. history, taking approximately 8,000 lives.[16]

Tornadoes

Like hurricanes, tornadoes are atmospheric vortices, or violently rotating columns of air, with an eye at the center, but the two types of storms have little else in common. Scientists believe the eye of a tornado is relatively calm with downward moving air; however, instruments that could measure conditions within the eye cannot survive the tornado's winds long enough to provide data. Other differences distinguish tornadoes from hurricanes. Whereas hurricanes form over water and die out when they move inland, tornadoes typically form over land. (A tornado that forms over water is called a waterspout.) While hurricanes measure hundreds of miles across, tornadoes are much smaller in diameter, normally measuring hundreds of yards across. A hurricane's life span typically lasts for days, whereas a typical tornado dies out in a matter of minutes. Though smaller in diameter, tornadoes produce winds that are usually much faster than those produced by hurricanes. A Category 5 hurricane has wind speeds of 156 MPH or greater. In contrast, the strongest tornadoes have wind speeds exceeding 200 MPH, with some of the strongest tornadoes possibly reaching speeds as high as 300 miles per hour.[17] Despite their greater wind speeds, tornadoes tend to cause far less damage than hurricanes, due in part to their smaller diameter and shorter life span.

Tornadoes form most often in thunderstorms and move across the central United States in the spring and summer. Though other countries experience tornadoes, the United States tops the list with an average of more than 1,000 each year. Most of these tornadoes occur in either Florida or Tornado Alley, an area in the central United States reaching roughly from central Texas northward to Iowa. In comparison, Canada ranks second with an average of 100 tornadoes each year.[18]

The Enhanced Fujita Scale (EF Scale) provides a means of classifying the intensity of tornadoes. The scale assigns a class of EF0 to EF5 to denote the destructive force of the storm. EF0 through EF2 denote weak systems, which last only a few minutes and account for few deaths. Stronger tornadoes with a classification of EF3 to EF4 have winds of 100 to 205 MPH. Tornadoes in this range account for around 30 percent of tornado deaths. The most dangerous storms, EF5 tornadoes, account for 70 percent of deaths. These storms can be more than a mile wide and last for more than an hour, with wind gusts exceeding 200 miles per hour.

Drought

A drought occurs when an extended period of below-average rainfall results in water shortages. Droughts happen around the world, with impacts varying from region to region. Drought impacts include reduced or destroyed crop productivity, increased fire hazards, lower water levels in lakes and rivers, increased mortality of farm animals, and damage to wildlife and fish habitats. Over an extended period, these impacts may cause significant economic damage or, in the most severe cases, threaten the lives of large numbers of people. About 1.1 billion people live in drought-prone areas of the world, with 419 million living in the areas most likely to experience droughts.[19]

Floods

Unusually high rainfall, sometimes coupled with factors such as recent snow accumulation, can raise the water levels of lakes and rivers above their normal boundaries. Levees, walls typically constructed of earth to regulate water levels, may give way under the stress, flooding the surrounding floodplains. Flooding can also result from the storm surge of a hurricane or a tsunami or from damage to levees or dams during an earthquake. The worst floods in history have claimed millions of lives. More than 2 billion people live in flood-prone areas of the world.[20] These areas have relatively high concentrations of agricultural activity, increasing the threat of economic damages.

In 2010, floods in Pakistan submerged more than one-fifth of the total land area of the country. According to the United Nations, the floods severely affected more than 20 million people, about an eighth of the population; nearly 2,000 people died and more than 1 million homes were destroyed in the flooding, which began in July 2010 following unusually heavy monsoon rains.

Earthquakes

An earthquake is a sudden shaking or trembling of the Earth's crust. Some quakes result from volcanic activity, but most result from a break in the Earth's crust, often in response to shifting tectonic plates. The Earth's surface consists of a collection of plates that constantly move, changing shape and size. The plates normally move at imperceptible speeds, often less than an inch per year, about the speed that fingernails grow, but sometimes the plates rupture, shifting suddenly and violently. Geologic faults occur where plates collide or slide past each other, sometimes miles beneath Earth's surface.

Effects of the ground's shaking cause the greatest danger from an earthquake, which can severely damage buildings and threaten the lives of people. Soil liquefaction can occur when groundwater mixes with sand, causing the ground to become very soft. In such circumstances, buildings may sink or tip. Ground displacement can also cause significant damage to structures along the fault as the two plates move in different directions or one moves upward as the other moves down. For example, during the 1906 earthquake along the San Andreas Fault in California, one section of road slid 20 feet from its previous position. Tsunamis can also result from earthquakes. An additional hazard from earthquakes occurs as electrical lines and gas pipes break, posing the risk of fire.

Scientists measure the magnitude, or destructive power, of an earthquake using the Richter scale. Magnitudes of 2.0 or less occur frequently but usually go unnoticed by people. At magnitude 4.5, equipment around the world can detect the shaking, and some damage may occur in populated areas. "Great earthquakes" have magnitudes larger than 8.0. A great earthquake occurs somewhere in the world every 1.5 years on average.

The most destructive events typically occur along subduction zones, places where one plate is forced under another. Sometimes called megathrust earthquakes, these events can exceed a magnitude of 9.0. The Great Chilean Earthquake in 1960, a 9.5 on the Richter scale, generated tsunamis that killed people as far away as Japan.

Tsunamis

When an underwater disturbance such as an earthquake, a landslide, or a volcanic eruption displaces a large amount of water, a series of enormous waves called tsunamis results. A tsunami can travel across the ocean at the speed of a jet airplane and smash into the shore with waves as high as 100 feet. Scientific speakers and writers often refer to tsunamis as seismic sea waves. Sometimes tsunamis are mistakenly called tidal waves, but in fact they have nothing to do with tides. A recent tsunami disaster occurred on March 11, 2011, when a magnitude 9.0 earthquake 250 miles (400 km) northeast of Tokyo sent a 23-foot wave smashing into the coast of Japan. The earthquake and its aftershocks threatened nuclear power plants along the coast; at least 17,000 people were left dead or missing, and hundreds of thousands lost their homes.[21]

Volcanoes

A volcano is a vent in the Earth's surface from which molten rock, hot rock, or steam may emerge. Often, a volcano is formed of rock that previously ejected from the vent.

When a volcano erupts, a number of hazards may result. Lava flows, glowing rivers of molten rock, rush down the mountain. Burning pyroclastic flows of hot ash and rock fragments speed down the mountain. In addition to these fiery hazards, melting ice or lakes pushed from their banks create large mudflows, or lahars, that rush down river channels. Lahars can cause devastating floods and can bury entire villages under several feet of mud. Besides these dangers, earthquakes or landslides during eruptions can trigger tsunamis.

In rare cases, a volcano's explosion is so violent that it causes the volcano to collapse into a crater called a caldera. For instance, the eruption of Thera in the 15th century b.c.e. caused most of the modern-day island of Santorini to collapse into the Mediterranean Sea. Scholars suspect that this event led to the fall of the Minoan civilization, which existed in the Bronze Age on the island of Crete. Some scholars speculate that Thera's eruption is the basis of the legend of Atlantis, a mythical island that sank into the ocean in a single day.

Just as scientists rank the strength of an earthquake on the Richter scale, they rank the strength of a volcano by using the Volcanic Explosivity Index (VEI). The VEI, introduced by Chris Newhall and Steve Self in 1982, classifies a volcanic eruption according to its magnitude, from 0 to 9. Scientists use the VEI to rank the event by observing characteristics such as the height of the plume (a mixture of gas and particles that shoots upward) and the volume of ash ejected from the volcano. As an example of this system, consider the 1980 eruption of Mount St. Helens in Washington State. That eruption registered as 5 on the VEI. In contrast, the largest volcanic eruption in recorded history was that of Mount Tambora in 1815, which registered a VEI of 9. The powdery dust of its debris entered the stratosphere and circled Earth, causing the so-called Year Without a Summer by blocking sunlight.

Limnic Eruption

In rare circumstances, gases such as carbon dioxide from volcanoes or other sources become trapped at the bottom of a deep lake. When the amount of gas becomes great enough, it rushes to the surface, creating a limnic eruption. The only recorded instances of such eruptions occurred at two lakes in Cameroon, a republic in western Africa. The first eruption occurred on August 15, 1984. On this day, gas escaping from Lake Monoun killed about 40 people. Two years later on August 21, 1986, gas erupted from Lake Nyos, killing 1,700 people.[22]

Landslides and Avalanches

When rocks, debris, and earth on a mountain or hill give way and move down the slope, a landslide occurs. Usually initiated by periods of heavy rainfall, these debris flows typically travel at speeds of 30 to 50 MPH. In the worst circumstances, speeds can reach 100 MPH, yielding a landslide that destroys nearly everything in its path. Each year, between 25 and 50 people lose their lives in landslides, with economic damages estimated between $1 and $2 billion.[23] The largest landslide in recorded history occurred during the 1980 eruption of Mount St. Helens. This landslide buried 24 square miles of land. It surged more than 13 miles down the North Fork of the Toutle River, filling the valley with 150 feet of rock and debris.

An avalanche is the rapid flow of snow down a slope, usually in mountainous terrain. A slab avalanche—a massive sheet of snow that breaks off all at once and then crashes down the slope—typically flows at 60 to 80 MPH, while wetter avalanches, which tend to start at one point and then spread out, travel more slowly, at approximately 20 MPH. The fastest avalanches are formed of powdered snow mixed with air and can move at speeds of 200 MPH or greater. Large avalanches can pull in air, water, rocks, small vegetation, and even trees, creating a powerful and destructive force. On February 9, 1999, in the Alps of France, an avalanche that sped down the slope of Montagne de Peclerey destroyed 17 chalets in the hamlets of Le Tour and Montroc. This massive wall of snow measuring 50 feet high by nearly 1,000 feet wide killed 10 people and severely injured five others.[24] Each year, approximately 150 people die in avalanches worldwide.[25] In the United States, most avalanches occur during the winter and early spring months of December to April.

Wildfires

Wildfires are uncontrolled fires that blaze in forests, plains, and other wilderness areas. The source of the fire may be natural, such as a lightning strike, or man-made, such as a smoldering campfire or a blaze deliberately set by an arsonist. Sometimes, firefighters set prescribed fires to manage fire threats in a controlled manner. If a wildfire approaches a populated area, it poses various threats. It can destroy homes and buildings. It can produce thick clouds of smoke that travel many miles, causing eye irritation and breathing difficulties for people with respiratory problems.

Despite the threats to people and their property, wildfires can have positive impacts on the forests in which they occur. Many plants and trees can survive a fire and even thrive. Ponderosa pines, whose bark is fire resistant, increase their growth rate in the years after a fire has swept through the area. Fire triggers or increases the release of seeds in lodgepole and jack pines, and it triggers flowering or fruiting in some herbs and shrubs. Fire burns away accumulated dead leaves, preparing the ground for germinating seeds. Fire also releases nutrients in the soil. These are just a few examples of how wildfire can be beneficial as well as destructive.

Despite the benefits of fire in some ecosystems, wildfires that threaten populated areas are clear dangers. Each year, the United States spends billions of dollars fighting wildfires and rebuilding after them. The year 2007 was an especially costly year, with wildfires burning 9.3 million acres[26] and causing $1,413 million in property and crop damages.[27] More recently, in 2009, wildfires burned 5.9 million acres of land.[28] While this figure is less than the five-year average of 7.7 million acres burned per year,[29] experts predict that fire seasons in the United States will get longer and bigger as a result of climate change, including shorter and wetter winters, warmer and drier summers, and an estimated two more decades of the current drought cycle that is affecting areas of the Southeast, Southwest, and West. During the period of 2010–14, experts expect to see wildfires burn an average of 10–12 million acres per year.[30]

In 2010, west and central Russia experienced hundreds of wildfires caused by unprecedented temperatures in the region. The fires cost an estimated US$15 billion in damages.

Extreme Weather

The main hazards of winter weather are due to extremely cold temperatures. One hazard of winter weather is hypothermia. This condition occurs when the body becomes so cold that it begins losing heat faster than it can warm itself. Symptoms include shivering, slurred speech, stumbling, confusion, very low energy, weak pulse, shallow breathing, and progressive loss of consciousness. If hypothermia is left untreated, it can lead to death.

Another hazard is frostbite. When exposed to extreme cold, body parts farthest from the heart—usually the hands, feet, nose, legs, and ears—can suffer frostbite, a condition in which blood flow to this body part slows or stops. Frostbitten skin becomes hard or waxy to the touch, and it may itch, burn, or become numb. Once thawed, the area is red and painful. In extreme cases, the affected areas require amputation.

While hypothermia and frostbite are serious conditions, the greatest source of danger in the winter months is heavy snow. A blizzard is a heavy snowstorm with winds more than 36 MPH. A blizzard can cripple an entire region, closing shipping routes, stranding motorists, blocking airport runways, and disrupting emergency services. Lesser storms may bring heavy accumulations of ice, which can knock down power lines and make bridges and overpasses treacherous. Additionally, high winds may produce a wind chill much colder that the actual temperature. As the wind increases, it removes heat from the body at an increasing rate, driving body temperature down and potentially leading to hypothermia or frostbite.

At the opposite temperature extreme are heat waves, which can be as deadly a hazard as winter weather. Under normal conditions, the human body dissipates heat through the regulation of blood circulation, perspiration, and panting. Temperatures above 90 degrees and high relative humidity tax the body's ability to dissipate heat effectively. When the body cannot cool itself adequately or can no longer compensate for salts and fluids lost through perspiration, heat-related illnesses such as heat cramps, heat exhaustion, and heat stroke can occur.

On average, 1,500 people in the United States die each year from excessive heat,[31] but severe heat waves can spike this number significantly higher in any given year. In 1980, a heat wave in the midwestern United States killed at least 1,250 people,[32] with some experts placing the number of deaths as high as 5,000.[33] In the summer of that year, Americans paid $1.3 billion more for electricity than they had the previous summer, and the higher demand for electrical power exceeded the supply, causing rolling blackouts. In 1995, a Chicago heat wave killed 522 people in just one month.[34]

An El Niño event causes extremes of both wet and dry weather, though in different regions, that can result in disasters, including flooding, drought, and wildfire. El Niño is a weather phenomenon characterized by unusually warm temperatures in the equatorial Pacific Ocean that may result in increased rainfall across the southern United States and in Peru while causing drought in the West Pacific, such as in Indonesia, Australia, the Philippines, southeastern Africa, and Brazil. The El Niño of 1997–98 contributed to record high temperatures, raising the "estimated average surface temperature for land and sea worldwide [by] 0.8°F."[35] As the meteorologist Katrina Glebushko explains, other results of the 1997–98 El Niño included droughts in "the Western Pacific Islands and Indonesia as well as in Mexico and Central America. In Indonesia drought caused uncontrollable forest fires and floods, while warm weather led to a bad fisheries season in Peru, and extreme rainfall and mud slides in southern California. Corals in the Pacific Ocean were bleached by warmer than average water, and shipping through the Panama Canal was restricted by below-average rainfall."[36] El Niño conditions typically recur every three to six years and last nine months to a year.

Plants and Animals

Plant diseases caused by bacteria, viruses, molds, and fungi can destroy entire harvests, devastating the food supplies of communities and regions and leading to economic loss, food shortages, and famine. The Irish Potato Famine of 1845–50 resulted from the destruction of Ireland's staple potato crop by late blight, a disease that causes lesions to develop on the plant and the tuber to rot rapidly. Late blight spores can survive from year to year in infected tubers, and the spores can be carried by wind for miles. Infected crops must be destroyed completely to eradicate the blight. Since potatoes are typically propagated with cuttings from tubers, to destroy a crop completely means to destroy the means of planting future crops from that stock. A completely new and uninfected source of tubers would be required.

Like plant diseases, animals can bring disaster to a community or larger region. Infestations by locusts or other insects can destroy crops and bring famine to hundreds or even millions of people. According to the United Nations World Food Programme (WFP), locust infestation is one of the most significant threats to food security in Madagascar, a country where "chronic food insecurity affects 65 percent of the population, with an 8 percent increase during the lean season."[37] Fleas and rats can spread diseases like the bubonic plague (colloquially called the Black Death) that killed millions of people in Asia and Europe in the Middle Ages. Mosquitoes can carry yellow fever, as they did during the yellow fever epidemic in New Orleans in 1905. Indeed, examples of epidemics and pandemics attributable to bacteria, viruses, and their carriers are so numerous that the category is often treated as its own class of disaster, linked to though separate from natural disasters attributable to geological processes.

INTERNATIONAL HISTORY
Mass Extinctions

The worst natural disasters that have occurred on Earth caused worldwide mass extinctions, and they happened millions of years ago. By studying changes in fossil assemblages in rock or sediment layers, scientists have identified events when many species of animals or plants on land and in the sea suffered large reductions of population or completely disappeared. Based on evidence preserved in the layers of the Earth, scientists have proposed different explanations for different mass extinction events. They include meteorite strikes, volcanism, changes in sea level, ice ages, and abrupt global climate change.

The most recent decimation of species occurred 65 million years ago when more than two-thirds of all species on Earth, including the dinosaurs, perished in a global catastrophe. The cause of this mass extinction event is preserved in a layer of sediment that marks the end of the Cretaceous (K) period and the beginning of the Tertiary (T) period in geologic time. The Cretaceous–Tertiary boundary is commonly referred to as the K–T.

Scientists who have studied the K–T boundary have found important clues to explain the cause of this mass extinction event. They observed that this layer of sediment contains unusually high concentrations of iridium, a heavy metallic element that is rare in Earth's crust but commonly found in meteorites. In the 1990s, geophysical researchers imaged a huge crater buried beneath the sediments of the Gulf of Mexico and the Yucatán Peninsula, which they identified as the impact site of a massive meteorite. They believe that when it smashed into Earth 65 million years ago, the meteorite created a fireball and blew an expansive cloud of dust and debris into the sky. The dust encircled the globe, creating a harsh environment in which most species could not survive.

In relatively more recent times, during the period of 800–1000 C.E., a series of intense droughts probably caused the collapse of the Maya civilization in Mesoamerica, in the area of modern-day Honduras and Mexico's Yucatán Peninsula. At the height of their civilization, the Maya numbered an estimated 15 million, and they thrived in huge cities linked by trade routes. Their towering pyramids still stand. The civilization reached its peak around 800–900 C.E. and then suddenly collapsed, leaving deserted cities and abandoned trade routes. Scientists speculate that the civilization died in famines, probably caused by a combination of natural drought and deforestation.

Historical

Mass extinctions resulting from natural hazards are disasters on the largest scale, but history is riddled with disasters that, although they are on a comparatively smaller scale, stand out as some of the deadliest natural events in recorded history. Natural disasters that set records as the deadliest or greatest of their type in recorded history through the 19th century include the Mediterranean earthquake of 1201, the Shaanxi earthquake of 1556, the eruption of Tambora in 1815, the 1839 cyclone in Coringa, India, the New Madrid, Missouri, earthquakes of 1811–12, and the Irish Potato Famine of 1845–50. Though certainly not a comprehensive summary of the greatest natural disasters before 1900, these representative events show the staggering degree of death and destruction that various types of natural disasters can bring.

The deadliest earthquake in Earth's recorded history occurred in the eastern Mediterranean on July 5, 1201. An estimated 1.1 million[38] people lost their lives, mostly in Egypt and Syria. Nearly as deadly was the earthquake in Shaanxi Province in central China in 1556. In late January or early February, an earthquake shook this region along the Wei River, killing an estimated 830,000 people.[39]

The deadliest volcano eruption in recorded history is the eruption of Tambora on Sumbawa, an Indonesian island, in April 1815. The resulting ashfall, tsunami, disease, and starvation killed 92,000 people.[40] This disaster actually includes two eruptions that occurred within a few days of each other. First, on April 5, a thunderous eruption sent ash into the air; the sound of the eruption was heard up to 870 miles away. Then, during April 10 and 11, Tambora erupted again, ejecting approximately 1.7 million tons of debris into the air. A cloud of volcanic ash circled the Earth, blocking the sun and lowering global temperatures, causing the Year Without a Summer.

In November 1839, Coringa, India, experienced natural disasters that permanently crippled the city. A cyclone caused a massive 40-foot storm surge that crashed through the harbor and into the streets of the city. Forty vessels were destroyed in the harbor, while 300,000 people in the city lost their lives. Coringa had been hit by cyclones before. In 1789, a cyclone spawned three storm surges that killed 20,000 people.[41] Afterward, survivors rebuilt the city over many years. After the 1839 cyclone, however, the city was never completely rebuilt.

The New Madrid, Missouri, earthquakes of 1811–12 were the greatest natural disaster in U.S. history at that point. Scientists estimate their magnitudes to be between 7.2–8.1 on the Richter scale, although this scale had not been developed at that time. The series of earthquakes began early on December 16, 1811, when a powerful earthquake struck the small town of New Madrid in the Mississippi River valley in what was then the Louisiana Territory. The force was enough to tear down trees, topple chimneys, and send people running from their houses. A few hours later, another earthquake struck New Madrid, causing the ground to ripple like water and splitting the Earth with long fissures. Entire buildings toppled into the holes, and sulfurous fumes rose out of the holes. During the three-month period between December 16, 1811, and March 15, 1812, more than 200 earthquakes occurred along the fault line. They affected 50,000 square miles in the United States and parts of Canada. Since the population in and around New Madrid was small—less than 3,000 people—the death toll was low. Six people lost their lives in the area, and perhaps 100 more lost their lives on the Mississippi River.[42]

During the Irish Potato Famine of 1845–50, Ireland lost one-fourth of its entire population. About half this loss—approximately 1 million people—resulted from starvation, typhoid, typhus, scurvy, and other health-related issues. In addition, an estimated 1–1.5 million people emigrated to escape the disaster; most went to America.[43] The famine was sparked by the potato blight, a disease that struck crops. Even before the famine, however, the populace of Ireland suffered from poverty caused by heavy taxation by the British and local landlords. Thus, the historical record of the famine is mired in political and economic controversy. By the 1840s, one-third of the Irish were subsisting mainly on potatoes. Then the blight wiped out this dietary staple for three successive years. People died by the tens of thousands of outright hunger, of diseases from which they were too weak to recover, or of exposure when they were evicted from their homes for being unable to pay rent. Ireland was permanently changed by the disaster, and still today the nation's history is divided into the pre-Famine years and the post-Famine years.

1900 to the Present

Recent history includes its share of record-setting natural disasters. During the past century or so, China suffered two of the country's deadliest earthquakes ever recorded, East Africa and the Sahel nations suffered more than 1 million famine deaths, the United States experienced two hurricanes that each broke records as the country's most expensive, Myanmar suffered one of the world's worst recorded disasters (Cyclone Nargis), and a tsunami in the Indian Ocean changed the way the world responds to tsunamis. These natural disasters are representative of those that have shaped economies, societies, and the understanding of the natural world in modern times.

Within the last century, China has experienced two separate earthquakes that rank among the deadliest in China's history. The earlier of the two occurred on December 16, 1920, on the Haiyuan Fault in north-central China. This magnitude 7.8 earthquake caused the deaths of 200,000 people. Destruction spread across seven provinces and regions. In Haiyuan County, more than 73,000 people were killed, and more than 30,000 more were killed in Guyuan County. In Xiji County, a landslide buried the entire village of

Sujiahe. About 125 miles of surface faulting occurred, and rivers became dammed or changed course.[44]

The 1976 earthquake in Tangshan, China, was even more destructive, killing or injuring more than 1 million people. The earthquake struck on July 27, 1976, about 87 miles southeast of Beijing. It obliterated Tangshan, a city of 1 million people. First, a magnitude 7.5 earthquake occurred, and 15 hours later a major aftershock occurred. Additional aftershocks in the 5.0–5.5 range hit the region as well. Concrete buildings crumbled and collapsed, bridges crumpled, and railroad tracks bent. Damage and fatalities were reported as far away as Beijing, a densely populated industrial city. In all, more than 240,000 people lost their lives, and 799,000 more were injured.[45]

Unlike rapid-onset hazards, drought and famine can cause disasters that unfold over years or recur over decades. In East Africa and the Sahel, famine has wracked the population since 1983. Though famine's death toll peaked between 1984 and 1986, deaths from starvation and related health failures continue into the 21st century. More than 1 million people have died, and the death toll continues to grow. For instance, in 2010, half of Niger's population of 15 million people faced moderate or severe malnutrition due to grain shortages, according to the country's government.[46] A combination of natural and human elements combined to fuel the famine problem, factors such as poor societies dependent on farming for survival, desertification of agricultural land bordering the Sahara, decreased annual rainfall, shrinking water sources, and civil war.

In 1992, Hurricane Andrew caused $25 billion of property damage in the southern Florida peninsula and south-central Louisiana.[47] At that time, it was the most expensive natural disaster in U.S. history (to be surpassed by the $81 billion in damages caused by Hurricane Katrina in 2005).[48] Dade County, Florida, was the hardest hit. Here, Andrew destroyed more than 25,500 homes and damaged more than 100,000 others. An estimated 250,000 people were left temporarily homeless, and 40 people lost their lives. Ed Rappaport of the National Hurricane Center (NHC) noted that "a combination of good hurricane preparedness and evacuation programs likely helped minimize the loss of life."[49]

On December 26, 2004, a magnitude 9.0 earthquake off the western coast of Sumatra caused a massive tsunami in the Indian Ocean. The U.S. Geological Survey (USGS) reported this to be the world's third largest earthquake since 1900. It ripped along nearly 750 miles of the Sunda Trench subduction zone, forcing a piece of the ocean floor to snap suddenly upward as high as 50 feet. People reported feeling the shaking as far away as India and Sri Lanka. The megathrust earthquake triggered an enormous tsunami, measuring 100 feet at its highest, that swept across the ocean to destroy structures and entire cities in Indonesia, Thailand, Sri Lanka, India, and the Republic of Maldives. The earthquake and tsunami killed an estimated 227,898 people. Approximately 1.7 million people in 14 countries in South Asia and East Africa were displaced by the disaster. Economic losses from the tsunami totaled $10 billion.

At the time of the disaster, no organization held the responsibility for issuing tsunami warnings for the Indian Ocean, and none of the countries struck by the tsunami had evacuation plans in place. Following the disaster, the Intergovernmental Oceanographic Commission (IOC), part of the United Nations Educational, Scientific and Cultural Organization (UNESCO), began working with United Nations member states bordering the Indian Ocean to establish an Indian Ocean Tsunami Early Warning System. A network of coastal and deep-ocean stations was set in place to detect potential tsunamis in the Indian Ocean and transmit data. Through 2010, the Pacific Tsunami Warning Center and the Japan Meteorological Agency were responsible for compiling data from the Indian Ocean and issuing tsunami watches, with these duties scheduled to transfer to Regional Tsunami Watch Providers in Indian Ocean countries in 2011.[50]

On May 2, 2008, Cyclone Nargis struck southern Myanmar in one of the world's worst recorded disasters. The cyclone, whose wind speeds reached 121 MPH, drove a huge storm surge inland. The winds and saltwater surge destroyed a huge portion of the fertile Irrawaddy Delta. When the storm had passed, the entire lower portion of the delta was underwater, and 1 million people were homeless.[51] The salty water contaminated a million acres of rice paddies, and winds destroyed 700,000 homes. Three-fourths of the livestock died, and half the fishing fleet sank. Yangon, the nation's main city, suffered heavy damage and loss of life as well. Finding and tallying the storm's victims stretched on as the government only gradually realized the extent of the destruction. One year later, the number of dead was estimated to be 85,000 people; however, an additional 54,000 people were still listed as missing.[52]

RESPONDING TO INTERNATIONAL NATURAL DISASTERS
The Disaster Risk Index

Natural disasters affect countries around the world. Often, the nature and extent of the damage are closely linked to the level of the nation's development. Nations that are more developed typically have greater resources for identifying risk and exposure and for managing and mitigating the effects of disasters. These nations tend to have buildings and infrastructures that are more disaster resistant. At the same time, rapidly growing urban areas may create greater exposure. Less-developed countries fare less well, with buildings and infrastructures not suited to withstand natural disasters and fewer systems in place to predict and monitor disasters or respond to their impact. Both developed and undeveloped countries, therefore, need tools for understanding and managing natural disasters and for reducing their damage.

In September 2000, the General Assembly of the United Nations adopted the Millennium Declaration. The resolution lists eight Millennium Development Goals (MDGs) to achieve worldwide by the year 2015. Chief among them is ending world poverty and hunger. The goal is to reduce by half, between 1990 and 2015, the proportion of people who earn less than $1 a day and in the same time period to reduce by half the proportion of people who suffer from hunger. The United Nations Development Programme (UNDP) concluded that natural disasters are a "significant threat"[53] to the achievement of the MDGs. Natural disasters can suddenly devastate a region, wiping out vital economic infrastructures including the means of earning an income, buying or raising food, continuing an education, and so on. According to the UNDP, "While humanitarian action to mitigate the impact of disasters will always be vitally important, the global community is facing a critical challenge: How to better anticipate—and then manage and reduce—disaster risk by integrating the potential threat into its planning and policies."[54]

In response to this global challenge, the UNDP devised the Disaster Risk Index (DRI). Published in 2004, the DRI "measures the relative vulnerability of countries to three key natural hazards—earthquake, tropical cyclone, and flood; identifies development factors that contribute to risk; and shows in quantitative terms, just how the effects of disasters can be either reduced or exacerbated by policy choices."[55]

Risk Factors and International Development Goals

The policies of a government directly affect the vulnerability of its population to natural hazards. When a country disregards factors that increase risk, a natural disaster can have a devastating effect on a large portion of its citizens. Over time, these risks accumulate, placing an ever-increasing number of people in harm's way. For instance, as poverty increases, a person's ability to acquire nourishment diminishes, raising the individual's vulnerability to famine. To complicate matters, this person may reside in a city with inadequate building codes or an area prone to flooding and, due to constraints of poverty, possess little means of surviving an earthquake or a tsunami.

The United Nations and other international organizations work closely with developing countries to achieve long-term development goals, including nutrition, education, gender equality, health care, and environmental sustainability. During the creation and implementation of projects to achieve these goals, one priority is to identify and reduce risk factors that increase the vulnerability of a specific population to natural hazards. Such risk factors include extreme poverty and hunger, the lack of primary education, poor health of mothers and young children, HIV/AIDS and other diseases, corruption, and substandard building practices.

EXTREME POVERTY AND HUNGER

DRI research statistically links poverty with human vulnerability to natural disasters. For people who lack basic nutrition or financial resources, the shock of a disaster can be overwhelming. Flooding or drought often destroys meager sources of nutrition, causing food reserves to deplete quickly. For any given country, eliminating disaster risk often goes hand in hand with reducing poverty.

The process of implementing these changes often occurs slowly due to complex political situations. Developing countries with low gross domestic products (GDPs) often depend entirely on international aid to address poverty and hunger. UNDP facilitates ongoing programs to address these needs, but a natural disaster can quickly overwhelm the international resources stationed within a country. In order to address the heightened need, additional personnel, equipment, and supplies must be acquired and transported to the disaster site, a process that can often be less than timely. In 1984, for example, in the emergency food program for Burkina Faso, Mali, Mauritania, Senegal, and Somalia, the average time to approve, procure, and transport food exceeded six months.[56]

LACK OF PRIMARY EDUCATION

Increasing the education levels within nations is one of the major goals of disaster management. Literacy enables individuals to become more involved in their communities and in decisions that can ultimately reduce risk during a catastrophic event. Significant disasters can have lasting effects on the educational system of an affected region or country. For instance, when schools are destroyed in a disaster, the means of obtaining education is severely limited or eliminated outright. Similarly, when disaster strikes, education may take a low priority in comparison to meeting immediate needs for food, water, shelter, and money. People's struggles turn solely to survival. As a result, vulnerability to disaster increases a situation that creates the possibility of a vicious downward cycle if multiple disasters occur over a short period of time.

Introduction

POOR HEALTH OF MOTHERS AND YOUNG CHILDREN

A natural disaster can have a greater negative effect on children under five years of age than on older individuals. In addition to the increased sensitivity to injury, the risks to nutrition, drinking water, and sanitation make young children particularly vulnerable. The death of parents or the loss of household income can be particularly damaging to the psychological and physical health of these children. In the same way, maternal health, the health of pregnant women or those caring for young children, can also be adversely affected. As the strain from the disaster drains household resources, women and girls, in many cases, are the last to receive what they need to survive.

HIV/AIDS, MALARIA, AND OTHER DISEASES

During a disaster, people are often cut off from much-needed medical assistance, but when the person already suffers from a medical condition such as AIDS or another chronic disease, the stress can become overwhelming. Medicine may become unavailable, access to doctors or medical facilities may be cut off, or emergency responders may be unavailable. Consequently, the vulnerability of this group of people increases significantly. In addition, loss of clean drinking water and sanitation increases the risk of other diseases such as malaria, raising the risk of an epidemic.

CORRUPTION

Humanitarian operations during a natural disaster often involve complex distribution networks to deliver relief supplies. Food must be transported, warehoused, and separated into manageable quantities as it moves from large shipping depots to local distribution centers to individuals in need. Relief workers must establish these networks swiftly and make ever-changing management decisions as fieldworkers submit assessments of the situation. If not carefully monitored, this hectic environment presents opportunities for financial and nonfinancial corruption to occur.

The amount of money devoted to a relief operation can be sizable. For example, in the three months following the January 2010 earthquake in Haiti, relief funding from the United States Agency for International Development (USAID) and the U.S. Department of Defense (DoD) exceeded a billion dollars.[57] With such enormous amounts of money involved, relief operations must be closely monitored. Common forms of financial fraud include kickbacks, bribes, fraud, and extortion, activities that can occur at an organizational level, such as in bid-rigging, or the manipulation of the bidding process for relief supplies. Another common practice involves the removal and sale of portions of supplies during transportation. Moreover, financial corruption can occur at the individual level when fraudulent claims are used to request disaster relief. For instance, in the aftermaths of Hurricanes Katrina and Rita, FEMA paid nearly $20 million to thousands of individuals who filed duplicate claims for damages to the same house.[58]

Other forms of corruption relate to the total dependence of disaster victims on relief workers. Nonfinancial fraud involves activities such as diverting relief supplies, conditional release of rations, preferred treatment of family and friends, and coercion or intimidation of staff to overlook or participate in other forms of corruption. One instance of nonfinancial corruption involved extensive sexual exploitation related to the distribution of relief supplies. A 2002 investigation by the United Nations High Commissioner for Refugees (UNHCR) implicated 40 agencies and 67 individuals working in Liberia, Guinea, and Sierra Leone.[59] To counter this type of corruption, aid agencies have joined with human rights organizations such as Transparency International to identify areas of risk and implement zero-tolerance policies designed to eliminate corruption in humanitarian operations.[60]

SUBSTANDARD BUILDING PRACTICES

Model building codes provide standards to ensure that newly constructed commercial and residential structures can safely withstand a natural disaster. Developed countries require construction projects to conform to government standards such as the International Building Code (IBC) in the United States or the Eurocode in Europe. These codes mandate that buildings be able to survive the physical forces of earthquake, resist the winds of a hurricane, and slow the spread of fires. Building codes also restrict the types of buildings that can be constructed in flood plains or mandate that levees provide adequate protection against flooding.

Building codes are formulated to protect the safety of a building's occupants, but they do not necessarily remove the risk of damage to the building; consequently, a disaster that results in no fatalities may nevertheless cause significant financial damage to structures. The adoption of building codes may account for the decrease in deaths from natural disasters in recent decades despite the fact that economic damages continue to rise. From 1900 to 2003, natural disasters killed 62 million people worldwide, but 85 percent of those deaths occurred before 1950. While deaths have declined, financial damages rose from $93 billion in the 1960s to $778 billion in the 1990s[61] to more than $885 billion between 2000 and 2010.[62]

For nondeveloped countries, building practices seldom follow an established standard. For one reason, countries with widespread poverty do not have the resources to develop and enforce a model building code. The residents of these countries typically use readily available materials for construction without considering the structure's vulnerability to a natural disaster.

For instance, the 2010 earthquake in Haiti, a magnitude 7.0 quake, devastated Port-au-Prince, the country's capital. A study by the Organization of American States (OAS) found that many of the buildings could not have survived an earthquake of even half that magnitude.[63]

Measuring Risk

The DRI measures the risk of death during a natural disaster and gives a means of comparing the relative risks of different countries to natural disasters. It takes into account both the exposure and the vulnerability faced by citizens of each country using data from three hazard types—earthquakes, tropical cyclones, and floods—that account for 39 percent of deaths from natural disasters worldwide. A DRI using data from droughts and famines, which account for 55 percent of deaths from natural disaster worldwide, is still under development. A multi-hazard DRI that combines the risks of all hazards in an area is also being studied.[64]

EARTHQUAKE HAZARD

Between the years 2000 and 2010, earthquakes caused 697,907 deaths worldwide,[65] including the deaths from the January 2010 earthquake in Haiti. Only two of the earthquake deaths in that 10-year period occurred in the United States. The deadliest years in this period were 2004, with 228,802 deaths, and 2010, with 226,895. These are the years in which the Sumatra earthquake and the Haiti earthquake happened, respectively. Around the world, 130 million people are exposed to earthquake risk, with the greatest exposed populations living in Japan, Indonesia, and the Philippines.[66] The DRI study asserts that vulnerability to earthquakes increases as cities grow and that countries with predominantly large cities have the highest risk.

TROPICAL CYCLONE HAZARD

Cyclones, also called hurricanes or typhoons, potentially endanger 119 million people each year worldwide. Between 1980 and 2000, 251,384 deaths were associated with tropical cyclones worldwide. Countries found to have the highest exposure were China, India, the Philippines, Japan, and Bangladesh. Since a large portion of Bangladesh's population lives along the coast, it has a much greater vulnerability to such storms than other countries. More than 60 percent of the deaths recorded during the study period occurred in that country.[67] Some island countries, such as Haiti and the Dominican Republic, also have considerable risks from hurricanes, while neighboring Cuba, a more developed country, has a lower ranking. Countries with predominantly rural populations tend to have significantly higher vulnerability to tropical cyclones.

FLOOD HAZARD

According to the DRI, each year an estimated 190 million people in 90 different countries are at risk from flooding hazards. Between 1980 and 2000, a total of 170,010 deaths worldwide resulted from flooding.[68] Countries with the greatest vulnerability include India, Bangladesh, Pakistan, and China. In these countries, large portions of the population live in river floodplains or low-lying coastal areas. Sparsely populated rural areas, where preparation, warning systems, and health coverage are weak or nonexistent, were found to be the most vulnerable areas.

DROUGHT HAZARD

Although the DRI does not officially cover droughts and famine, the study includes some research in the area. Between the years 1980 and 2000, 832,544 deaths were linked to drought, with droughts in Ethiopia, Somalia, and Mozambique together accounting for the most deaths.[69] The difficulties in assessing risk posed by drought come from the speed at which the disaster occurs. Instead of a rapid onset, as with the other hazards studied, droughts develop slowly and last over a number of years. For example, the drought associated with the dust bowl in the United States lasted six years, from 1934 to 1941. With 220 million[70] people annually exposed to drought, the development of a more precise analysis is a clear need.

MULTI-HAZARD RISK

Another area that the UNDP has slated for further study within the DRI concerns countries where multiple hazards can affect the same population. For instance, Mozambique has dangers from cyclones, floods, and drought. Consequently, this country is one of the most dangerous countries in the world in terms of natural disasters. Other countries with significant risk from more than one hazard are Ethiopia, Malawi, and the Philippines.

International Response to Natural Disasters

National governments shoulder the responsibility for responding with assistance and aid when a natural disaster occurs. For developed countries that have prepared for catastrophic events, a disaster may significantly strain local authorities, requiring the mobilization of national resources. In this case, international organizations play a lesser role or have no involvement at all. However, for undeveloped countries and governments lacking the resources to respond adequately, the international community often plays a major role in planning for disasters and responding when they occur.

COORDINATING NATIONAL AND INTERNATIONAL EFFORTS

The Office for the Coordination of Humanitarian Affairs (OCHA), which is a branch of the United Nations, appoints an Emergency Relief Coordinator (ERC) who is responsible for directing the response to a natural disaster. Other UN branches may become involved as the situation unfolds. Referred to as emergency tools, these teams have specific training and responsibilities and deploy to the disaster site to provide assistance such as real-time alerts concerning approaching disasters, humanitarian aid, coordination of other relief organizations, civil defense, search and rescue teams, environmental damage assessment, and emergency funding for relief efforts.

Local government policies determine the tools and resources available to international response teams. UN search and rescue teams establish an On-Site Operations Coordination Centre (OSOCC) to centralize the management of all assistance provided to local authorities and to assess the need for additional resources. International relief teams arriving in the country receive critical information about the disaster, local policy limitations, and assignments through the OSOCC. Limitations placed on UN teams working within a country vary greatly. The deployment of relief teams and supplies often uses military resources such as aircraft, ships, and trucks. Some local governments prohibit the use of foreign military personnel or equipment within their borders, which may delay relief efforts while teams acquire suitable civilian transportation. In extreme cases, a government may refuse to allow international aid to enter the country. For example, in 2008, Cyclone Nargis caused 100,000 deaths and left millions of people without food in Myanmar, but the government of Myanmar refused to allow UN relief teams into the country.[71] U.S. and French naval ships loaded with relief supplies remained anchored offshore for two weeks before withdrawing.[72]

EMERGENCY RELIEF TEAMS

The Inter-Agency Standing Committee (IASC) of the United Nations manages the early warning unit, which is the primary means of coordinating humanitarian efforts related to natural disasters. Some of the primary goals of this organization are to develop humanitarian policies, allocate responsibilities among the various agencies that can provide assistance, provide an ethical framework for all providers of humanitarian aid, identify areas where plans or resources could fall short during a disaster, and resolve disputes between agencies that provide humanitarian assistance. In practice, the International Committee of the Red Cross and other large relief organizations outside of the United Nations are included in the decision-making process. When OCHA deems a disaster to be imminent, a team from this unit can be deployed from New York or a regional OCHA office to assist with contingency planning and other preparations needed to support humanitarian efforts.

When a disaster response moves to the relief phase, a United Nations Disaster Assessment and Coordination (UNDAC) team may deploy to the affected region to assist the government or UN personnel already in the area. UNDAC teams are designed for rapid deployment, sometimes on notice as short as 12 hours, to anywhere in the world. Each team is comprised of staff with expertise in disaster management and has predetermined procedures for dealing with sudden-onset emergencies. For some disaster-prone nations, such as India, the United Nations maintains a standing UNDAC team to allow quicker response to developing emergencies.

Search and rescue teams perform one of the most important tasks in the aftermath of a disaster. In 1991, the International Search and Rescue Advisory Group (INSARAG) was formed as a result of initiatives that followed the 1988 earthquake in the Armenian Soviet Socialist Republic. This disaster killed at least 25,000 people and marked the first humanitarian efforts within the Soviet Union by the United States and other Western nations.[73] The role of INSARAG is to examine the methodology and coordination of search and rescue teams responding to earthquakes around the world. Each year, this organization hosts regional search and rescue simulations and training courses to develop the skills of first responders in earthquake-prone regions.

In the direct aftermath of a disaster, another key issue is providing medical services to the injured and dealing with potential complications such as the spread of diseases. The World Health Organization (WHO) plays a significant role in mobilizing emergency personnel to deal with the immediate medical needs and the risks of outbreaks during a disaster when local and national health systems become overloaded. In preparation for disasters, WHO trains medical professionals to operate in the hectic and ill-defined aftermath of a disaster with skills such as rapid diagnosis, containment, communications, emergency surgery, and the creation of temporary medical facilities. WHO also ensures that these professionals have the support and materials necessary to perform key tasks during emergency responses to disasters.

Disaster-response personnel must also assess and manage potential health-related complications. For example, the presence of open wounds and abrasions increases the likelihood of tetanus infections. The loss of access to clean drinking water raises the risk of contracting waterborne illnesses such as diarrheal diseases, hepatitis, and leptospirosis. Overcrowding in shelters increases the chances that diseases such as measles, meningitis, and acute respiratory infections will transmit more readily. Another potential complication is an increase in vector-borne diseases. These are illnesses such as malaria and dengue fever that are transmitted through a secondary factor, such as increased mosquito populations or high humidity.

United Nations Humanitarian Civil-Military Coordination (UN-CMCoord) directs the complicated issues of using military personnel during a disaster. Military teams provided by almost all UN members are available for deployment, but a variety of factors such as a country's policies concerning foreign militaries, the location of the disaster, and current diplomatic relationships between countries can place significant limitations on what teams can actually be used. Most commonly, the deployment of air support provides transportation for emergency personnel and supplies, particularly during the initial surge phase of response, when needs are the greatest and access by other means of transportation may be cut off. Other services that military personnel may perform during a disaster include medical treatment, search and rescue, communications, engineering, power supply and distribution, and sea transportation.

EMERGENCY FUNDING SOURCES

During natural disasters and other humanitarian emergencies, funds must be made available to international aid organizations in a timely fashion. Otherwise, relief efforts may be disrupted. Approved by the UN in 2005, the Central Emergency Response Fund (CERF) provides grants of up to $450 million and loans of up to $50 million to agencies engaged in disaster relief. One of the primary purposes of the fund is to provide immediate cash via a loan to agencies that have received pledges of funds for lifesaving activities but which have not yet received the monies. This arrangement provides a means of rapid cash flow while donor pledges are being transferred.

The World Bank provides another source of funding in time of disaster in the form of emergency loans or grants. Since 1984, World Bank lending related to natural disasters has grown from just over $500 million to over $3 billion in years with significant disasters. As of 2006, the World Bank maintained a portfolio of 528 projects related to natural disasters, representing 9.4 percent of all the loans made by the bank since 1984.[74]

INFORMATION MANAGEMENT SERVICES

OCHA works closely with organizations that track potential and developing emergency situations. The IASC Humanitarian Early Warning Service (HEWS) provides a global source for early warnings and forecasts of natural hazards online at hewsweb.org. Maps and graphics that present validated information in nontechnical terms provide real-time data regarding droughts, floods, storms, locusts, volcanoes, earthquakes, El Niño, and other hazards. Launched in 1996, this UN Web site, ReliefWeb, provides documents, maps, news, and all necessary resources for the humanitarian relief community. Reaching approximately 70,000 subscribers, ReliefWeb provides around-the-clock updates, posting approximately 150 maps and documents each day and maintaining a database of close to 300,000 maps and documents going back to 1981.

The Global Disaster Alert and Coordination System (GDACS) collects and organizes valuable information from around the world, reporting precise details used to facilitate the levels of response needed by other organizations. As quickly as possible, this system provides disaster alert and impact estimation, timely alerts to organizations that can provide humanitarian aid, coordination of relief activities through targeted discussions and periodic summaries, customized satellite images and maps available via the Web, and an integrated feed from ReliefWeb, news sources, and hewsweb.org.

During a disaster, the timely dispatch of specialized personnel and teams significantly influences the impact of the emergency. In addition, relief supplies and equipment must be located quickly. In 1991, the UN created the Central Register of Disaster Management Capacities (Central Register) to track such resources. The Central Register tracks three categories of contact persons—local officials needed to coordinate customs issues, disaster response personnel, and major donors that provide emergency humanitarian assistance. In addition, five directories of disaster management assets are maintained: search and rescue teams, military and civil defense resources, emergency stockpiles of relief supplies, rosters of disaster management expertise, and a register of advanced technologies for disaster response.

VICTIM IDENTIFICATION

Following a natural disaster, identifying the wounded and dead becomes a priority. INTERPOL, the world's largest international police organization, developed the Disaster Victim Identification Guide (DVI Guide) to provide guidelines for INTERPOL member countries in identifying disaster victims. Each member country is encouraged to assemble and train its own DVI team, but if a disaster occurs in a country that does not have such a team, it can request that INTERPOL send a team from another nation. One hundred and eighty-eight countries from around the world are members of INTERPOL, including the United States.

Within the DVI team, specialists accomplish the widespread duties of finding disaster victims, both alive and dead, and of gathering and protecting evidence necessary for their identification. In this regard, the DVI team members perform the traditional police work of securing an area, establishing a search grid, collecting fingerprints and forensic evidence, and so on. However, in this case, the "crime" is the natural disaster.

The Recovery and Evidence Collection Team recovers the bodies of those who have died. This team is also responsible for collecting and preserving evidence at the scene, including personal property such as wallets, purses, jewelry, suitcases, and so on that can help identify victims.

The Ante Mortem Team, or AM Team, collects information that will help identify living victims, prepares missing persons reports, and notifies authorities of their identification of victims. Similarly the Post Mortem Team (PM Team) gathers dental, medical, and forensic data to identify victims who have died. The team includes professionals in the fields of fingerprint analysis, forensic pathology, forensic odontology, and DNA analysis.

The Reconciliation Team compares the files that the AM and PM Teams produce, looking for matches. For instance, the AM Team may fill out a missing persons report, including identifying details of the person. The PM Team may fill out a report on a deceased victim who matches that description. The Reconciliation Team matches these reports and forwards the data to the Identification Board for evaluation and final decisions.

A natural disaster is devastating to victims' physical and mental health. Likewise, the professionals who identify the victims—including living persons, deceased persons, and body parts—may suffer physically or psychologically due to the nature of their work. For this reason, doctors and psychologists who make up the Care and Counseling Team provide medical and psychological treatment to DVI team members.

---

[1] USGS Earthquake Hazards Program. "Earthquakes with 1,000 or More Deaths Since 1900." Available online. URL: http://earthquake.usgs.gov/regional/world/world_deaths.php. Accessed November 17, 2009.

[2] Maxx Dilley et al. *Natural Disaster Hotspots: A Global Risk Analysis.* Washington, D.C.: World Bank, 2005, p. 26.

[3] ———. *Natural Disaster Hotspots,* p. 1.

[4] ———. *Natural Disaster Hotspots,* p. 4.

[5] ———. *Natural Disaster Hotspots,* p. 4.

[6] Steve Nix. "Cedar Fire Disaster—San Diego County, California, Late October 2003." Available online. URL: http://forestry.about.com/od/forestfire/ss/top_fires_na.htm. Accessed May 4, 2010.

[7] Richard Melson. "Bengal Famine of 1770." Available online. URL: http://www.cambridge-forecast.org/MIDDLEEAST/BENGAL.html. Accessed May 4, 2010.

[8] James Glanz. "Geothermal Power." Available online. URL: http://topics.nytimes.com/top/news/business/energy environment/geothermal-power/index.html. Accessed May 4, 2010.

[9] ———. "Geothermal Drilling Safeguards Imposed." Available online. URL: http://www.nytimes.com/2010/01/16/science/earth/16alta.html?_r=1& ref=james_glanz. Accessed September 13, 2010.

[10] International Strategy for Disaster Reduction. "Disaster Statistics: Impact: Killed." Available online. URL: http://www.unisdr.org/disaster-statistics/pdf/isdr-disaster-statistics-impact.pdf. Accessed November 16, 2009.

[11] P. Peduzzi et al. "Assessing Global Exposure and Vulnerability Towards Natural Hazards: The Disaster Risk Index." *Natural Hazards and Earth System Sciences* 9 (7/17/09): 1,149. Available online. URL: http://www.nat-hazards-earth-syst-sci.net/9/1149/2009/nhess-9-1149-2009.pdf. Accessed November 16, 2009.

[12] United Nations Office for the Coordination of Humanitarian Affairs. "Mozambique: A Success Story." Available online. URL: http://ochaonline.un.org/News/InFocus/InternallyDisplacedPeopleIDPs/ MozambiqueASuccessStory/tabid/5142/language/en-US/Default.aspx. Accessed November 17, 2009.

[13] United States Geological Survey. "Magnitude 7.0—Haiti Region." Available online. URL: http://earthquake.usgs.gov/earthquakes/recenteqsww/Quakes/us2010rja6.php #summary. Accessed March 14, 2010.

[14] Guardian.co.uk. "Chile Earthquake Death Toll Likely to Rise as Rescue Effort Moves to Coast." Available online. URL: http://www.guardian.co.uk/world/2010/mar/02/earthquake-chile-death-toll-looting. Accessed March 14, 2010.

[15] World Meteorological Organization. "Natural Hazards." Available online. URL: http://www.wmo.int/pages/themes/hazards/index_en.html. Accessed November 17, 2009.

[16] United States Geological Society. "Hurricane Hazards—a National Threat." Available online. URL: http://pubs.usgs.gov/fs/2005/3121/2005-3121.pdf. Accessed November 17, 2009.

[17] NOAA National Severe Storms Laboratory. "A Severe Weather Primer: Questions and Answers about Tornadoes." Available online. URL: http://www.nssl.noaa.gov/primer/tornado/tor_faq.shtml. Accessed May 5, 2010.

[18] NOAA Satellite and Information Service. "U.S. Tornado Climatology." Available online. URL: http://www.ncdc.noaa.gov/oa/climate/severeweather/tornadoes.html. Accessed November 17, 2009.

[19] Dilley et al. *Natural Disaster Hotspots*, p. 35.

[20] ———. *Natural Disaster Hotspots*, pp. 35, 43.

[21] CNN. "Japan Raises Nuclear Alert Level." Available online. URL: http://www.cnn.com/2011/WORLD/asiapcf/03/18/japan.disaster/index.html?hpt=T2#. Accessed March 18, 2011.

[22] Youxue Zhang. "Cracking the Killer Lakes of Cameroon." *Geoscience News* (July 1996). Ann Arbor: Department of Geological Sciences, University of Michigan, p. 3.

[23] United States Geological Survey. "Landslide Hazards Program." Available online. URL: http://landslides.usgs.gov/learning/faq/?PHPSESSID=5dj39virf3uute9gf0fs3l1122. Accessed November 17, 2009.

[24] Andrew S. Goudie. *The Nature of the Environment*. Available online. URL: http://www.blackwellpublishing.com/goudie. Accessed May 5, 2010.

[25] National Snow and Ice Data Center. "Avalanche Awareness." Available online. URL: http://nsidc.org/snow/avalanche. Accessed May 5, 2010.

[26] National Oceanic and Atmospheric Administration. "U.S. Wildfire Annual 2007." Available online. URL: http://www.ncdc.noaa.gov/sotc/?report=fire&year=2007&month=13. Accessed May 5, 2010.

[27] NOAA Economics. "Economic Costs: Annual U.S. Fire Weather Summary." Available online. URL: http://www.economics.noaa.gov/?goal=weather&file=events/fire. Accessed May 5, 2010.

[28] National Oceanic and Atmospheric Administration. "State of the Climate: U.S. Wildfire 2009." Available online. URL: http://www.ncdc.noaa.gov/sotc/?report=fire&year=2009&month=13&submitted=Get+Report. Accessed May 5, 2010.

[29] ———. "State of the Climate: U.S. Wildfire 2009.

[30] Fire Executive Council of the National Association of State Foresters. *Quadrennial Fire Review 2009*. Available online. URL: http://www.nifc.gov/QFR/QFR2009Final.pdf. Accessed May 5, 2010.

[31] National Oceanic and Atmospheric Administration. "Heat Wave: a Major Summer Killer." Available online. URL: http://www.noaawatch.gov/themes/heat.php. Accessed May 17, 2010.

[32] ———. "Heat Wave: a Major Summer Killer."

[33] Christopher R. Adams. "Impacts of Temperature Extremes." Available online. URL: http://sciencepolicy.colorado.edu/socasp/weather1/adams.html. Accessed May 5, 2010.

[34] ———. "Impacts of Temperature Extremes."

[35] Katrina Glebushko. "The El Niño Phenomenon: From Understanding to Predicting." Available online. URL: http://www.csa.com/discoveryguides/prednino/review.pdf. Accessed May 6, 2010.

[36] ———. "The El Niño Phenomenon."

[37] World Food Programme. "Madagascar: Response to Recurrent Natural Disasters and Seasonal Food Insecurity." Available online. URL: http://www.wfp.org/content/response-recurrent-natural-disasters-and-seasonal-food-insecurity. Accessed May 5, 2010.

[38] Steven Dutch. "Top Ten Earthquakes and Volcanic Eruptions." Available online. URL: http://www.uwgb.edu/dutchs/PLATETEC/TOPTEN.HTM. Accessed November 17, 2009.

[39] ———. "Top Ten Earthquakes and Volcanic Eruptions."

[40] ———. "Top Ten Earthquakes and Volcanic Eruptions."

[41] Emergency and Disaster Management, Inc. "Cyclones." Available online. URL: http://www.emergency-management.net/cyclone.htm. Accessed November 17, 2009.

[42] Lee Davis. *Natural Disasters, Revised Edition*. New York: Facts On File, 1992, p. 95.

[43] Irish Famine Curriculum Committee. "The Great Irish Famine." Available online. URL: http://www.nde.state.ne.us/SS/Irish/Irish_pf.html. Accessed November 17, 2009.

[44] United States Geological Survey. "Historic Earthquakes: Haiyuan, Ningxia (Ninghsia), China." Available online. URL: http://earthquake.usgs.gov/earthquakes/world/events/1920_12_16.php. Accessed November 17, 2009.

[45] ———. "Historic Earthquakes: Tangshan, China." Available online. URL: http://earthquake.usgs.gov/earthquakes/world/events/1976_07_27.php. Accessed November 17, 2009.

[46] Mark John. "'Strong Risk' of 2010 Famine in Africa's Sahel." Available online. URL: http://www.alertnet.org/thenews/newsdesk/LDE60Q14F.htm. Accessed May 6, 2010.

[47] Ed Rappaport. "Hurricane Andrew, 16–28 August 1992." Available online. URL: http://www.nhc.noaa.gov/1992andrew.html. Accessed November 6, 2009.

[48] U.S. Department of Health and Human Services. "Hurricane Katrina." Available online. URL: http://www.hhs.gov/disasters/emergency/naturaldisasters/hurricanes/katrina /index.html. Accessed November 17, 2009.

[49] Rappaport. "Hurricane Andrew."

[50] Reuters AlertNet. "How the Indian Ocean Tsunami Warning System Works." Available online. URL: http://www.alertnet.org/db/an_art/59567/2009/09/28-122125-1.htm. Accessed May 6, 2010.

[51] Associated Press. "UN Officials: Myanmar Cyclone a 'Major, Major Disaster.'" Available online. URL: http://www.livescience.com/environment/080506-ap-myanmar-cylcone.html. Accessed November 6, 2009.

[52] New York Times. "Cyclone Nargis." Available online. URL: http://topics.nytimes.com/topics/news/international/countriesandterritories /myanmar/cyclone_nargis/index.html. Accessed November 17, 2009.

[53] Bureau for Crisis Prevention and Recovery. "Foreword." In Reducing Disaster Risk: A Challenge for Development. Available online. URL: http://www.undp.org/cpr/disred/english/publications/rdr.htm. Accessed November 16, 2009.

[54] ———. "Foreword."

[55] ———. "Foreword."

[56] Famine in Africa: Improving U.S. Response Time for Emergency Relief. (April 1986). Washington D.C.: United States General Accounting Office, pp. 8–9.

[57] USAID. "Haiti—Earthquake." Available online. URL: http://www.usaid.gov/our_work/humanitarian_assistance/disaster_assistance /countries/haiti/template/fs_sr/fy2010/haiti_eq_fs50_04-16-2010.pdf. Accessed May 17, 2010.

[58] Government Accountability Office. "Continued Findings of Fraud, Waste, and Abuse." Available online. URL: http://www.gao.gov/highlights/d07300high.pdf. Accessed May 17, 2010.

[59] BBC News. "Child Refugee Sex Scandal." Available online. URL: http://news.bbc.co.uk/2/hi/africa/1842512.stm. Accessed May 18, 2010.

[60] Transparency International. "Preventing Corruption in Humanitarian Operations." Available online. URL: http://www.transparency.org/content/download/49759/795776/ Humanitarian_Handbook_cd_version.pdf. Accessed May 18, 2010.

[61] U.S. News and World Report. "Why Natural Disasters Are More Expensive—but Less Deadly." Available online. URL: http://www.usnews.com/money/business-economy/

articles/2010/03/24/why-natural-disasters-are-more-expensivebut-less-deadly.html. Accessed May 19, 2010.

[62] EM-DAT: The OFDA/CRED International Disaster Database. "Advanced Search." Available online. URL: http://www.emdat.be/advanced-search. Accessed May 20, 2010.

[63] CNN World. "Problems with Haiti Building Standards Outlined." Available online. URL: http://www.cnn.com/2010/WORLD/americas/01/13/haiti.construction/index.html. Accessed May 20, 2010.

[64] Andrew Maskrey et al. *Reducing Disaster Risk.* New York: John S. Swift Co., 2004, p. 29.

[65] USGS Earthquake Hazards Program. "Number of Earthquakes Worldwide for 2000–2011." Available online. URL: http://earthquake.usgs.gov/earthquakes/eqarchives/year/eqstats. php. Accessed May 6, 2010.

[66] Maskrey et al. *Reducing Disaster Risk,* p. 34.

[67] ———. *Reducing Disaster Risk,* p. 36.

[68] ———. *Reducing Disaster Risk,* p. 40.

[69] ———. *Reducing Disaster Risk,* p. 47.

[70] ———. *Reducing Disaster Risk,* p. 49.

[71] *New York Daily News.* "Myanmar Government Refuses U.S. Aid." Available online. URL: http://www.nydailynews.com/news/national/2008/05/08/2008-05-08_myanmar_ government_re fuses_us_aid.html. Accessed May 7, 2010.

[72] *New York Times.* "Cyclone Nargis." Available online. URL: http://topics.nytimes.com/ topics/news/international/countriesandterritories/myanmar/cyclone_nargis/index.html. Accessed May 7, 2010.

[73] David Brand and Ann Blackman. "Soviet Union When the Earth Shook." Available online. URL: http://www.time.com/time/magazine/article/0,9171,956559,00.html. Accessed November 17, 2009.

[74] World Bank Independent Evaluation Group. *Hazards of Nature, Risks to Development: An IEG Evaluation of World Bank Assistance for Natural Disasters.* Washington, D.C.: World Bank, 2006, p. 11.

2

Focus on the United States

Each year in the United States, natural disasters cause hundreds of deaths and billions of dollars in economic losses. In 2005, the United States was one of the top three countries hit by natural disasters, with 16 of these events occurring. Only China and India suffered more natural disasters that year, with 31 events and 30 events, respectively.[1] That year on August 28, Hurricane Katrina hit the southern coast of the United States. By the end of the storm, more than 1,800 people had lost their lives and more than $81 billion in damages had occurred.[2]

No U.S. state or territory is free of the risk of natural disasters. For example, according to the U.S. Geological Survey (USGS), landslides occur in every state. Each year, they cause 25–50 deaths and $3.5 billion in damages. While the frequent wildfires in California receive a great deal of media coverage, wildfires occur throughout the country. In 2004, 40 states suffered wildfires that burned a total of 8 million acres.[3] Besides threats of earth and fire, water disasters pose hazards nationwide. More than half of the country's population lives within 50 miles of a coast. Many of these coastal areas, especially along the Atlantic and Gulf coasts, face the threat of hurricane. The mainland's west coast, Hawaii, Alaska, and island territories in the Caribbean and Pacific face the threat of tsunami. An average of 1,000 tornadoes strike the United States each year, with approximately 20 of those being violent.[4] A tornado can occur at any hour of the day or night and at any time of year if conditions are right. While most tornadoes occur in Tornado Alley in the southern plains and in Florida, tornadoes have occurred in each state in the country. When it comes to earthquakes, Alaska faces the nation's greatest hazard; however, earthquakes have occurred throughout the United States. In the past 30 years, all states except Connecticut, Delaware, Florida, North Dakota, and Wisconsin have experienced at least one earthquake with a magnitude of 3.5 or greater.[5]

In an effort to minimize the deaths and damages associated with natural disasters, a number of U.S. organizations work to predict disasters, monitor disasters as they occur, and reduce future disaster risk. Their tools include computer networks, weather- and Earth- and ocean-observing satellites, specialized sensors that measure deformation rates and vibrations of the Earth, and ocean-observing systems that detect tsunamis and monitor tides and wave amplitudes. Strategies such as these cannot prevent disasters from occurring, but they can help mitigate deaths and damages.

NATURAL DISASTERS IN U.S. HISTORY

Throughout U.S. history, natural disasters have challenged its citizens to overcome destruction and devastation on the small and the large scale, from the efforts of individuals in local communities to the efforts of the powerful federal machine known as the Federal Emergency Management Agency (FEMA). Some of the noteworthy natural disasters to strike the nation include the Peshtigo Fire of 1871, which burned more than a million acres in northeastern Wisconsin and upper Michigan and killed an estimated 1,200 to 2,400 people. It is the deadliest fire in recorded U.S. history. The Blizzard of 1888 brought 40–50 inches of snow to the New York City region, including New York City and parts of New Jersey, Massachusetts, and Connecticut. The storm killed more than 400 people, grounded or wrecked more than 200 ships,[6] and caused $25 million in damages.[7] The Great Flood of 1993 brought major or record-breaking flooding to nine states: North Dakota, South Dakota, Nebraska, Kansas, Minnesota, Iowa, Missouri, Wisconsin, and Illinois. The disaster killed 50 people and caused $15 billion in damages.[8]

Natural disasters such as these have contributed a hefty dose of death and destruction to the nation, but they have also contributed in positive ways. Each disaster has aided our understanding of Earth's processes and the possible outcomes of them. So, too, have these disasters nudged or downright forced the nation to cope, rebuild, and better prepare for future natural hazards. A closer examination of a representative group of disasters can help illustrate the nation's history of disaster and response, including examinations of the Johnstown Flood of 1889, the Galveston hurricane of 1900, the 1906 San Francisco earthquake, the dust bowl of 1934–41, the eruption of Mount St. Helens in 1980, and Hurricane Katrina in 2005. These events represent a range of types of disasters that have struck the nation and illustrate the many different controversies and responses that followed in the wake of these catastrophes.

Johnstown Flood, 1889

On May 31, 1889, floodwaters burst the South Fork Dam above Johnstown, Pennsylvania. The Conemaugh River rushed forward in the form of a 150-foot-high wall of water. It slammed through several villages before reaching Johnstown an hour later. In the space of an hour, 2,500 people lost their lives and thousands of others became homeless.[9]

The earthen dam had been built 14 miles upstream of Johnstown in the 1820s as part of a system of canals that would allow barges to haul coal, iron, and limestone from mines and quarries near Johnstown to Pittsburgh. The South Fork Dam was 272 feet thick at its base, narrowing to 10 feet thick at the top. It stretched 931 feet wide and had a spillway along the east side that was 72 feet wide by nine feet deep. At the time, the South Fork Dam was the world's largest earthen dam, and it held back the largest artificial lake in the world. Ironically, by the time the dam and canal system were complete, they had been superseded by railway transport routes. Deemed expendable by the state of Pennsylvania, the dam was sold in 1879 to Benjamin F. Ruff, who started a hunting and fishing club on the site. The dam fell into disrepair.

In the spring of 1889, the precursors of disaster clicked into place. In April, more than a foot of snow melted, saturating the ground in a 12,000-square-mile region around the Conemaugh River. Then, in late May, a storm dumped heavy rains on the soggy earth. Waterways throughout the region were full to overflowing, including the river and lake above Johnstown. On May 31 at 3:10 P.M., the dam burst, sending 4.5 billion gallons of water roaring into the valley.

The wall of water rushed past the village of South Fork, built on high ground. It crashed through Mineral Point, destroying the buildings and killing 16 people. As the water rocketed along, it carried the debris of its own destruction: trees, lumber, animals, and the bodies of people. The water crashed through the Pennsylvania Railroad yard, washing away 33 locomotives, 315 freight cars, and 18 passenger cars. Next the water hit Woodvale, where it destroyed more than 1,000 buildings and houses and killed nearly 1,000 people. From there, the water crashed through the Gautier Works, a steel mill that was part of the Cambria Iron Company, completely destroying it.

Just after 4:00 P.M., the churning mass of water and debris hit Johnstown. The destruction was total. Ten minutes later, more than 1,000 people had lost their lives, and the buildings of the town were ripped apart. The floodwaters came to a crashing stop just past Johnstown, against a 32-foot stone railroad bridge. Masses of wood from buildings, trees, animals, people (some still alive), and freight from railroad cars jumbled together, and fires broke out in the mess. Some people were pulled to safety, but others perished in that final jumble of water, debris, and fire.

This particular flood was a natural disaster that had significant aid from human actions and negligence. When Benjamin F. Ruff bought the dam and its reservoir in 1879, he built the South Fork Fishing and Hunting Club on the real estate. He lowered the top of the dam by 20 feet in order to put a road through, and he built a clubhouse where wealthy folks who had purchased $2,000 memberships could socialize. He built luxury cottages where they could stay while enjoying their outdoor pursuits. Despite these expensive developments, the dam itself received shoddy repair and maintenance, consisting of patching holes with underbrush and removal altogether of the discharge pipes. This, together with the lowered crest and other maintenance shortcuts, created a clear flood hazard that, though evident, went unaddressed. Floods that did occur prompted Daniel J. Morrell, president of Cambria Iron Company in Johnstown, to hire an engineer to inspect the dam. In the face of the engineer's damning report, Ruff nevertheless assured everyone that they were in no danger from the dam. Nine years later, the dam burst and took its deadly toll.

Immediately following the disaster, volunteers and supplies poured into the valley. Throughout the difficult summer of 1889, $3 million in money, medical supplies, clothing, and other materials was put to use in the cleanup and rebuilding of the disaster area. Clara Barton, who had founded the American Red Cross eight years earlier, distributed half a million dollars in money and supplies and organized the construction of temporary housing for residents who had lost their houses. While the efforts of that summer addressed the worst of the damage, cleanup of the Johnstown area took years. For months, bodies continued to be discovered. Most people placed the blame for the disaster on the South Fork Fishing and Hunting Club and its negligence in maintaining the dam. Despite popular opinion, neither Benjamin F. Ruff nor his club were ever held legally accountable for the flood.

Galveston Hurricane of 1900

Galveston, Texas, is a city located 50 miles southeast of Houston on Galveston Island in the Gulf of Mexico. In 1900, Galveston was the fourth wealthiest city in the United States[10] and had a population of more than 40,000 people. On September 8, 1900, a powerful tropical hurricane struck, with wind speeds that reached 120 miles per hour (MPH) before destroying the weather bureau's anemometer, making additional measurements impossible. By today's standards, the storm would have been at least a Category 3 storm on the Saffir-Simpson Hurricane Wind Scale. By the time most people under-stood the danger they were in, the storm had knocked out two bridges to the mainland and swamped a causeway. As a result, residents were trapped in the city. In Galveston, the storm killed 6,000 people and injured another 6,000. Historians place the storm's total death toll at twice that, at 10,000–12,000 deaths, ranking the hurricane as the deadliest natural disaster in U.S. his-tory.[11] Tens of thousands of people lost their homes, and more than half of Galveston's taxable property was destroyed.[12] In all, the storm caused $30 million in economic damages.[13]

The storm caught the city off guard. In 1900, the United States did not have a system of detecting and tracking storms at sea. Between August 30 and September 3, the U.S. Weather Bureau had tracked the hurricane as it hit the islands of Antigua, Haiti, and Cuba in the Caribbean Sea. Once the hurricane left Cuba, however, the Weather Bureau was unable to track it. The bureau predicted that the storm would hit the U.S. coast east of Galveston.

The brothers Isaac and Joseph Cline operated the Galveston branch of the U.S. Weather Bureau. On September 7, the day before the hurricane struck Galveston, the Clines concluded from wind directions that the hurri-cane was headed toward the city, not to the east of it. The next morning, however, the U.S. Weather Bureau sent word that the hurricane would not hit land. By now, the city was already flooding, but people went to work as usual. Others went down to the coast to watch the huge waves crash onto shore. Despite the official Weather Bureau prediction, Dr. Isaac Cline tele-graphed warnings of the hurricane to cities all along the Gulf Coast.

By afternoon, half of Galveston was flooded. That evening, fierce winds tore the roofs off buildings and hurled debris through the air. The Weather Bureau's anemometer registered 120 MPH winds just before the storm ripped it from its position and blew it away. By now, people were desperate to find safety, but the city was already cut off from the mainland. Many people were killed by falling and flying debris as they waded through the streets. Others drowned or were buried under collapsing buildings. After the storm, looters moved in swiftly, worsening the damage. Dr. Isaac Cline sur-vived the storm, and the U.S. Weather Bureau appointed him to direct the nation's first scientific study of hurricanes. During the storm, Cline had dem-onstrated that the bureau's understanding of hurricanes and its ability to anticipate a hurricane's path were far from proficient. Just before noon on the day of the storm, the U.S. Weather Bureau had predicted that the hurricane would stay out at sea and had directed Cline to issue wind warnings accord-ingly. Instead of changing the wind warnings as he had been directed to do, Cline wired hurricane warnings to towns along the coast. His last wire of the day, just before the storm tore down the telegraph wires, was to the U.S. Weather Bureau, stating that the storm had struck Galveston.

Despite the overwhelming devastation the hurricane inflicted on the island, residents of Galveston sprang into action on the day after the storm. The historian David G. McComb reports that church bells called people to worship the morning after the storm. By the end of that first day, Mayor Walter C. Jones had established the Central Relief Committee. During that first week, the city managed to restore telegraph and water services, and workers had begun laying new lines for telephone service. By the third week, the city had electric trolleys running again, and ships had begun operation again in the harbor.[14] As part of the relief effort, the American Red Cross founder Clara Barton arrived with a contingent of volunteers. In what was to be her last disaster relief mission, the 78-year-old helped to construct an orphanage for young victims of the storm and to acquire lumber for the rebuilding of survivors' homes. Other relief groups arrived from Houston. Besides the enormous efforts by ordinary citizens to mend and rebuild, engineers tackled the question of how to prepare for and mitigate a similar disaster in the future. In 1901, the city of Galveston hired civil engineers Alfred Noble, Henry M. Robert, and H. C. Ripley to determine how best to protect the city. On their recommendations, the city began the Herculean task of raising the city 17 feet higher at the seawall (yet to be constructed) and higher in town as needed. Over the next decade, workers raised utilities higher—sewer pipes, water pipes, gas lines—and jacked up structures, from homes to office buildings, across 500 city blocks. Some structures were raised a few inches while others were raised as high as 11 feet. Section by section, the ground level was built up by pumping in a slurry of sand and water dredged from the ship channel. In addition, the city began construction of the seawall itself. Between 1902 and 1904, Galveston built the first section of seawall, and the U.S. Army Corps of Engineers added to the seawall in 1905. Over the following 56 years, the seawall grew longer as workers added to it, section by section, until in 1961 it was more than 10 miles long. The cost of raising the sea level of the island and building the seawall is estimated to be $16 million.[15]

1906 San Francisco Earthquake

On April 18, 1906, an earthquake along the San Andreas Fault brought San Francisco, California, to a fiery ruin. The first tremors of the quake occurred at 5:13 A.M. and lasted for 40 seconds. The second set of tremors lasted a terrifying 75 seconds. The earthquake registered 8.3 on the Richter scale, although later this number was revised to 7.8. All over the city, fires broke out, either from earthquake damage to gas lines or from attempts by fire-fighters to stop fires by using dynamite to create a backfire. An estimated 700 people died (some estimates place the death toll as high as 3,000[16]), 500 city blocks were destroyed, and $400 million to $500 million in damages occurred.[17]

The destruction shocked and horrified people. All over the city, buildings toppled, crumpled, shifted, and collapsed. People were crushed by buildings, trapped in cracks in the Earth, and burned by the waves of fire that rushed through the city. Countless looters were shot, adding to the death toll. In Chinatown, thousands of rats teemed out of burning buildings and underground tunnels. Many of these rodents carried bubonic plague. Over the next year, more than 150 cases of bubonic plague due to rat bites were reported.

The fire burned for three days before firefighters were able to create a successful backfire to stop it. Assisting the firefighters were military troops from the nearby Presidio garrison. General Frederick Funston had declared martial law, ordering the troops to patrol the city, where they fought fires and shot looters. In declaring martial law, General Funston had acted without proper authority. The city's mayor, Eugene Schmitz, also acted without authority when, after the declaration of martial law, he authorized the general's action. Their actions joined countless breakdowns of law and order.

Most banks had lost their deposits to looters and fire, but the Bank of Italy had saved $80,000 of its deposits. From this fund, people of the city obtained loans to begin the long process of rebuilding. The financial toll of reconstruction, however, would far exceed this figure. In less than two years, the cost of reconstruction reached $90 million.[18] According to the Multidisciplinary Center for Earthquake Engineering Research (MCEER), "San Francisco received approximately $9 million in relief from individuals, cities, states, the federal government, and other countries. . . . China and Japan made significant donations of about $250,000 each, but racial prejudice kept these funds from reaching Chinese- and Japanese-American quake victims."[19]

The discrimination against Asian residents of the city was not limited to the distribution of relief funds. Chinese-Americans were unwelcome in relief camps and consequently were moved from camp to camp. In addition, during rebuilding the city attempted—unsuccessfully—to seize the valuable real estate occupied by Chinatown and relocate this cultural mecca.

More than 225,000 of the city's 400,000 residents were homeless following the disaster, and the provision of shelter and basic care were top priorities of relief workers. The disaster had crossed all social and economic boundaries, and rich and poor alike awaited assistance. The chief providers of relief efforts were the San Francisco Red Cross and Relief Corporation and the U.S.

Army, who offered medical treatment and organized soup kitchens. The army set up 21 refugee camps filled with tents. Later, the tents were replaced with shacks, and still later the residents of the shacks were offered the opportunity to buy the shacks and move the dwelling to a home site. Rebuilding the city was not only costly, but the process brought a spotlight to bear on the question of building codes and city planning for emergencies. According to MCEER, "After the earthquake, concerned citizens, government officials, and business leaders argued the need to develop meaningful building codes and construction processes—the dialogue yielded nothing substantive. The insurance industry struggled mightily to ascertain damages and to fulfill policies. The losses were unrivaled. . . . [T]he 1906 Earthquake and Fires forever altered the way insurance and building industries conducted business."[20]

The disaster produced an urgent concern in the American scientific community for achieving a more advanced understanding of seismology. According to the USGS, "The 1906 earthquake marked the dawn of modern scientific study of the San Andreas Fault system in California. . . . When the 1906 earthquake struck on April 18, nearly all scientists in California began to assemble observations of the earthquake and its effects." One outcome was the California governor's establishment of the State Earthquake Investigation Commission. "Thus, the first integrated, government-commissioned scientific investigation into earthquakes in the U.S. was launched. . . . The Commission's final report (published in 1908, and now commonly referred to as the Lawson report) was an exhaustive compilation of detailed reports from more than twenty contributing scientists on the earthquake's damage, the movement on the San Andreas Fault, the seismograph records of the earthquake from around the world, and the underlying geology in northern California. . . . To this day, the report remains a document of the highest regard among seismologists, geologists and engineers—a benchmark for future, integrated investigations into the effects of earthquakes in the U.S."[21]

Dust Bowl of 1934–1941

The dust bowl is the name given to the years of drought and famine in Kansas, Oklahoma, Texas, Arkansas, and Missouri in the 1930s that affected an estimated 50 million to 100 million acres of land.[22] More than 300,000 farmers and their families were driven off their land by the combined effects of the drought and the economic hardships of the Great Depression.[23] The dust bowl, spanning years, caused the deaths of thousands of people. Never before had the United States seen a disaster like this one on its own soil.

The disaster gathered force from a handful of factors, including drought and poor land management practices. According to a report published by the

Earth Institute at Columbia University, "Much of the Plains had been plowed up in the decades before the 1930s as wheat cropping expanded west. Alas, while natural prairie grasses can survive a drought the wheat that was planted could not and, when the precipitation fell, it shriveled and died exposing bare earth to the winds. This was the ultimate cause of the wind erosion and terrible dust storms that hit the Plains in the 1930s."[24]

Faced with crop failures and financial ruin, farmers sought assistance from the U.S. government. In 1933, the Emergency Farm Mortgage Act allotted $200 million for refinancing mortgages to help farmers facing foreclosure, and the Farm Credit Act of 1933 established 12 district banks and numerous local credit associations to offer short-term loans for agricultural production and low-interest rates to farmers who faced foreclosure. According to a 1937 report, 21 percent of rural families in the Great Plains at that time were receiving government assistance. The financial assistance provided by the government throughout the dust bowl is estimated to be as high as $1 billion.[25]

December 1934 was a turning point. For six months, winds raked across the land, drying up every drop of moisture, emptying ponds and stock tanks, and whipping away any remaining seeds. Massive clouds of dust rose up and blocked the sun, moving across the land. Cattle died from suffocation, their noses full of dirt. People died from breathing in too much dirt. Unable to grow crops and with their cattle dying, farmers lost their incomes. When they could not pay the mortgages on their houses and land, banks took the property. As if all that was not bad enough, the 1930s was the time of the Great Depression in the United States, when people across the nation lost money in the stock market and suffered lowered wages and lost jobs.

In desperation, farmers gathered their families and left the ruined Midwest plains. Known as Okies, they went to California hoping to find work in cotton fields, orchards, and farms there. Few found jobs, and most faced resentment or hostility from Californians, who were in financial straits themselves. The Okies lived in shacks and tents and trucks along roadsides and in empty lots, with nowhere to go and no way to earn a living. By the time steady rain began to soften and heal the land in the Midwest, thousands of people had died.

Even before the drought and dust storms were over, the U.S. government took steps to mitigate the disaster. In 1934, as part of the New Deal, President Franklin D. Roosevelt signed the Taylor Grazing Act, enabling the government to take up to 140 million acres of federally owned land out of the public domain and turn the acreage into grazing land in an attempt to halt the damage of overuse. In 1935, Congress established the Soil Conservation Service (SCS) in the Department of Agriculture (moving and renaming the Soil Erosion Service that had been formed in 1932 in the Department of the Interior). Led by Hugh Hammond Bennett, a soil conservation pioneer, the SCS developed conservation programs to retain topsoil and reduce damage to the land. For example, the Soil Conservation and Domestic Allotment Act of 1936 enabled the government to pay farmers to reduce the acreages they farmed in order to conserve soil and stop erosion. The act educated farmers about new farming methods such as strip cropping, terracing, crop rotation, contour plowing, and cover crops that would help prevent topsoil from being blown away. These farming methods are still in use today. In 1994, under President Bill Clinton, the SCS was renamed the Natural Resource Conservation Service (NRCS).

Eruption of Mount St. Helens, 1980

Mount St. Helens is an active stratovolcano located in Washington State, in the Cascade Range. A stratovolcano is a tall cone-shaped mountain formed of layers, called strata, of hardened lava and volcanic ash. Stratovolcanoes are associated with subduction zones, places where one plate is forced under another.

On March 16, 1980, a series of small earthquakes shook Mount St. Helens. Over the next 11 days, hundreds more small earthquakes followed. On March 27, steam blasted through the volcano's ice cap, and over the next week, due to steam explosions, the summit's crater widened to 1,300 feet across. The earthquakes continued—more than 10,000 occurred by May 17.[26] On the north-facing side of the volcano, a huge bulge formed, a sign that magma was rising inside the volcano.

At 8:32 A.M. on May 18, a magnitude 5.1 earthquake shook Mount St. Helens. Seconds later, the volcano erupted, blasting out a column of ash that reached 15 miles into the atmosphere within 15 minutes. Less than an hour later, a second eruption blasted another column of debris into the air. The 520-million-ton cloud of ash spread across the United States in three days. It circled the earth in 15 days.

With the first eruption, the mountain's summit and the massive bulge on its side dropped away, creating the largest landslide in Earth's recorded history. The landslide depressurized the volcano's magma system, resulting in explosions. Rock, ash, volcanic gas, and steam rocketed upward and outward at speeds that reached 300 MPH. Then rock and ash rained down in the blast area. The second eruption occurred within the hour.

During the afternoon, avalanches of hot ash, pumice, and gas—called pyroclastic flows—gushed out of the crater at 50 to 80 MPH. The glaciers on the sides of the volcano that had not blasted away now began melting. As water flowed down the mountainside, it formed a slurry of mud and rocks called a lahar. Several lahars raced down the mountainside that afternoon. They destroyed roads, bridges, houses, and trees.

The May 18, 1980, eruptions of Mount St. Helens caused the deaths of 57 people and countless wildlife, including 7,000 big game animals and 12 million salmon fingerlings in hatcheries. Twenty-seven bridges and nearly 200 homes were destroyed. In forests, 4 billion board feet of timber was blasted down.[27] In Washington State, economic damages resulting from the eruption totaled an estimated $860 million, with $450 million of that attributed to the loss of standing timber and between $40 million and $100 million of the total attributed to agricultural losses. In the neighboring state of Oregon, the port of Portland estimated a loss of $5 million in lost revenues due to the closure of the Columbia River shipping channel for several months following the eruption.[28]

Hurricane Katrina, 2005

On August 29, 2005, Hurricane Katrina hit the southern coast of the United States. It was the third largest hurricane to make landfall in recorded U.S. history, with hurricane force winds reaching 75 miles from the storm's center.[29] With a death toll of more than 1,800 people, it is one of the deadliest hurricanes in U.S. history.[30] With more then $80 billion in estimated damages, it is the costliest hurricane in U.S. history.[31] New Orleans, the city hit hardest by the storm, lost 25 percent of its population, and the racial and socioeconomic makeup of its population have changed significantly as a result of the disaster.

On August 23, a tropical storm formed in the Bahamas, and the next day it was named Tropical Storm Katrina. On August 25, Katrina reached hurricane status as a Category 1 storm on the Saffir-Simpson Hurricane Wind Scale. Shortly thereafter, it made its first landfall, on the southeastern coast of Florida. Over the next six hours, it crossed the peninsula on a southwestern path before entering the Gulf of Mexico.

As Katrina moved northwest across the Gulf, coastal areas braced for its landfall. On August 26, the Louisiana governor Kathleen Blanco declared a state of emergency and activated the Louisiana National Guard. The next day, President George W. Bush declared a federal state of emergency for the state, and the New Orleans mayor C. Ray Nagin declared a state of emergency for his city. Evacuation of New Orleans residents began, with many residents crowding into the Louisiana Superdome, a sports facility, for shelter. By August 30, the Superdome was severely overcrowded with 20,000 people, most of them from poor and African-American communities in the city. As later reports of the storm would demonstrate, the disaster's toll "was disproportionately borne by the region's African-American community, by people who rented their homes, and by the poor and unemployed."[32]

Hurricane Katrina's strength grew rapidly. By August 27, it had strengthened to a Category 3 hurricane with wind speeds of 115 MPH. The next day it jumped to a Category 5 hurricane with winds exceeding 160 MPH. Hurricane force winds extended nearly 100 miles from the center of the storm. On August 29, the storm weakened to an upper-level Category 3 storm, still capable of violent damage. At 6:10 A.M., Katrina made landfall with sustained winds at an estimated 125 MPH. Storm surge was 25 feet high, the highest in recorded U.S. history.

Levees and floodwalls that protected New Orleans began to breach, or fail. At 6:30 A.M., levees along the Intracoastal Waterway failed, flooding eastern New Orleans. Shortly after, water overflowed the floodwalls of the Industrial Canal, a 5.5-mile waterway that connects the Mississippi River to the Intracoastal Waterway. More water flowed into New Orleans, flooding the Lower Ninth Ward and other parts of St. Bernard's Parish. At 9:45 A.M., the 17th Street Canal breached. A mass of water from Lake Pontchartrain rushed into New Orleans.

At 10:00 A.M., Hurricane Katrina made landfall again, this time near the mouth of the Pearl River at the Louisiana-Mississippi border. The storm's hurricane force winds still extended nearly 100 miles from the hurricane's eye. Finally as Katrina moved north and then northeast into Mississippi, it weakened to a Category 1 hurricane.

The destruction of the massive storm shocked the nation. Storm surge—as high as 28 feet in western Mississippi and 15 feet in western Alabama—caused major or catastrophic damage along the coastlines of Louisiana, Mississippi, and Alabama.[33] The storm spawned 43 tornadoes, including 20 in Georgia and others in Mississippi, Alabama, and Florida.[34] On the Gulf Coast from Louisiana to Alabama, 1.2 million people had been under evacuation order during the storm.[35] People in five states lost their lives, most of them in Louisiana, but some in Mississippi, Florida, Alabama, and Georgia.

The city of New Orleans suffered catastrophic damage. Eighty percent of the city was flooded,[36] and breaches in levees continued to occur after Katrina's passing. Residents clung to balconies and rooftops, and the bodies of those who had drowned floated in floodwaters. Looters spread through the city. The storm had ripped the roof off the Superdome, where people had gathered for safety, but it was not until September 1 that the first busloads of evacuees left the Superdome. After the storm hundreds of thousands of local residents were left unemployed. The National Hurricane Center called Katrina "one of the most devastating natural disasters in United States history."[37]

As information and data about the disaster and its victims were gathered, one fact became glaringly apparent: the poor, the unemployed, and African Americans had suffered the brunt of the storm's devastation. John R. Logan, professor of sociology at Brown University, prepared a report explaining the disparity of suffering by race, housing tenure, and poverty and unemployment status. Logan states that "[d]amaged areas were 45.8 percent black, compared to 25.4 percent in undamaged areas" and that "45.7 percent of homes in damaged areas were occupied by renters, compared to 30.9 percent in undamaged communities." He also reports that "20.9 percent of households had incomes below the poverty line in damaged areas, compared to 15.3 percent in undamaged areas, 7.6 percent of persons in the labor force were unemployed in damaged areas (before the storm), compared to 6.0 percent in undamaged areas."[38] The racial and economic statistics that Logan and other researchers collected illuminate the fact that, while all storm victims had suffered unjustly, poor and black people bore the added injustice of being, by dint of race and/or economic status, in the wrong place at the wrong time with few or no options for adequate response. Further studies of disaster response would shed light on how this had come to be, and the federal government drew the greatest share of blame.

The U.S. government's response to the disaster and FEMA's response in particular drew sharp criticism from across the nation. The view was—and is—widely held that the government responded slowly, inadequately, and sometimes ineptly. Debarshi Chaudhuri of MIT explains that when Katrina struck, the Department of Homeland Security mobilized 1,000 workers but "[i]n an effort to organize the response, FEMA also asked that no firefighters or ambulance crew respond to areas hit by Hurricane Katrina without being first mobilized by local and state authorities, a declaration that undoubtedly slowed response to the disaster. . . . FEMA seemed almost unwilling to accept help from non-government organizations. For example, the American Red Cross was not allowed into New Orleans following the disaster and was unable to supplement the government's response."[39]

In the face of public outcry against FEMA's performance, the FEMA director Michael Brown was relieved of his duties nine days after the hurricane struck. A congressional report entitled *Failure of Initiative* was released six months later and describes the U.S. government's response as a "failure of leadership" and concludes that "the preparation for and response to Hurricane Katrina should disturb all Americans."[40] Specific areas of criticism include, among others, the inadequate levee system that led to flooding in New Orleans, the slow and incomplete evacuation of people, the slow arrival of relief supplies and equipment, confusion and delay caused by "unsubstantiated rumors and uncritically repeated press reports," the "collapse of local law enforcement," and the failure of "government at all levels" to react "more effectively" to a hazard that had been "predicted with unprecedented timeliness and accuracy."[41]

NATURAL HAZARDS IN THE UNITED STATES

During the 20th century, the population of the United States increased from 77 million in 1900 to 281 million in 2000. The first decade of the 21st century saw this figure increase by more than 7 million. With the upsurge in population and increase in physical infrastructures associated with that population comes increased vulnerability to natural hazards. While natural disasters of the past century are not notably more severe than natural disasters of the more distant past, the loss of life and economic damages resulting from natural disasters have increased. According to Munich-Re, one of the world's largest reinsurers, natural disasters cost the global insurance industry around $22 billion in 2009.[42] In contrast, in the 1950s, "the average cost of catastrophic events was a mere $3.9 billion."[43] According to some experts, "Natural disasters may be getting more expensive because more people are building more expensive infrastructures in areas that are prone to natural disasters, like coastal areas, fire-prone forests, steep mountain slopes, and riverbanks."[44]

Earthquakes

Forty-one U.S. states and territories are at moderate to high risk for earthquakes.[45] During the 20-year period of 1975–95, only four states did not have an earthquake. These states are Iowa, Wisconsin, Florida, and North Dakota. The latter two states have the fewest earthquakes in the entire United States.

Most earthquakes, and volcanoes too, occur along the boundaries of tectonic plates. The Earth has 12 major plates, including the North America Plate covering North America and northeast Siberia. As well, the Earth has numerous smaller plates, including the Juan de Fuca Plate, located between the North America Plate to its east and the Pacific Plate on its west. The San Andreas Fault forms a portion of the tectonic boundary between the Pacific Plate and the North America Plate. This fault system runs more than 800 miles in length through California. The plates on either side of the San Andreas Fault are moving in the same direction to the northwest, but at different rates. In approximately 15 million years, the cities of Los Angeles, which is on the faster-moving Pacific Plate, and San Francisco, located on the slower North America Plate, will be next to each other.

In the United States, damaging earthquakes occur most frequently in California. Each year in southern California, around 10,000 earthquakes occur. Most are too small for people to feel. A few hundred per year are greater than magnitude 3.0, and 15 to 20 are greater than magnitude 4.0. Alaska is one of the most seismically active regions in the world. More earthquakes occur in Alaska than in any other state, and most of the nation's largest earthquakes occur there. The largest recorded earthquake in the United States was a 9.2 quake that occurred at Prince William Sound, Alaska, on March 28, 1964. This earthquake was nearly as powerful as the largest recorded earthquake in the world, a 9.5 earthquake that struck Chile on May 22, 1960. Alaska experiences a magnitude 7.0 earthquake once a year, on average, and a magnitude 8.0 or greater earthquake every 14 years. Most occur in uninhabited areas.

In Utah, the Wasatch Range runs north and south, marking an active fault zone. Nearly 75 percent of Utah's population lives near the Wasatch Fault, a 240-mile-long zone made up of several segments, each of which is capable of producing up to a magnitude 7.5 earthquake. During the past 6,000 years, a magnitude 6.5 or greater earthquake has occurred along this fault about once every 350 years. According to the USGS, it has been 350 years since the last earthquake of this magnitude occurred on this fault. An earthquake on this fault is due at any time.

Tornadoes

More than 1,000 tornadoes are reported in the United States each year.[46] As tornado detection systems improve, scientists expect even more tornadoes to be reported yearly. Tornadoes occur most frequently in the central plains, east of the Rocky Mountains and west of the Appalachian Mountains. But no state is immune to tornadoes. They roar across all types of landscapes, including mountains, valleys, plains, swamps, deserts, gullies, and rivers. Tornadoes are reported most often in the spring and summer, but they can occur at any time of year under the right conditions. Tornadoes kill an average of 60 people per year, mostly as a result of flying or falling debris.

Tornado season is a term used to refer to the time of year when the most tornadoes occur. On the Gulf Coast, tornado season is in the early spring. In the southern plains, it is during May and early June. In the northern plains and the upper Midwest, tornado season is in June or July. Tornado Alley is an informal name given to the area of the United States that experiences the most numerous tornadoes, year in and year out. In 2009, in Texas, for example, 108 tornadoes were reported—and this was the state's least active year in a decade.[47] The most violent tornadoes are consistently reported in eastern South Dakota, Nebraska, Kansas, Oklahoma, northern Texas, and eastern Colorado. According to the National Severe Storms Laboratory (NSSL), the greatest number of violent tornadoes start in south-central Oklahoma.

Scientists measure the strength of a tornado on the Enhanced Fujita Scale (EF Scale), based on the degree of damage and the intensity of the wind gusts, among other factors. The EF Scale was implemented in the United States on February 1, 2007. Before that an older version of the scale, called the Fujita Scale, was used. The Enhanced Fujita Scale ranks tornadoes as shown in the chart below.

Enhanced Fujita Scale		
Scale	MPH (3-second gust)	Expected Damage
EF0	65–85	light damage
EF1	86–110	moderate damage
EF2	111–135	considerable damage
EF3	136–165	severe damage
EF4	166–200	devastating damage
EF5	more than 200	incredible damage

An EF0 tornado, for instance, may break branches or knock out windows with flying debris. Most tornadoes in the United States—74 percent of them[48]—are weak tornadoes in the EF0 or EF1 range. An EF2 tornado will damage buildings by ripping roofs completely off and knocking down exterior walls. Violent tornadoes—those in the EF3 to EF5 range—lift and toss vehicles, level houses at the foundations, and uproot trees. Only 2 percent of reported tornadoes each year are EF4 or EF5.[49] Tornadoes in the EF3 to EF5 range cause 67 percent of tornado deaths each year.[50]

Around 7 percent of the U.S. population lives in mobile homes, and these homes are the most vulnerable to tornado damage. They offer occupants little protection even if the home is tied down properly. Nearly half of tornado fatalities in the country occur in mobile homes.[51]

Hurricanes

Each year, an average of 10 tropical storms form in the Atlantic Ocean, Caribbean Sea, or Gulf of Mexico, and six of these usually turn into hurricanes. More often than not, hurricanes stay at sea. Every three years, an average of five hurricanes makes landfall in the United States. Of these five, two are usually ranked a Category 3 or above on the Saffir-Simpson Scale, meaning that the hurricane has winds of 111 MPH or greater.

In a typical hurricane year, between 50 and 100 people lose their lives as a result of a hurricane. These deaths occur anywhere from Texas to Maine, depending on the hurricane's landfall. Notable exceptions occur. In 2005, Hurricane Katrina caused the deaths of an estimated 1,800 people.

More than half of the U.S. population lives within 50 miles of a coast. The densely populated coastlines of the Atlantic and the Gulf Coast are at particular risk from storm surges that accompany hurricanes. A storm surge is water pushed toward the shore by winds of the storm. This dome of water can be as high as 20 feet and can stretch 50 to 100 miles wide. The storm surge combines with the regular tide to create a hurricane storm tide. The mean water level can increase 15 feet or more. Besides that, wind can create waves on top of the storm tide, adding more height. The result can be severe flooding in coastal areas. The Gulf Coast and Atlantic coastlines lie less than 10 feet above sea level. As a result, these areas are at great risk for storm surge flooding.

Homes and buildings in flood plains are at risk for flooding due to the many inches—sometimes nearly a foot—of rain that a hurricane can dump in just a few hours.

Volcanoes

There are 169 active volcanoes in the United States.[52] More than 40 of these are in Alaska,[53] in a volcanic chain that stretches more than 1,500 miles along the Aleutian Islands, Alaska Peninsula, and Cook Inlet. On average, one or two of these volcanoes erupts each year. Since the 1500s, when volcanic eruptions in Alaska began to be recorded, more than 230 eruptions have occurred. Today, all Alaskan volcanoes are monitored by satellite, and 27 are monitored with seismic networks.

Volcanic eruptions in Alaska are a risk to flying jet aircraft in the North Pacific. The eruption of Mount Redoubt on December 15, 1989, is a case in point. The explosion sent a column of ash 40,000 feet into the air, and the ash—consisting of rock fragments less than 1/2 inch in diameter—was carried north by strong winds. One hundred and fifty miles away from the volcano, a 747 jetliner carrying 244 people was en route from Amsterdam to Anchorage. Ninety minutes after the explosion the jetliner descended into the ash cloud, which filled the aircraft with grit and gas. All four engines failed, and the jet fell nearly 12,000 feet. When the plane was a few thousand feet from the ground, the engines restarted and the plane landed safely in Anchorage.

In Washington State, Mount St. Helens is the most active volcano, with more eruptions than any other volcano in the Cascade Range in the last 4,000 years. This volcano has erupted with powerful explosions of pumice and ash (such as the May 1980 eruption), as well as with slow, continuous extrusions of viscous lava and with eruptions of fluid lava.

Other active volcanoes in Washington include Mount Rainier, Mount Baker, and Glacier Peak. The USGS has stated that Mount Rainier is "one of our Nation's most dangerous volcanoes."[54] This 14,410-foot-high volcano is located in Mount Rainier National Park just outside Seattle. At its feet, suburban communities spread out, built on rock and earth hurtled down the mountainside in countless previous eruptions. Many of these communities sit square on the projected paths of future lahars that can accompany eruptions. Around 80,000 people and their homes are at risk in the lahar-hazard zones. Major highways and utilities also cross the zones. Nearly 2 million tourists a year visit the park. All of these are in danger of the debris flows, pyroclastic flows, lahars, and lava flows that could occur with an eruption. Moreover, the USGS has stated that Mount Rainier could erupt with little or no warning, and that the risk of a large lahar reaching these lowlands is about a one in 10 chance during the average person's life span.[55]

In eastern California, the Long Valley Caldera and the Mono-Inyo Craters volcanic chain form a long volcanic system that has been active in recent decades. Twelve eruptions have occurred in the past 1,000 years, and scientists believe that future eruptions are certain. After four magnitude 6 earthquakes occurred in May 1980, scientists found that the central part of Long Valley Caldera was slowly rising. Rising ground, together with recent earthquakes, commonly preceded volcanic eruptions.

Other active volcanic areas in California include those in the Lassen Volcanic National Park area in the northern part of the state, in the Cascade Range. Here, Lassen Peak is one of the largest lava domes on Earth.

The Hawaiian Islands are part of a chain of volcanic islands that has been 70 million years in the making. Each island of Hawaii is formed of one or more volcanoes that first erupted on the seafloor and built itself higher and higher until it pushed above sea level. Most of the active volcanoes in the Hawaiian Islands are on the island of Hawaii, which is formed of five volcanoes. Here, residents face numerous hazards, including lava flows, explosive eruptions, volcanic smog, earthquakes, and tsunamis. Most Hawaiian volcanoes erupt with lava flow as opposed to violent explosions. Still, large flows of fluid lava can travel great distances quickly.

Mauna Loa, the world's largest active volcano, last erupted in 1984. The eruption covered 16 square miles with lava in just three weeks' time. Like

Mauna Loa, Kīlauea has erupted numerous times in the past century. In 1990, lava from Kīlauea flowed into the coastal community of Kalapana, destroying and partly burying most of the town and burying two nearby towns, Kaimū and Kaimū Bay, under more than 50 feet of lava. This lava flow was part of Kīlauea's ongoing eruption, begun in 1983. Repeated lava flows have buried miles of highway, destroyed more than 180 homes, and destroyed a visitor center in Hawai'i Volcanoes National Park. Scientists warn that future damage to residential areas is likely.

In Wyoming, Yellowstone is one of the world's most active volcano systems, though it is not currently a violent system. In the past few million years, Yellowstone has produced several massive eruptions as well as numerous smaller eruptions and steam explosions. However, no eruptions of lava or ash have occurred in several thousand years. Scientists predict that over the next few hundred years, volcanic hazards in the area will be limited to geysers and hot springs, occasional steam explosions, and moderate to large earthquakes. At the Yellowstone Volcano Observatory, scientists continually monitor volcanic activity in the region.

Tsunamis

In the United States, the vulnerability to tsunamis lies mainly in the Pacific states—Alaska, Hawaii, Washington, Oregon, and northern California—and the U.S. Caribbean islands. As long ago as the year 900, a tsunami struck the Puget Sound area in northwestern Washington. As recently as 1964, a magnitude 9.2 earthquake in Prince William Sound, Alaska, caused a tsunami that struck the Pacific coasts of Alaska, Washington, Oregon, and northern California. One hundred and ten people lost their lives, including some as far away as Crescent City, California.[56]

Four years before that, in 1960, a magnitude 9.5 earthquake occurred off the coast of Chile, in South America. It caused a tsunami that raced ashore along 500 miles of the Chilean coast. The tsunami also swept across the Pacific and struck the Hawaiian Islands, killing 61 people and causing $24 million in damage. Buildings were smashed and swept from their foundations, and debris-laden waves bent over the metal poles of parking meters.[57]

In 1946, a magnitude 7.3 earthquake in the Aleutian Islands of Alaska generated a Pacific-wide tsunami. The wave killed more than 165 people and caused an estimated $26 million in damage. Much of the death and destruction occurred in the Hawaiian Islands, where more than 100 people lost their lives. The run-up, or height of the water on shore above sea level, reached nearly 40 feet on the island of Hawaii.[58]

The island of Puerto Rico is situated with the Caribbean Sea on its south and the Atlantic Ocean on its north. Bunce Fault, a fault system similar to the San Andreas Fault in California, occurs in very deep water near the island and poses a significant tsunami threat. In 1918, a magnitude 7.5 earthquake just off the island's northwestern coast caused a tsunami that struck the island. Damage from the earthquake and tsunami totaled $4 million, and 116 people lost their lives.[59]

In the past century, tsunamis that have struck the U.S. Pacific coast have been distant tsunamis, meaning they were generated by earthquakes in the Pacific basin. In addition to being at risk of distant tsunamis, the Pacific Northwest is at risk of local tsunamis. The Pacific Northwest is the area where the Juan de Fuca Plate, an oceanic tectonic plate, is being subducted under the North America Plate, a continental plate. This is called the Cascadia subduction zone. Earthquakes that occur along this fault line may generate local tsunamis. In recorded history, only one major earthquake has occurred in this zone; however, the USGS warns that the area may be "poised between major earthquakes"[60] and is therefore a potential threat.

Extreme Weather

Each year in the United States, extreme weather events such as droughts, floods, heat waves, and blizzards account for many billions of dollars of economic damage and loss of life. In 2009, parts of the Southwest, Great Plains, and southern Texas suffered drought conditions for much of the year, with damages estimated at $5 billion.[61] That same year, flooding in the United States caused $1 million in damages, a figure much lower than the $6.9 million in flood damages in 2008.[62] In July 2006, a heat wave brought more than 50 record-breaking high temperatures to areas of the central and western United States,[63] killing 22 people in 10 states.[64] In 2009, extreme winter weather (winter storms, ice, and avalanches) caused more than $1.5 billion in damages and killed 35 people.[65]

Experts predict that weather disasters, including extreme weather events, will increase in the coming years from climate change. According to a report of NASA's Earth Observatory, "Global climate models do predict that a warmer climate could lead to higher sea levels and coastal flooding, more intense storms, deadly heat waves, and more extreme flood-drought cycles in the twenty-first century."[66] Consequences of these disasters can include higher rates of fatality, such as by increased deaths in heat waves; increased economic damages and losses, such as by lost revenues in winter resort areas experiencing early spring; and human displacement, both temporary and permanent, such as by flooding.

FORECASTING NATURAL DISASTERS

Natural disasters cannot be prevented from occurring. However, in an effort to reduce the deaths and damages that occur, scientists have developed extensive technology for forecasting when and where a disaster might strike and for monitoring disasters in progress.

Earthquakes

The USGS continually monitors earthquake activity in the United States and Puerto Rico. Despite extensive research, scientists have not discovered a reliable way to predict earthquakes. Based on past earthquake activity, the location and extent of faults, plate tectonics, and other factors, the USGS creates and regularly updates National Seismic Hazard Maps. These maps show risk areas and shaking levels of earthquakes that could occur in the United States. The maps are used in the design of buildings, bridges, highways, and utilities that can withstand earthquakes. The maps are also used to help create and update building codes and to help insurance companies set rates.

Though the exact time and place of an earthquake cannot be forecast, the possibility of an earthquake's occurring on a given day is 100 percent. In fact, several hundred earthquakes, most of which register only on sensitive equipment, occur each week. Every year, a few hundred of these are large enough to cause significant damage, though these typically occur in unpopulated areas.

In an attempt to more narrowly define the probability of where damaging earthquakes could occur, a team of scientists and engineers called the Working Group on California Earthquake Probabilities (WGCEP) created the Uniform California Earthquake Rupture Forecast. In the group's statewide forecast, issued in April 2008, they predicted the chance of a magnitude 6.7 or greater earthquake as 63 percent in the San Francisco region and 67 percent in the Los Angeles area over the next 30 years. Such an earthquake would cause damages similar to the 1989 Loma Prieta earthquake near San Francisco, which had a magnitude of 6.9 and caused an estimated $6 billion in damages. Over the same period, the forecast places a 2 percent likelihood of San Francisco's experiencing a magnitude 8.0 earthquake similar to the earthquake that devastated the area in 1906.[67] WGCEP was the fifth such forecasting group in California since 1988, but it was the first group to make predictions statewide as opposed to specific faults or regions only. It is likely that a similar working group will be formed in coming years to build upon existing forecasts using new and updated methodologies.[68]

For the Cascadia subduction zone off the coasts of Washington and Oregon, the USGS forecasts a 4.5 percent chance of a megathrust earthquake over the next 50 years.[69] Such an earthquake would have a magnitude similar to the 2004 earthquake in the Indian Ocean off the coast of Sumatra. It would cause a massive tsunami that would strike the coasts of Washington, Oregon, California, and Alaska. The wave would travel across the Pacific Ocean and potentially endanger the eastern coasts of Asia and Australia and the western coast of South America.

Volcanoes

Much like earthquake prediction, the exact prediction of volcanic eruptions is impossible. However, volcanoes typically exhibit many warning signs before a major event occurs. By watching closely for changes on the mountain, scientists form educated estimates of the likelihood of an eruption, and they issue an alert level for each volcano in the United States. In a non-eruptive state, a volcano rated "normal" poses little danger. As the volcano becomes more active, the alert level rises to "advisory" and then to "watch," notifying residents to prepare for an eruption. The final level, "warning," signifies that a dangerous eruption is either occurring or imminent.

To determine the current state of a volcano, scientists use a variety of monitoring techniques. As molten rock, or magma, and gases flow beneath the mountain's surface, rocks fracture and vibrate, causing swarms of small earthquakes. Seismometers placed around a volcano measure these vibrations. Volcanologists interpret the minute tremors as debris flows, rock falls, glacier movement, or tectonic earthquakes, depending on the type of signature recorded by the equipment. Scientists also monitor the levels of gases being released by the volcano, especially carbon dioxide and sulfur dioxide. Changes in the composition of the gases escaping the volcano may result from the movement or pressurization of magma changes within the volcano. The most thorough method of gas sampling requires a visit to the volcano, which may be too dangerous. In some cases, equipment is installed that continuously monitors gas emissions.

Extreme Weather

The National Weather Service (NWS) Storm Prediction Center (SPC) monitors the United States for severe weather, such as tornadoes. This agency continuously watches changes in temperature and wind patterns that could lead to the four ingredients needed for a tornado to form—moisture, instability, lift, and wind shear. Forecasters compile observations from satellites, weather stations, balloons, aircraft, and radar to create the weather outlook. Using a variety of tools, from hand-drawn analyses to sophisticated computer programs, the SPC issues warnings for areas where severe weather is likely. The SPC also monitors heavy rainfall that could cause flooding, heavy snowfall, and fire danger levels.

The NWS National Hurricane Center (NHC) is responsible for monitoring hurricanes that could strike the United States. When a tropical cyclone forms in the Atlantic Ocean or the Pacific Ocean, the NHC continuously tracks the storm, using information collected from satellites, buoys, reconnaissance aircraft, and radar to predict its path. As the storm progresses, the NHC issues warnings for areas likely to be impacted by high winds, storm surge, tornadoes, or flooding.

DISASTER PREPAREDNESS AND RESPONSE

Preparing for potential catastrophes requires a cooperative effort between the people living in a hazardous area and local, state, and national governments. The levels of responsibility required for each depend on the type of hazards in the area. A large part of the governmental role in disaster preparedness involves education programs that teach people how to prepare for an emergency and what to do when one occurs.

Personal Preparedness

Some preparations people need to take are similar for most disasters. Families in disaster-prone areas need to create a disaster kit containing flashlights, extra batteries, emergency food, extra water, a first aid kit, and essential medications. During a disaster, large portions of a city or region may lose electricity or communications and roads may become impassable. Such outages may last for days or even weeks. In addition, outages can happen as the result of wind storms, flooding, and winter weather. Though these events present few dangers, residents must be prepared to weather the storm when they occur.

Surviving more serious natural disasters requires additional preparation. People living in areas where earthquakes are likely to occur need to secure pictures, mirrors, canned goods, light fixtures, or any other household items that could cause injury during an earthquake. In areas likely to be struck by a hurricane, residents must secure their properties with storm shutters or plywood to protect against high winds. People living near volcanoes or coastal areas where tsunamis or hurricanes may occur must be ready to evacuate. While evacuations may be advised or required days or sometimes months ahead of a disaster, the warning might also come suddenly, giving only minutes to move to safety.

Emergency Response

When major natural disasters occur, personal preparedness only goes so far. After following evacuation orders, many people may find themselves stranded. Other victims may be trapped by floodwaters, buried under debris, or in need of medical care. Emergency workers such as firefighters, emergency medical staff, police officers, city workers, and local volunteers go to work immediately to provide assistance to those in need. If necessary, the governor of the state may declare a state of emergency and direct additional resources, including the state's National Guard, to assist in the relief effort.

When relief measures exceed the abilities of local and state governments, the governor of the state requests a major disaster declaration. In response, FEMA dispatches a team that performs a damage assessment. From the information gathered by this assessment, the president of the United States issues a presidential major disaster declaration. Though this process may sound complex, it can take only a few hours to implement for a major disaster.

Once a presidential major disaster declaration is issued, victims gain access to long-term recovery programs provided by the federal government. Emergency housing may be provided for up to 18 months. Victims may receive grants to cover personal property losses, transportation, medical needs, or funeral expenses. Uninsured homeowners and renters gain access to low-interest loans from the Small Business Administration (SBA).

For major disasters, the FEMA Public Assistance (PA) Grant Program typically pays 75 percent of the costs associated with debris removal, emergency protective measures, and the repair or replacement of public facilities such as schools.[70] In the 10-year period from 1999 to 2008, FEMA handled an average of 50 major disasters per year[71] and spent approximately $58 million per disaster for public assistance.[72] In 2009, FEMA declared 59 major disasters for catastrophes ranging from severe winter storms to flooding, tornadoes, and landslides and including a major disaster in American Samoa, which had suffered an earthquake, tsunami, and flooding in September.[73]

Government Agencies for Preparedness and Response

Disaster preparedness and response in the United States is a complex blend of efforts from diverse agencies. Governmental departments and agencies play a huge role in this arena. The government's Web site DisasterAssistance.gov serves as a portal for coordinating the disaster assistance of 17 governmental entities, including such diverse agencies as the Department of Agriculture, the Department of Homeland Security, the Social Security Administration, and the Department of Transportation. In the wake of a disaster, victims can go to this centralized source to apply for and receive assistance from one or more of these agencies.

At the forefront of the nation's response to and management of disasters, both natural and human-caused, is FEMA, the agency that coordinates the federal government's responsibilities in disaster preparedness, prevention, mitigation, response, and recovery. FEMA is an agency of the Department of Homeland Security. FEMA was founded in 1979, but the roots of the agency reach back to the Congressional Act of 1803, legislation that provided assistance to a New Hampshire town that had suffered extensive damage from a fire. For the next century, the government legislated disaster assistance on an ad hoc basis in response to hurricanes, earthquakes, floods, and other natural disasters. The 1930s saw the formalization of some federal disaster aid through the Reconstruction Finance Corporation, the Bureau of Public Roads, the U.S. Army Corps of Engineers, and other agencies. Early in the 1960s, most federal disaster assistance was brought under the Department of Housing and Urban Development (HUD). Within HUD, the Federal Disaster Assistance Administration was established. This agency cut its teeth on a series of natural disasters in the coming years, including Hurricane Carla (1961), the Alaskan Earthquake (1964), Hurricane Betsy (1965), Hurricane Camille (1969), the San Fernando Earthquake in California (1971), and Hurricane Agnes (1972). In 1974, the Disaster Relief Act provided for the presidential declaration of disaster.

Despite the creation of the Federal Disaster Assistance Administration, federal response to disasters still originated from a diverse array of agencies. According to FEMA's Web site, "When hazards associated with nuclear power plants and the transportation of hazardous substances were added to natural disasters, more than 100 federal agencies were involved in some aspect of disasters, hazards and emergencies. Many parallel programs and policies existed at the state and local level, compounding the complexity of federal disaster relief efforts."[74] As a result of the increasing complexity of federal disaster response, President Jimmy Carter issued Executive Order 12127, establishing FEMA and gathering many federal responsibilities pertaining to disaster response within this agency.

FEMA emerged in the national spotlight as it managed the government's responses to environmental disasters such as the toxic waste contamination of Love Canal, a neighborhood in Niagara Falls, New York, revealed in 1976, and the meltdown at the Three Mile Island nuclear power plant in 1979. Natural disasters in FEMA's early years include the Loma Prieta Earthquake in 1989 and Hurricane Andrew in 1992. In 2001, the terrorist attacks of September 11 brought national attention to the link between governmental response to disaster—in this case a human-caused disaster—and the nation's security. Consequently, two years later "FEMA joined 22 other federal agencies, programs and offices in becoming the Department of Homeland Security."[75] Operating as an agency within the Department of Homeland Security, FEMA received its most recent improvement in the wake of the fallout of FEMA's response to Hurricane Katrina in 2005. In 2006, President George W. Bush signed the Post-Katrina Emergency Reform Act, which "significantly reorganized FEMA, provided it substantial new authority to remedy gaps that became apparent in the response to Hurricane Katrina in August 2005, . . . and included a more robust preparedness mission for FEMA."[76] Today, FEMA continues to operate with the vision "A Nation Prepared."

[1] International Strategy for Disaster Reduction. "2005 Disasters in Numbers." Available online. URL: http://www.unisdr.org/disaster-statistics/pdf/2005-disaster-in-numbers.pdf. Accessed November 20, 2009.

[2] U.S. Department of Health and Human Services. "Hurricane Katrina." Available online. URL: http://www.hhs.gov/disasters/emergency/naturaldisasters/hurricanes/katrina /index. html. Accessed November 20, 2009.

[3] United States Geological Survey. "Natural Hazards—A National Threat." Available online. URL: http://pubs.usgs.gov/fs/2007/3009/2007-3009.pdf. Accessed November 20, 2009.

[4] National Climatic Data Center. "U.S. Tornado Climatology." Available online. URL: http:// www.ncdc.noaa.gov/oa/climate/severeweather/tornadoes.html. Accessed May 10, 2010.

[5] United States Geological Survey. "Earthquake Hazards Program." Available online. URL: http://earthquake.usgs.gov/earthquakes/states/last_earthquake.php#30years. Accessed May 10, 2010.

[6] National Climatic Data Center. "The Big One! A Review of the March 12–14, 1993 'Storm of the Century' [with comparisons to the Blizzard of 1888]." Available online. URL: http:// www1.ncdc.noaa.gov/pub/data/blizzard/blizz.txt. Accessed May 10, 2010.

[7] Borgna Brunner. "The Great White Hurricane: The Blizzard of 1888: March 11–March 14, 1888." Available online. URL: http://www.infoplease.com/spot/blizzard1.html. Accessed May 10, 2010.

[8] Lee W. Larson. "The Great USA Flood of 1993." Available online. URL: http://www.nwrfc. noaa.gov/floods/papers/oh_2/great.htm. Accessed May 10, 2010.

[9] Lee Davis. *Natural Disasters, Revised Edition.* New York: Facts On File, 1992, p. 183.

[10] ———. *Natural Disasters,* p. 288.

[11] Michael A. Smith. "Post-Storm Rebuilding Considered 'Galveston's Finest Hour.'" Available online. URL: http://www.1900storm.com/rebuilding/index.lasso. Accessed May 12, 2010.

[12] Davis. *Natural Disasters,* p. 288.

[13] Smith. "Post-Storm Rebuilding."

[14] ———. "Post-Storm Rebuilding."

[15] ———. "Post-Storm Rebuilding."

[16] Donald Hyndman and David Hyndman. *Natural Hazards and Disasters,* 2nd ed. Florence, Ky.: Brooks Cole, 2008, p. 89.

[17] Davis. *Natural Disasters,* p. 92.

[18] Multidisciplinary Center for Earthquake Engineering Research. "The Great 1906 Earthquake and Fires of San Francisco." Available online. URL: http://mceer.buffalo.edu/1906_Earthquake/san-francisco-earthquake.asp. Accessed May 12, 2010.

[19] ———. "The Great 1906 Earthquake and Fires of San Francisco."

[20] ———. "The Great 1906 Earthquake and Fires of San Francisco."

[21] United States Geological Survey. "1906 Marked the Dawn of the Scientific Revolution." Available online. URL: http://earthquake.usgs.gov/regional/nca/1906/18april/revolution.php. Accessed May 12, 2010.

[22] NOAA National Weather Service Forecast Office. "What Is Meant by the Term Drought?" Available online. URL: http://www.wrh.noaa.gov/fgz/science/drought.php?wfo=fgz. Accessed May 10, 2010.

[23] Davis. *Natural Disasters,* p. 125.

[24] Ben Cook, Ron Miller, and Richard Seager. "Did Dust Storms Make the Dust Bowl Drought Worse?" Available online. URL: http://www.ldeo.columbia.edu/res/div/ocp/drought/dust_storms.shtml. Accessed May 10, 2010.

[25] National Drought Mitigation Center. "What Is Drought?" Available online. URL: http://www.drought.unl.edu/whatis/dustbowl.htm. Accessed May 10, 2010.

[26] Steven R. Brantley and Bobbie Myers. "Mount St. Helens—From the 1980 Eruption to 2000." Available online. URL: http://pubs.usgs.gov/fs/2000/fs036-00/. Accessed November 20, 2009.

[27] ———. "Mount St. Helens."

[28] United States Geological Survey. "Economic Impact of the May 18, 1980 Eruption." Available online. URL: http://vulcan.wr.usgs.gov/Volcanoes/MSH/May18/description_economic_impact.html. Accessed May 10, 2010.

[29] The Weather Channel. "Katrina's Statistics Tell Story of Its Wrath." Available online. URL: http://www.weather.com/newscenter/topstories/060829katrinastats.html. Accessed November 20, 2009.

[30] ———. "Katrina's Statistics."

[31] Kathy Finn. "Made Homeless by Katrina—and Now FEMA." Available online. URL: http://www.alertnet.org/thenews/newsdesk/N19459965.htm. Accessed November 20, 2009.

[32] John R. Logan. "The Impact of Katrina: Race and Class in Storm-Damaged Neighborhoods." Available online. URL: http://www.s4.brown.edu/katrina/report.pdf. Accessed May 13, 2010.

[33] The Weather Channel. "Katrina's Statistics."

[34] ———. "Katrina's Statistics."

[35] ———. "Katrina's Statistics."

[36] Finn, "Made Homeless."

[37] Richard D. Knabb, Jamie R. Rhome, and Daniel P. Brown. "Tropical Cyclone Report: Hurricane Katrina, 23–30 August 2005." Available online. URL: http://www.nhc.noaa.gov/pdf/TCR-AL122005_Katrina.pdf. Accessed November 20, 2009.

[38] Logan. "The Impact of Katrina."

[39] Debarshi Chaudhuri. "Government: Response to Katrina." Available online. URL: http://web.mit.edu/12.000/www/m2010/finalwebsite/katrina/government/government-response.html. Accessed May 13, 2010.

[40] United States Congressional House. *A Failure of Initiative*. 109th Congress, 2nd session. H. Rpt. 109–377. Washington, D.C.: GPO, 2006. Available online. URL: http://www.gpoaccess.gov/serialset/creports/katrina.html. Accessed May 13, 2010.

[41] ———. *A Failure of Initiative*.

[42] Ulrike Dauer. "Munich-Re Sees Climate Related Losses Mounting." Available online. URL: http://online.wsj.com/article/SB10001424052748703510304574625931956804434.html. Accessed May 11, 2010.

[43] Holli Riebeek. "The Rising Cost of Natural Hazards." Available online. URL: http://earthobservatory.nasa.gov/Features/RisingCost/printall.php. Accessed May 11, 2010.

[44] ———. "The Rising Cost of Natural Hazards."

[45] National Disaster Education Coalition. "Earthquake." In *Talking about Disaster: Guide for Standard Messages*. Washington, D.C.: NDEC, 1999.

[46] ———. "Tornado." In *Talking about Disaster: Guide for Standard Messages*. Washington, D.C.: NDEC, 1999.

[47] National Climatic Data Center. "State of the Climate: Tornadoes: Annual 2009." Available online. URL: http://lwf.ncdc.noaa.gov/sotc/?report=tornadoes&year=2009&month=13&submitted=Get+Report. Accessed May 10, 2010.

[48] NOAA National Severe Storms Laboratory. "A Severe Weather Primer: Questions and Answers about Tornadoes." Available online. URL: http://www.nssl.noaa.gov/primer/tornado/tor_damage.html. Accessed November 20, 2009.

[49] ———. "A Severe Weather Primer: Questions and Answers about Tornadoes."

[50] ———. "A Severe Weather Primer: Questions and Answers about Tornadoes."

[51] ———. "A Severe Weather Primer: Questions and Answers about Tornadoes."

[52] United States Geological Survey. "Volcano Hazards Program." Available online. URL: http://volcanoes.usgs.gov/. Accessed November 20, 2009.

[53] ———. "Historically Active Volcanoes in Alaska: A Quick Reference." Available online. URL: http://pubs.usgs.gov/fs/2000/fs118-00/fs118-00.pdf. Accessed November 20, 2009.

[54] ———. "Mount Rainier: Living Safely with a Volcano in Your Backyard." Available online. URL: http://pubs.usgs.gov/fs/2008/3062/fs2008-3062.pdf. Accessed November 20, 2009.

[55] ———. "Mount Rainier: Living Safely with a Volcano in Your Backyard."

[56] ———. "Tsunami Hazards—A National Threat." Available online. URL: http://pubs.usgs.gov/fs/2006/3023/2006-3023.pdf. Accessed November 20, 2009.

[57] University of Washington Department of Earth and Space Sciences. "1960 Chilean Tsunami." Available online. URL: http://www.geophys.washington.edu/tsunami/general/historic/chilean60.html. Accessed November 20, 2009.

[58] ———. "1946 Aleutian Tsunami." Available online. URL: http://www.geophys.washington.edu/tsunami/general/historic/aleutian46.html. Accessed November 20, 2009.

[59] University of Southern California Tsunami Research Center. "1918 Puerto Rico Tsunami." Available online. URL: http://www.usc.edu/dept/tsunamis/caribbean/webpages/1918prindex.html. Accessed November 20, 2009.

[60] United States Geological Survey. "Local Tsunamis in the Pacific Northwest." Available online. URL: http://walrus.wr.usgs.gov/tsunami/cascadia.html. Accessed November 20, 2009.

[61] National Climatic Data Center. "Billion Dollar U.S. Disasters." Available online. URL: http://www.ncdc.noaa.gov/oa/reports/billionz.html. Accessed May 17, 2010.

[62] National Weather Service. "Hydrologic Information Center." Available online. URL: http://www.weather.gov/hic/flood_stats/Flood_loss_time_series.shtml. Accessed May 17, 2010.

[63] National Oceanic and Atmospheric Administration. "Summer's Peak Has Arrived: Caution: Deadly Heat Wave Reaches East Coast." Available online. URL: http://www.noaanews.noaa.gov/stories2006/s2674.htm. Accessed May 17, 2010.

[64] National Weather Service. "Largest Power Outage in the History of St. Louis." Available online. URL: http://www.crh.noaa.gov/lsx/?n=july_2006. Accessed May 17, 2010.

[65] ———. "Summary of Natural Hazard Statistics for 2009 in the United States." Available online. URL: http://www.weather.gov/os/hazstats/sum09.pdf. Accessed May 17, 2010.

[66] Riebeek. "The Rising Cost of Natural Hazards."

[67] United States Geological Survey. "Forecasting California's Earthquakes—What Can We Expect in the Next 30 Years?" Available online. URL: http://pubs.usgs.gov/fs/2008/3027/fs2008-3027.pdf. Accessed November 20, 2009.

[68] National Earthquake Hazards Reduction Program. "Forecasting What's Coming Up from Down Below: The Uniform California Earthquake Rupture Forecast." Available online. URL: http://www.nehrp.gov/pdf/SeismicWavesAug09.pdf. Accessed May 10, 2010.

[69] United States Geological Survey. "Near-Term Probability of a Great Earthquake on the Cascadia Subduction Zone." Available online. URL: http://earthquake.usgs.gov/hazards/about/workshops/PacNWworkshoptalks/AdamsCascCondProbUSGS06.pdf. Accessed November 20, 2009.

[70] Federal Emergency Management Agency. "Public Assistance Grant Program." Available online. URL: http://www.fema.gov/government/grant/pa/index.shtm. Accessed November 20, 2009.

[71] ———. "Number of Declarations Per Calendar Year since 1999." Available online. URL: http://www.fema.gov/government/grant/pa/stat1.shtm. Accessed November 20, 2009.

[72] ———. "Average Total Obligations by Year and by Declaration." Available online. URL: http://www.fema.gov/government/grant/pa/stat2.shtm. Accessed November 20, 2009.

[73] ———. "2009 Federal Disaster Declarations." Available online. URL: http://www.fema.gov/news/disasters.fema?year=2009. Accessed May 10, 2010.

[74] ——. "FEMA History." Available online. URL: http://www.fema.gov/about/history.shtm. Accessed May 11, 2010.

[75] ——. "FEMA History."

[76] ——. "FEMA History."

3

Global Perspectives

Each year, natural disasters around the globe cause an average of $98.9 million in economic damages and kill an average of 93,177 people.[1] These disasters range from mild to horrific and no country in the world is immune; rather, each country or region possesses its own set of vulnerabilities, a unique conglomeration of environmental, political, and economic factors that exposes a place and its population to natural hazards that may lead to disaster. In some years, a disaster demands worldwide notice due to the magnitude of its destruction of property and environment or the overwhelming number of its victims, both dead and alive. These are worst-case scenarios come true, and in these years the economic impact and loss of life spike far higher than the average, forcing a scramble for more effective preparedness and counterstrategies. Each year, the world is more prepared for natural disaster than it was the year before but still the "big ones" catch most people off guard.

The case studies that follow illustrate how environment, politics, and economics can combine to produce catastrophes that not only command the world's notice but also contribute to global understanding of disaster preparedness, response, and recovery. Floods in China have proven over millennia that some disasters can be expected but not averted. The 1984–85 famine in Ethiopia shocked developed nations into donating millions of dollars of aid in one of the most popularized instances of foreign aid. The 1985 volcanic eruption in Colombia inspired the creation of the world's first volcano crisis response team. And finally, the 2004–05 Indian Ocean tsunami caused the world to ask why there had been no tsunami warning system for that region. While each of these disasters originated in nature, each played out against the backdrop of a country or region's environment, culture, and history. Catapulted into the global spotlight, each disaster expanded to include factors of media coverage, foreign aid, and the learning curve of the human race.

SUMATRA-ANDAMAN EARTHQUAKE AND TSUNAMI, 2004–2005

On December 26, 2004, a magnitude 9.0 earthquake off the west coast of northern Sumatra, an Indonesian island, caused massive tsunamis in the Indian Ocean. Entire cities in Indonesia were destroyed, and Thailand, Sri Lanka, India, and the Republic of Maldives suffered significant damages. The U.S. Geological Survey (USGS) reports this to be the third largest earthquake in the world since 1900, behind the 1960 earthquake in Chile and the 1964 earthquake in Prince William Sound, Alaska. However, the destruction wrought by the two stronger earthquakes pales in comparison to the death and damages caused by the 2004 Sumatra earthquake and tsunami, which killed between 226,000 and 300,000 people, displaced approximately 1.5 million people in 14 countries in South Asia and East Africa, and caused billions of dollars in damages. The magnitude of this disaster led to the creation of the Indian Ocean Tsunami Warning System.

The Earthquake

On the eastern side of the Indian Ocean, the Indonesian island of Sumatra lies about midway between India and Australia. Along the western coast of Sumatra, 36 volcanoes rise from the mountains, part of an arc of 150 volcanoes within Indonesia. These volcanoes form the western edge of the Ring of Fire, a stretch of unusually high volcanic and tectonic activity in the Pacific Basin. The term *tectonic* refers to the forces and movements of the Earth's plates as they interact to change the surface of the world. Moving no faster than a fingernail grows, the forces normally can be measured only with scientific instruments, but sometimes the plates shift suddenly and erratically, causing an earthquake.

Just off the coast of Sumatra, two large tectonic plates converge along a stretch of the ocean floor known as the Sunda Trench. The Sunda Trench plunges to one of the deepest parts of the Indian Ocean, the point at which the Indian Plate slowly forces itself underneath the Burma Microplate (also called the Burma Plate) at a rate of just under two and a half inches per year. This downward movement, or subduction, has created a trench hundreds of miles long and several miles deeper than the surrounding ocean floor. Also called a subduction zone, this area has the potential to release powerful megathrust earthquakes. A megathrust earthquake occurs when the plate that rests on top of the subduction zone, in this case the Burma Microplate, suddenly thrusts upward.

On December 26, 2004, at 7:58 A.M. local time, a megathrust earthquake began off the western coast of Sumatra. Scientists located the epicenter at approximately 155 miles south southeast of Banda Aceh, the capital city of the Aceh Province in northern Sumatra, Indonesia. The earthquake ripped along nearly 750 miles of the Sunda Trench subduction zone, forcing a piece of the ocean floor to snap suddenly upward as high as 50 feet. The energy released by the event was equivalent to 475,000 tons of TNT, or 23,000 times larger than the atomic bomb dropped on Nagasaki.[2] The earthquake occurred 18 miles underwater and had a magnitude of at least 9.0. In comparison to earthquakes since 1900, only four have occurred with similar severity—the 1960 Chile earthquake at 9.5, the 1964 Prince William Sound, Alaska, earthquake at 9.2, the 1957 Andreanof Islands earthquake at 9.1, and the 1952 Kamchatka earthquake at 9.0. All of these earthquakes occurred along underwater subduction zones.[3]

On the islands closest to the epicenter, the shaking caused the collapse of many buildings. Some mid-rise reinforced concrete structures collapsed under the stress, roads cracked, and people fell to the ground. People reported feeling the shaking as far away as India and Sri Lanka.

Though the earthquake was massive, it directly caused only a fraction of the damages associated with the entire disaster event. Most certainly, the shaking decimated nearby islands, but its influence on the ocean was the more significant factor. When the ocean floor shifted, it displaced all of the water above it. As though a rock had been thrown into a pond, the water rippled away from the point of disturbance. In this case, the earthquake set trillions of tons of water in motion. As the water approached the coasts of nearby landmasses, the ocean floor forced the water upward. Scientists call this a run-up, in the sense that the wave "runs up" as it comes ashore. Technically, a run-up is the vertical distance the seawater travels up the beach before it recedes into the ocean.

In most cases, the run-up from an earthquake occurs without notice. Just three days before the Sumatran earthquake, an earthquake of magnitude 8.1 off the coast of Australia caused five measurable run-ups peaking at eight inches. In contrast, the Sumatra earthquake caused 997 measurable changes in ocean levels around the globe. For the areas nearest the earthquake, the water transformed into huge waves traveling at hundreds of miles per hour. These waves are called a tsunami, a Japanese word meaning "a wave in the harbor." Just minutes after the Sumatra earthquake, the resulting tsunami reached a maximum height of 167 feet off the northwest coast of Sumatra.[4] Along the coast of Sumatra, the tallest waves reached 100 feet, with an average height of 33 feet. To the northeast, the tsunami struck Thailand and Myanmar; Thailand saw waves 13 to 26 feet tall. After two hours, the waves reached India and Sri Lanka, where the westward-moving water peaked at 40 feet on the coasts of Sri Lanka.[5] The waves continued across the Indian Ocean until finally claiming the last lives in Somalia, Kenya, and Tanzania more than seven hours later.

During the next few months, more than 100 aftershocks (subsequent earthquakes that are related to a major earthquake) occurred. The most significant was an 8.7 magnitude earthquake on March 28, 2005, that caused further damage to buildings and triggered a much smaller tsunami.

Tsunami Damage Assessments

While it is impossible to determine the exact number of people who lost their lives in the Sumatra tsunami, estimates place the number of dead and missing at between 226,000[6] and 229,866, with some estimates reaching nearly 300,000 fatalities.[7] The Centre for Research on the Epidemiology of Disasters (CRED), in collaboration with the World Health Organization (WHO), has maintained an Emergency Events Database EM-DAT since 1988. EM-DAT's International Disaster Database states a total of 226,408 fatalities. In addition, 1.5 million people were forced out of their homes by the disaster. According to EM-DAT, 12 countries reported fatalities, including Indonesia (165,708 deaths), Sri Lanka (35,399 deaths), India (16,389 deaths), Thailand (8,345 deaths), Somalia (298 deaths), Maldives (102 deaths), Malaysia (80 deaths), Myanmar (70 deaths), Tanzania (10 deaths), Seychelles (3 deaths), Bangladesh (2 deaths), and Kenya (1 death). The figures include the deaths of foreign tourists from 37 other countries.[8]

The earthquake that sparked the disaster caused most of its damage on the islands nearest its epicenter, the location where the earthquake occurred. The distance of populated areas from the subduction zone limited earthquake damage to some extent. On Andaman Island, Nicobar Island, and Car Nicobar Island, the shaking damaged buildings, cracked roads, and knocked people to the ground. Tremors from the earthquake were felt by people in countries around the northern Indian Ocean, including India, Sri Lanka, Thailand, Malaysia, Myanmar, and Bangladesh.

The staggering death toll of the Sumatra disaster results from the tsunami that followed the earthquake. With no early warning system for tsunamis in place at the time, people were caught off guard. They had little or no time to attempt escape. Buildings and other structures provided little, if any, protection from the massive waves. Most of these structures were unable to withstand high winds, much less the force of a tsunami.

In Sri Lanka and Thailand, for instance, buildings of wood, masonry, and concrete lined the beaches. Due to reduced Coriolis force (a term that

71

describes how the velocity of Earth's rotation changes with latitude), countries within 10 degrees north or south of the equator do not experience tropical cyclones and storm surges. Both Sri Lanka and Thailand are within this area. Since these locations had seldom experienced high winds, builders had not designed structures to withstand significant loads. The most common building technique in Sri Lanka was unreinforced masonry, and typical Thai architecture was bamboo framing with thatched roofing. Both types of structures are particularly susceptible to collapse when struck by a tsunami.

Economic losses from the Sumatra tsunami totaled $10 billion. The countries of Indonesia, Thailand, Sri Lanka, and India together accounted for 75 percent of the damages. The Republic of Maldives experienced the worst economic impact, with losses amounting to 45 percent of the country's gross domestic product (GDP), the total of all goods and services produced within the country in a year's time. Though Indonesia experienced the highest costs at $4.5 billion in economic losses, the overall damages lowered GDP by only 0.2 percent in 2005.[9]

The low level of insurance coverage contributed to the crisis in some countries. In many cases, insurance policies included no coverage for damages caused by earthquakes or tsunamis. Of the $10 billion in total damages, insurers covered only $1.3 billion. In Indonesia, policies covered only 12 percent of damaged properties. A similar situation existed in India. In Sri Lanka, only 2 percent of the affected population held property insurance. The only country with a high proportion of insurance coverage that experienced significant damages was Thailand, where insurance policies covered about 50 percent of damaged properties.[10]

INDONESIA

In Indonesia, nearly all casualties and damages from the Sumatra tsunami occurred in the Nanggroe Aceh Darussalam Province in northern Sumatra. The Indonesian Ministry of Health reported 128,803 dead, 37,066 missing, and 155,327 in need of medical care. The tragedy devastated countless cities, leaving more than 750,000 homeless while it forced an additional 250,000 to live with relatives.[11] Indonesia's Ministry of Health reported that the nation's health care was in a "disaster situation." The tsunami "had caused many missing people, victims, health problems, and a mess in almost all the social and public facilities in this province and part of North Sumatra." In addition the report stated, "The large number of health personnel victims who died and [are] missing and . . . the damaged health facilities had made the function of health services for people [to] become disabled."[12]

Waves as high as 100 feet ripped across the coast of Sumatra. Of the city of Leupung's 10,000 residents, only a few hundred survived, and the city was completely destroyed. In Calang, only a third of a population estimated between 9,000 and 12,000 survived. In Meulaboh, a city of 120,000, seven waves killed 40,000 people, left 50,000 homeless, and destroyed port facilities and a significant portion of the town. Close to the epicenter, Simeulue Island lost only five people to the tsunami, most likely because the island rose as the seafloor lurched upward.[13]

The tsunami destroyed or heavily damaged almost all structures less than 33 feet above sea level along the 106-mile coast of the Aceh Province. Estimates place fatalities at 50 percent of the total population of this province. In most places, the water reached about a half mile inland, but in the densely populated and low-lying city of Banda Aceh, waves reached up to two and a half miles inland.[14]

THAILAND

About two hours after the earthquake, the tsunami struck the western coast of Thailand, causing widespread destruction in the coastal provinces of Ranong, Phang-Nga, Phuket, Krabi, Trang, and Satun. The Minister of Interior officially reported 5,395 dead, 8,457 injured, and 3,001 missing. The fishing industry suffered the loss of 3,307 fishing boats, 15,534 fish-breeding baskets, and 35,727 other fishing-related items that were damaged or swept out to sea. Many tourist resorts in Phang-Nga and Krabi experienced significant damage due to their close proximity to the shore. Tourists filled the beaches as the waves arrived, and the water swept many away. Foreigners accounted for 1,953 of the confirmed dead and 962 of those still missing.[15]

An area stretching along 15 miles of coastline, Khao Lak in Phang Nga Province experienced some of the heaviest damage. Everything from the five-star hotels down to the budget hostels lined the beach, making it a prime tourist destination. The largest number of casualties occurred in this area when the 45-foot wave pushed a half of a mile inland in under a minute. All of Khao Lak's hotels suffered heavy damages, except one hotel that was built high on a hill.

Farther south, the Phuket peninsula partially shielded the Phi Phi Islands, reducing the height of the waves that hit to 10 feet and 18 feet. The waves hit in succession from either side of the tiny island. The waves damaged approximately 70 percent of the island's buildings. As of December 2005, workers had recovered 850 bodies while another 1,300 remained missing.[16] One of the few warnings came from the owners of the Phi Phi Island Village Beach Resort and Spa, who witnessed unusually high tides as the tsunami struck the Phuket peninsula. Authorities had issued no public warning, but the owners had seen news of the earthquake near Sumatra and decided to evacuate the hotel's beaches as a precaution. The decision saved all but two guests.

BANGLADESH, MALAYSIA, AND MYANMAR

Though in close proximity to the earthquake, the countries to the north and south of Thailand escaped the catastrophic damages suffered by their neighbor. To the north in Bangladesh and Myanmar, the north-south orientation of the earthquake generated less significant waves traveling northward. Officials reported 59 dead, 3 missing, and 3,205 displaced in Myanmar.[17] Bangladesh weathered the waves with no official casualties, though two children reportedly drowned when a wave capsized a tourist boat.

Following the disaster, the International Federation of Red Cross and Red Crescent Societies created the Cyclone Preparedness Programme in Bangladesh to better prepare for storm surges from cyclones or tsunamis. Program activities have helped educate people in 11 cyclone-prone districts on how to be better prepared for these types of disasters.

INDIA, SRI LANKA, AND MALDIVES

Two hours and 20 minutes after the Sumatra earthquake, the waves began to hit the southern coast of the Indian peninsula, Sri Lanka, and the Republic of Maldives. On mainland India, authorities reported 8,835 dead and 86 missing.[18] Starting at 9:05 A.M., a series of three to five waves swept across the Indian coastline. On the far side of the peninsula, the east coast experienced minimal damage. The waves affected the low-lying southeast coast more severely, where the largest waves reached over 30 feet high. The event affected more than a million people and destroyed or damaged almost 150,000 homes.[19]

The first wave hit Sri Lanka at 9:10 A.M. Measuring less than five feet high, it did little damage. A damaged clock that rested among the debris on the beach in the Galle District had stopped at 9:20 A.M. as the second 32-foot wave struck that region. As the larger wave washed across the eastern coast of Sri Lanka, it severely damaged low-lying areas where "the only indication of a preexisting building was a remnant of a foundation. In some areas, there was a zone near the coast where all structures were completely destroyed...."[20] In the affected regions, the waves completely destroyed 49,983 out of 88,544 houses.[21]

Southeast of India and Sri Lanka, the Republic of Maldives consists of a chain of 198 populated islands in the Indian Ocean, spread over 500 miles. Averaging only five feet above sea level, the extremely low elevations of the Maldivian islands make them particularly susceptible to changes in sea levels. Waves from five to 15 feet began to strike the islands at 9:20 A.M. Official death tolls recorded 83 confirmed deaths and 25 missing, along with 1,300 injuries. Significant damage occurred on 39 islands. The waves completely destroyed 14 islands, where utilities were destroyed or freshwater sources became contaminated by seawater. The disaster adversely affected around a third of the country's population of 300,000, of which 7 percent was displaced from their original island of residence.[22] In the capital city of Male, a 12-foot seawall reduced damages. Without the wall, the tsunami could easily have destroyed half the city.

MADAGASCAR AND THE EASTERN AFRICAN COAST

Traveling at just under 400 miles per hour (MPH), the tsunami crossed the entire Indian Ocean, about 2,800 miles, in a little more than seven hours. Then it dealt the greatest amount of damage along the coast of Somalia, where it killed 300 people. An additional 100 people were missing.[23] The waves spread across 400 miles of the Somali coastline, causing extensive damage to houses, water sources, and fishing gear. Relief workers estimated that 18,000 households were significantly affected and in need of humanitarian assistance.[24]

Over the next two hours, the tsunami swept down the eastern coast of the African continent. Kenya reported two deaths and more than 200 boats lost. In Tanzania, the waves killed 10 people. Sheltered by Madagascar, Mozambique escaped significant damages, but the rising waters did affect South Africa, where port authorities reported that 20 people had drowned. Just off the coast of Africa, Madagascar also proved less vulnerable. Officials there reported two deaths, minor flooding, and a small number of fishing boats damaged or destroyed.[25]

Response

One of the most common responses to the Indian Ocean tsunami was the realization that it was the worst natural disaster in anyone's personal experience. Vanessa Gezari, a foreign correspondent who covered the aftermath of the tsunami, wrote, "In scope, the tsunami was beyond any natural disaster that I or any reporter had ever seen. Although I've worked in Afghanistan and other conflict areas, I had never seen as much concentrated death as I did in Sri Lanka and Indonesia in the three weeks after the wave."[26] In those first weeks in the disaster zones, the primary concerns were the continued survival of survivors, including the need for food, shelter, and medical aid, and finding, identifying, and burying the dead. Beyond these goals were the enormous challenge of reconstruction and the obvious need to establish a tsunami early warning system for Indian Ocean countries. These goals and challenges were met with varying degrees of success, from the frequent impossibility of identifying water-damaged bodies to the triumphant completion of the Indian Ocean Tsunami Warning System 18 months later.[27]

FOREIGN AID

By January 7, 2005, more than 40 countries offered pledges totaling more than $4 billion in relief funds and other types of assistance, and that number would continue to grow. Those that pledged $10 million or more included, in order from largest to smallest pledge, Australia, Germany, Japan, the United States, the United Kingdom, Canada, Sweden, Denmark, Spain, China, France, Spain, Saudi Arabia, the Netherlands, Switzerland, India, and Belgium.[28] In addition, private donors including corporations and individuals pledged hundreds of millions of dollars. When all was said and done, approximately $14 billion was pledged from foreign countries and their citizens in addition to $2.5 billion from tsunami-affected countries.[29] These figures include both short-term relief funding and long-term reconstruction funding. Of the $14 billion, 40 percent was pledged by the general public internationally.[30]

Nonmonetary assistance from foreign nations took many forms, including equipment, supplies, services, and personnel. Besides its pledge of $350 million, the United States sent a force of 20 navy ships. Seven of the ships carried equipment capable of purifying 90,000 gallons of drinking water per day. The groups included 200 engineers, nine search and rescue planes, six C-130 transport planes, and a hospital ship. Indonesian navy ships ferried supplies, Japan deployed medical personnel and search and rescue teams, India dispatched naval vessels loaded with food to Sri Lanka, and the Australian military sent more than 600 health workers and technicians.[31]

Medical aid, food, and safe drinking water were top priorities. Within a week of the tsunami, 50 international medical aid groups had arrived in the Aceh Province in Sumatra. The International Committee of the Red Cross supplied a 100-bed portable field hospital. In Banda Aceh, the World Food Programme created a tent city and readied itself to feed around 500,000 people over the following two weeks.[32] The WHO sent 200 personnel to affected regions, supporting them with the technical and specialized expertise of 50 departments.[33]

Coordinating the massive relief efforts to avoid chaos and to prevent duplication of efforts was an enormous task. One week after the tsunami, the Council on Foreign Relations explained, "The United Nations is in charge of the massive international relief effort, coordinating the donations and efforts of myriad organizations from around the world. It has taken over the work of a U.S.–led 'core group' of nations, including India and Australia, that was set up after the disaster to make sure aid efforts were not duplicated. Jan Egeland, the United Nations Undersecretary-General for Humanitarian Affairs, is heading the U.N. effort. The U.N.–affiliated World Health Organization

(WHO) is also coordinating field-workers, aid agencies, governments, and private health organizations from its new command center in Geneva."[34]

The first wave of disaster response (medical aid, food supplies, burials, and temporary shelters) occurred in the month following the disaster. By the end of January 2005, emergency services were ending and the second wave of disaster response—reconstruction—was beginning. For instance, Human Rights Watch reported that it "visited several affected villages in Tamil Nadu State and the union territory of Pondicherry [in India] in January 2005, a month after the tsunami. By the time of our visit, emergency operations were winding down, the relief camps had been closed, and the government had embarked on phase two rehabilitation efforts. The local administration had begun providing temporary shelters and restoring infrastructure. A detailed damage assessment had been started. Schools had reopened in many of the affected districts."[35]

Disaster relief efforts were not without breakdowns and failures. Key problems were the delay of vital deliveries due to factors such as damaged roads, bridges, shipping ports, and airport runways and the need to clear deliveries through customs before distribution. In some circumstances, there was competition for supplies. For instance, Human Rights Watch reported that it had received "consistent reports of discrimination based on caste status in the distribution of aid [in India]. While the [Indian] government changed some policies when it was made aware of discriminatory practices on the ground, the greatest source of trouble seemed to be discrimination against Dalits by other victims of the tsunami, notably the communities of fishermen, who view themselves as belonging to a higher caste."[36] Four years after the disaster, a workshop of the Tsunami Recovery Impact Assessment and Monitoring System (TRIAMS) reported, "The availability of emergency aid and the attendant pressure to spend it undermined the role of recovery planning. At the same time, the multitude of development actors rendered normal coordination mechanisms unworkable."[37]

After the tsunami, several international organizations joined together to monitor and assess relief efforts, focusing their attention on the four hardest-hit countries: Indonesia, the Maldives, Sri Lanka, and Thailand. These organizations are the United Nations Development Programme (UNDP), the WHO, and the International Federation of Red Cross and Red Crescent Societies. Together, they formed TRIAMS and over the years since the Indian Ocean disaster have held workshops and published reports about the disaster response. For example, the 2007 report points out that recovery is far from complete and discusses the need for "replanning of recovery activities." The 2009 report explains lessons learned in monitoring post-tsunami recovery in areas such as data collection, analysis, and utilization. The findings and lessons explained in these reports are vital contributions to the global learning curve in disaster response and recovery.

In the wake of the tsunami, more than 40 aid agencies, including the United Nations, governments, and nongovernment groups, formed the Tsunami Evaluation Committee (TEC) to examine how effectively tsunami aid had been distributed and under what constraints relief efforts had been carried out. In July 2006, TEC published their findings in "Funding the Tsunami Response." The report states that nine months after the disaster, 56 percent of the $5.8 billion pledged by 22 foreign countries and the European Commission had been delivered. Most of that 56 percent of paid money was earmarked for humanitarian aid. Specifically, nearly all (97 percent) of monies for humanitarian aid were delivered, contrasting to just 20 percent of reconstruction commitments delivered.[38] By July 2006, of the total $14 billion pledged, $11.6 billion had been committed or delivered,[39] prompting TEC to conclude that "government donor pledges are being honored so far."[40] The TEC did not issue further reports on countries' delivery of pledges.

DIGNIFIED RECOVERY OF BODIES

Certain characteristics of the tsunami complicated the process of identifying bodies so that the remains could be returned to relatives for burial. To accomplish the task, affected countries grouped specialists—including crime scene investigators, forensic pathologists, forensic dentists, fingerprint experts, DNA experts, and photographers—into Disaster Victim Identification teams. These teams followed standards published by the International Criminal Police Organization (INTERPOL) in the Disaster Victim Identification Guide (DVI Guide).

Workers used three official methods to identify those lost in the tragedy: fingerprints, DNA analysis, and dental records. Typically, specialists rely heavily on DNA matching, but the tropical climate and saturation by seawater caused many bodies to decay rapidly. Though rescue workers collected DNA samples from all victims, the rate of decay often degraded the accuracy of the DNA tests, making the data useless. The rapid decomposition also hindered the use of fingerprinting as a means of identification, often leaving dental records as the primary source of identification.

TSUNAMI WARNING SYSTEM

In the days after the disaster, the mounting death toll caused many to question why the loss of life had been so large. Why had there been no warning? Why had the coastal residents not been evacuated? At the time of the disaster, no organization held the responsibility for issuing tsunami warnings for the Indian Ocean, and none of the countries struck by the tsunami had evacuation plans in place.

Historically, the region has a long record of tsunamis, particularly along the coast of Sumatra. The last event occurred in 1861 when an earthquake with an estimated magnitude between 8.8 and 9.2 caused extensive damage. Other major events in 1797 and 1833 caused tsunamis that claimed tens of thousands of lives. However, these earthquakes happened far from the epicenter of the 2004 earthquake, which had not seen an earthquake of that magnitude for more than 500 years. The most recent earthquake in that area had a magnitude of 8.0 and caused a tsunami of only three feet in India. The only other tsunami recorded in the area occurred during the eruption of Krakatoa in 1883 when the volcano collapsed into the sea, generating a 130-foot wave that killed 36,000 people.[41]

Monitoring the Pacific Ocean, the Pacific Tsunami Warning System in Hawaii issued a Tsunami Information Bulletin just 15 minutes after detecting the Indian Ocean earthquake. The bulletin declared, "This earthquake is located outside the Pacific. No destructive tsunami threat exists based on historical earthquake and tsunami data." Forty-five minutes later, after the estimated magnitude of the earthquake was changed from 8.0 to 8.5, the center issued the following revision to the bulletin: "There is the possibility of a tsunami near the epicenter."[42] But since there were no instruments in the Indian Ocean to measure sea levels, officials from the center had no way to confirm the danger. Three hours passed before Reuters Internet wire service reported that a tsunami had struck Thailand and Sri Lanka.

The extent of the damages caused by this disaster prompted scientists to question whether a similar event could happen elsewhere in the world. The most likely areas for a tsunami of similar size are located along the subduction zone off the eastern coast of Japan, the Chilean trench along southwestern South America, and the Cascadia subduction zone that runs along the Pacific Northwest coast of the United States. These areas generate major tsunamis approximately every 500 years. The Pacific Tsunami Warning Center in Hawaii, established in 1949 as the Tsunami Warning Center, and the West Coast and Alaska Tsunami Warning Center in Palmer, Alaska, established in 1967, monitor all of these areas for tsunamis and issue warnings to all countries bordering the Pacific Ocean when potential threats exist.

In June 2005, the Intergovernmental Oceanographic Commission (IOC) began working with United Nations member states bordering the Indian Ocean to establish a Tsunami Early Warning System for Indian Ocean countries. Scientists deployed the first device of the Deep-ocean Assessment and Reporting of Tsunamis (DART®) system in December

2005, placing a buoy used to detect changes in sea level about halfway between Thailand and Sri Lanka. By 2008, the detection system consisted of 39 DART° buoys deployed by the U.S. National Oceanic and Atmospheric Administration (NOAA), one buoy owned by Chile, six stations owned by Australia, and two more stations owned by Indonesia and Thailand. Anchored in areas with a history of tsunami-generating events, these systems record sea-level estimations on a 15-minute interval and transmit the data to the tsunami warning centers via satellite once every six hours. The warning centers may also query a buoy and retrieve the data immediately if they suspect that an event, such as a large earthquake, has caused a tsunami. This system also includes seven buoys in the Atlantic Ocean used to detect tsunamis that could strike the eastern coast of the United States or islands in the Caribbean Sea. Through 2010, the Pacific Tsunami Warning Center and the Japan Meteorological Agency shared the responsibility for compiling data from the Indian Ocean and issuing tsunami watches, with these duties scheduled to transfer to Regional Tsunami Watch Providers in Indian Ocean countries in 2011.

RECONSTRUCTION

The United Nations Food and Agriculture Organization (FAO) set up programs to rehabilitate the agricultural, fishing, and forestry industries in rural coastal communities in seven countries struck by tsunami. In Indonesia, in the aftermath of the disaster, FAO administered 17 projects that benefited more than 370,000 people. Funding for these programs amounted to $20 million, mostly from charitable donations. These programs have helped reclaim almost 25,000 acres of farmland from areas inundated by the tsunami, providing almost 46,000 farmers with fertilizer, tools, seeds, and farming equipment. FAO has also funded programs to assist in the recovery of the fishing industry, providing boats and fishing gear, rehabilitating fish ponds, and establishing new fish hatcheries. FAO also employed 8,000 people to plant more than 200,000 seedlings where forests along the coast were destroyed.[43] FAO coordinated similar programs in the Maldives, Myanmar, Seychelles, Somalia, Sri Lanka, and Thailand.[44]

Reconstruction efforts continue in all countries struck by the disaster. In the hardest-hit area, Aceh Province in Sumatra, the United States Agency for International Development (USAID) and CHF International (originally Cooperative Housing Foundation) projects provided temporary shelter to around 3,000 uprooted families and constructed more than 1,000 permanent homes.[45] USAID anticipates reconstruction of the Aceh West Coast Highway, a main transportation route between the province and the rest of the island, to be completed in late 2010. For the 93-mile (150-km) project, 40 miles (65 km) of road had been opened to the public in mid-2010, and 16 new bridges had been completed, with six more under construction.[46]

WATER RESOURCE MANAGEMENT IN CHINA

China is one of the world's oldest civilizations. Throughout its long history, it has suffered repeated disastrous swings between flood and drought, making water management one of the nation's most pressing challenges. Flooding in China accounts for some of the most destructive natural disasters in Earth's recorded history. Millions of people have lost their lives or livelihoods as rivers inundated the farmlands surrounding their banks. Nearly once a decade, China experiences deadly floods in the middle and lower portions of its major rivers as a result of heavy seasonal rains, called monsoon rains, or typhoons striking the eastern coast. During the past 2,000 years, the Yangtze River (Changjiang)[47] has flooded more than 1,000 times. In 1887, flooding along the Yellow River (Huang He) killed nearly 2 million people. In 1931, flooding along the Yellow, Yangtze, and Huai Rivers caused the worst disaster in recorded history, killing 3.7 million people from drowning, disease, or starvation, and affecting more than 51 million, about one-fourth of China's population at the time. In ironic juxtaposition to the catastrophic flooding, periodic droughts exacerbate China's water-management problems. According to China's Ministry of Water Resources, "Since the founding of P[eople's] R[epublic of] China in 1949, more than 50 extraordinary floods and 17 widespreading severe droughts have hit China."[48]

China controls a substantial amount of water, ranking fifth in the world for total water resources; however, the country's large population (1.3 billion in 2010[49]) reduces the per capita share of water to a fraction of the world average. Location of water sources within the country contributes to this problem. While northern China contains 65 percent of the nation's farmland, it controls only 24 percent of the water. On a per capita basis, water share in the north reaches as low as one-tenth of the world average. The majority of the rainfall in the north, about 70 percent, occurs between June and September and must be managed carefully as the region produces roughly half of China's grain and has industries that output 45 percent of China's GDP.[50]

The People's Republic of China is enormous, with a land area of 3,705,407 square miles (9,596,961 sq km), and 15 major rivers, including the Yangtze, Yellow, and Amur (Heilongjiang), the three longest. These 15 rivers drain basins totaling more than 2,007,731 square miles (5,200,000 sq km) and produce a combined annual runoff averaging 489 cubic miles (2,039 cu km) of water. The largest rivers and flows are in southern China. Among them is the Yangtze, by far the most voluminous of the Chinese rivers, with a basin draining 698,266 square miles (1,808,500 sq km) in the southeast and an average annual runoff of 228 cubic miles (951 cu km).

This north-south divide in moisture distribution in China is attributable in part to Asian monsoons that move across parts of the Asian continent in summer. In far south China, the South Asian monsoon fueled by moist air from the Indian Ocean brings heavy rainfall to the provinces of Yunman and Guangxi in early summer. The East Asian monsoon carries moist air from the western Pacific Ocean and brings heavy rainfall in a belt across southern China, first to the Yangtze River valley and then up into the Yellow River basin. Monsoons can produce excessive precipitation and flooding in southern China, but much of the large Xinjiang Province in the northwest remains dry, protected by the Himalayas from these seasonal rains.

As illustrated in China's long history, all of the country's rivers respond to some degree to the seasonal monsoons, bringing life-sustaining water to the people and the land, but also death and disaster. Deluges precipitated by monsoons are one reason for flooding, but other natural processes contribute to this hazard. Consider the Yellow River. In its upper reaches, the Yellow River flows over a bed of loess, or silt carried by high winds into China from Siberia during the last ice age. The flow of the river sweeps silt downstream by the ton, filling its own channel bit by bit. The sediment carried downstream is about 1.6 billion tons yearly, about 60 times more sediment than the Mississippi River carries.[51] Around a third of it settles before the river reaches the sea. As a result the riverbed becomes suspended, or higher than the surrounding floodplain. Only levees hold the waters at bay.

Because of significant regular flooding in the south and periodic droughts in the north, water resource management is a vital issue in the lives and livelihoods of China's people and in the economy of the nation. According to China's Ministry of Water Resources, "Since 1990, the average loss resulting from floods has amounted to about 1.5 percent of GDP of the same period, and the average economic loss resulting from drought has been over 1 percent of GDP of the same period."[52] In an ongoing struggle to prevent or mitigate loss of life and economic damages, the Chinese leaders of today continue the monumental efforts of their predecessors to combat floods and droughts without inadvertently making these hazards worse.

History of Flooding in China

According to the *Shiji (Records of the Grand Historian)*, completed by the Han scholar Sima Qian in 100 B.C.E., a great flood destroyed villages and inundated croplands in the basin of the Yellow River at the time of the Xia. The mystical Xia dynasty, which Sima dates to 2270–1750 B.C.E., is said to have come before the Shang dynasty (1750–1040 B.C.E.), the first dynasty in Chinese history from which have been excavated artifacts and written records. The flood forced survivors to abandon their homes to live on hillsides or to relocate far away. In the myth, the government appointed a man named Gun to control the floods. When he failed, he was executed. After three years, however, his lifeless body showed no signs of decay. Then, from a cut on his body, out sprang Yu the Great, who became the focus of a great legend, on the scale of the Greek legends of the Trojan Horse or the 12 labors of Hercules.

According to legend, Yu contributed a significant technology to flood control along the Yellow River. Instead of building dikes and dams, the previous methods of flood control, he instructed his men to dredge riverbeds and to dig ditches and canals to divert floodwaters. He worked alongside the men, dredging all the rivers that fed the Yellow River. When he came to a mountain, he turned into a bear and dug a channel with his claws. After 13 years, he finally finished the task and freed the Yellow River basin of the evils of floods. He also boldly declared, "Whoever controls the Yellow River controls China." According to tradition his words came true. The Great Yu became emperor of China, founding the Xia dynasty.

Among the first significant water management projects in China's recorded history are canal constructions. The Han Canal, the first canal on record, linked the Yangtze River and the Huai River and was built during the later years of the Spring and Autumn Period (770–476 B.C.E.). The Hong Canal, built in 361 B.C.E., was the first canal to link the Yellow River and the Huai River and served as a major waterway for nearly 1,000 years. The Zhengguo Canal was constructed in 246 B.C.E. by Zheng Guo, a hydraulic engineer. The 90-mile-long (145 km) canal linked the Jingshui River and the Luoshui River and irrigated more than 98,000 acres of land in what is today's Shaanxi Province in east-central China. The resulting prosperity helped the Qin State defeat six other states to unify China under the Qin dynasty (221–206 B.C.E.).[53] During the Sui dynasty (581–618 C.E.), engineers completed construction of the longest canal in the world, the Grand Canal, whose 1,103-mile (1,775-km) length links the Yellow and the Yangtze river systems, following a north-south path from Beijing to Hangzhou. The canal linked north and south China and provided a ready means of transporting grain from the south to feed soldiers at China's northern border. Building these canals was backbreaking work that was never really complete, for to maintain the viability of the canals they had to be dredged of silt as it accrued from upstream.

Silt accrual was—and is—not just a problem in the canals but in the rivers as well. In 8 B.C.E., an engineer named Jiarang recommended that the

Yellow River be channelized, a technique that cuts channels to straighten the run of a river. The method increases the flow rate, and since the water runs faster, silting decreases. He also instituted programs to divert irrigation water into canals and basins to reduce flooding and projects to build higher levees. These strategies became the standard flood control methods in China for nearly 2,000 years.

Prior to 1949, the slow rising of the riverbed due to silting caused 26 major course changes in the Yellow River, sometimes moving the course of the river by hundreds of miles. On average, two levee breaches occurred every three years and one course change every 100 years. This unpredictable nature prevented any permanent bridges from being built over the Yellow River until 1905. Despite the early efforts to control the river every few years, the floods returned, sometime killing millions and prompting a second name for the river—China's Sorrow.

After the Communist Revolution in 1949, the new Chinese government consulted the Soviet Union for help in controlling floods along the river. The resulting plan was to build 46 new dams to establish points of control along the river. Dams upstream of the Great Bend region would be used to generate power, while those downstream of this region would be used to help irrigate farmland. Many of the projects experienced the problems with sediment and flooding that have plagued the Yellow River basin for thousands of years. One of the biggest dams among the 46 was the Sanmenxia Dam in Henan Province, at the place where the Yellow River passes through a rock-walled gorge. The design called for a 400-foot-high (122 m) dam that was two-thirds of a mile wide.[54] The Russian engineers assumed that afforestation (planting seeds or trees to change open land to forests) and soil conservation would eliminate the problem of silting, but this solution later proved to be ineffective. As builders completed the Sanmenxia Dam in 1960, newly printed banknotes carried the image of the dam as a symbol of triumph. But with each year that passed, problems mounted. Just four years after its completion, the dam had lost 40 percent of its storage capacity because silt deposits raised the bed of its reservoir. Sediment clogged the turbines used to generate electricity. Today, the reservoir holds only 10 percent of its original volume, and it generates electricity at only 25 megawatts instead of the projected 1,160-megawatt capacity. Some officials have called for the dam to be demolished.[55]

Between the 1950s and 1990s, specialized programs spurred modifications to flood control and mitigation along many of China's rivers. These programs covered three major areas of flood control: levees, reservoirs, and flood storage and retardation basins. First, levees were strengthened and new ones were built along more than 150,000 miles (241,402 km) of flood plains. They provide protection for 200 million people and 80 million acres of farmland. Second, more than 84,000 reservoirs provide major storage areas for floodwaters. Third, 100 flood storage and retarding basins provide additional areas to hold water during times of heavy rainfall.

Then, as China began a transition to a market-based economy, new methods of raising funds became available. These included loans, bonds, and stocks in addition to funding provided by the central government. Since the late 20th century, private and governmental sources have funded projects such as the levee system of the Shanghai Municipality, where design standards were strengthened to withstand 1,000-year floods—severe floods that happen, on average, once every 1,000 years. The Three Gorges Dam, the largest hydroelectric dam in the world, surpassing the Grand Coulee Dam in the United States, was also funded by bonds and loans from banks.

Current Situation

By the first decade of the 21st century, flood forecasting, flood warning, and water management in China were streamlined and organized within key governmental institutions. At the top level is the Ministry of Water Resources (MWR), founded in 1949 and charged with managing and protecting the nation's water resources, including such issues as flood control, irrigation, water conservation, and other policies. Within the MWR, seven River Basin Commissions regulate water resources along each major river. These agencies coordinate resources during floods and drought, mediate water disputes, manage water and soil loss in key areas of the river basins, and generally provide unified management of all the rivers, lakes, estuaries, and tidal flats within their jurisdictions.

Also under the auspice of MWR is the State Flood Control and Drought Relief Headquarters (SFCDRH), sometimes translated as the Flood Fighting and Drought Defying Headquarters. SFCDRH handles flood forecasting and warning, drawing data from a network of hydrological information systems, including hydrological stations that measure rainfall, water level, and river discharge rates. According to MWR, "There are about 34,000 hydrological or precipitation stations and more than 8,600 flood reporting stations built all over the country."[56] SFCDRH uses a computerized flood forecasting system that can incorporate real-time adjustments to increase the accuracy of predictions regarding imminent floods or floods in progress. The agency issues public warnings via radio, television, and newspapers when river levels exceed critical levels. Under the Water Law of the People's Republic of China (1988), SFCDRH has the authority to divert floodwaters with gates on rivers or retain floodwaters in flood-retarding basins as part of the flood control system.[57]

RECENT FLOODS

Floods continue to cause major disasters in China. With 189 million people living within potential flood areas and economic activity in potential flood areas accounting for more than 10 percent of GDP, the exposure to flood or drought conditions is high. During June through September of 1998, floods in the Yangtze Basin caused the deaths of 3,000–4,000 people, with the number varying among reports. An estimated 14 million people were made homeless, and the floods caused an economic loss of around $24 to $30 billion. Eleven million acres of cropland were destroyed. In 1999, flooding along the Yangtze River displaced almost 2 million people and caused the collapse of 400,000 houses. Estimates placed damages at $3 billion. In 2005, flooding along the Xijiang claimed several hundred lives. The State Flood Control and Drought Relief Headquarters reported flooding across 22 provinces and the displacement of hundreds of thousands of people in the worst-hit areas. China's flood-control authorities reported that "the worst flood on record was seen on the Mengjiang, a tributary of Xijiang River."[58] In May 2010, floods struck 13 provinces in southern China including Guangdong, Sichuan, and Zhejiang, damaging more than 80,000 homes and affecting more than 10 million people; 136 people were reported killed or missing.[59]

In response to the floods and droughts of recent decades and in acknowledgment of the disasters of centuries past, China's MWR deems flood and drought management to be "the major event . . . for the improvement of people's livelihood."[60] Mitigation measures including built-up embankments, reservoirs, and flood-retention areas have controlled most normal flooding. In 2010, according to MWR, "[M]ain river sections of major rivers and lakes are capable of withstanding 100-year-return floods; medium and small rivers are capable of preventing normal floods; flood defense capacity of major sea dykes has risen to the 50-year-return level. When medium level drought occurs, the industrial, agricultural and ecological water consumption will not be affected as water supply in both urban and rural areas are basically secured."[61] Despite the optimism of MWR's statement, silting in rivers and canals continues to be a problem. As reservoirs slowly lose their storage capacity and riverbeds steadily rise, the dangers of major flooding grow. In some areas, for instance, the Yellow River flows 32 feet above the surrounding plains, above the rooftops of nearby buildings. For this reason, the government continues to seek out methods of combating silt accrual in the Yellow and other rivers.

WATER SCARCITY

Since the 1950s, China has built up a large bureaucracy focused on two main goals. One goal is to manage flood mitigation systems that are designed to prevent or at least manage the devastating floods of China's major rivers. The other goal is to supply agricultural and industrial development with water resources effectively. In many respects, the nation succeeded in the implementation of these goals, especially the latter goal. However, the policies did not place emphasis on water conservation until the 1988 Water Law; consequently, the nation faced water-shortage issues in many areas. One of the chief causes of water scarcity in China is groundwater and surface water overdraft, or using water supplies faster than they can naturally replenish. Droughts contribute to water scarcity, as does pollution of freshwater supplies and climate change, which can increase the intensity of droughts and contribute to lessened rainfall in already arid regions.

Between 1970 and 1997, the water flow of the Yellow River slowed to zero discharge during the dry season, meaning that the river's flow dried up before reaching the ocean. For five to 10 days in the early 1970s, the river slowed to zero discharge, and between 1995 and 1997, the zero discharge peaked at 100–200 days. The lack of water in the lower reaches of the Yellow River caused significant economic and environmental issues. Decreases in surface water have led to an increased use of groundwater in northern China. By 1995, regions with limited surface water obtained nearly 80 percent of their needs from groundwater sources. As the use of ground sources increased, water table depth began to sink, sometimes severely. Along the Fuyang River basin in Hebei Province, the water table dropped from around 40 feet (12 m) in 1974 to a depth of 120 feet (37 m) in 1998. During the 1990s, more than 750 square miles (1,942 sq km) of coastal water tables dropped below sea level. In turn, more than 8,000 water wells became contaminated by seawater, and nearly 100,000 acres of farmland had to be abandoned.[62]

In 1997, the Yellow River Commission met to assess the issues causing the problem. They found water usage in the middle river during dry seasons to be excessive. A year later, the MWR granted the Yellow River Commission full authority over the entire river basin. Upon completion of a comprehensive plan to manage the river, the commission solved the zero discharge issue. Still, a 2003 report by the United States Department of Agriculture found that water usage along the Yellow River exceeded supply and that further action was needed to maintain the river in a sustainable manner.[63]

THE THREE GORGES DAM

Along the Yangtze, China's longest river and the third longest river in the world, 27,000 miles of levees line the river's banks. The river has overflowed its banks many times and caused great destruction and loss of life. Five major floods occurred along the Yangtze River in the last century. In 1931, floods affected 28 million people and killed 145,000. Four years later, flooding affected 10 million people and killed 142,000. Other major floods occurred in 1949 and 1954. Most recently, in 1998, floods struck major portions of the Yangtze Valley. Despite flood control efforts by the Chinese military, the flooding still affected more than 2 million people and killed 1,526.[64]

To halt the flooding of the Yangtze as well as to provide hydroelectric power and improve navigation on the river, the Chinese government built the Three Gorges Dam. On September 8, 2004, the dam successfully controlled the Yangtze's third largest flood in history while maintaining normal power generation.

Construction of the Three Gorges Dam began in 1994 and was completed in 2003. The dam rises 331 feet (101 m) high and is 1.3 miles (2 km) long. Its vast reservoir stretches for 410 miles (660 km). Chinese officials state that the cost of the project was $27.2 billion, while unofficial estimates put the price at two to three times that amount.[65] The Three Gorges Dam provides 100-year flood protection for 15 million people and 3.7 million acres of farmland. The dam also provides increased shipping capacity between the cities of Shanghai and Chongqing by allowing large ships to travel along the course of the Yangtze River.

Despite the benefits of the Three Gorges Dam, the project has attracted many criticisms. Official sources cite that 13 cities, 140 towns, and 1,350 villages along the densely populated Yangtze Valley were submerged by the project's reservoir. The relocation of the people who once lived in those places became the world's largest resettlement project, requiring the movement of a million people.[66]

The reservoir created other unexpected hazards along the banks of the Yangtze. Each year, the water level in the reservoir ranges from 475 to 575 feet. As the water rises and falls, it destabilizes slopes along the Yangtze Valley and creates erosion issues and the possibility of landslides. In the five months after the reservoir began operation, more than 150 dangerous geological events were recorded. Erosion problems are expected to cause the relocation of an additional 530,000 people. Silt levels at the mouth of the river dropped to a third of previous levels, allowing up to 1,000 acres of coastal wetlands to be eroded each year.[67] As a result, seawater has moved inland, destroying farmland and threatening freshwater supplies.

Large reservoirs such as the Three Gorges Dam may have increased the chance of reservoir-induced seismicity (RIS). Scientists theorize that RIS occurs when the massive weight of a reservoir, or water seeping into a fault under the reservoir, triggers an earthquake. RIS is still a debated phenome-non. Scientists continue to study a number of earthquakes that occurred near large reservoirs, seeking more insight into the potential risks of building dams in areas with seismic activity. In China, a 7.9 magnitude earthquake near the Zipingpu Dam in May 2008 is being studied as a possible RIS event. The earthquake killed approximately 80,000 people and damaged as many as 2,380 dams along 185 miles of the ruptured fault line.[68]

SOIL EROSION

In the last few decades, the Chinese government has closely studied causes and effects of soil erosion in western China, where the Yangtze River begins its flow. An increase in erosion in this area increases the amount of silt that the river carries downstream. Consequently, the costs of managing the middle and lower reaches of the river increase. In addition, the loss of topsoil to erosion results in poorer farming yields. In predominantly rural western China, the government identified deforestation, farming of steep slopes, and overgrazing as major causes of erosion. Erosion, in turn, is the major cause of disasters such as flooding and landslides. Studies by the FAO found soil erosion issues to be a serious threat to the longevity of land use in China.

To precisely identify problem regions, China compiled nationwide sta-tistics relating to the types, intensities, and localities of soil erosion in 1999. The program used satellite images, maps, and data from local monitoring stations to classify the degree of soil erosion as very slight, slight, moderate, strong, very strong, and ultra strong.

Erosion affects land in two ways. Sheet erosion occurs as rain hits the ground and dislodges particles of earth that the runoff carries away as silt. Linear erosion begins when rainfall exhausts the ground's ability to absorb water. On sloped ground the rain forms into small streams, or rills, that cut into the ground. Over time, the rills cut deeper into the ground and combine to form gullies. Once erosion reaches this point, corrective measures are difficult.

Based on data gathered in the soil erosion survey, additional studies have determined that the greatest concentration of soil erosion occurs on farmland with a slope between 10 and 25 degrees. By comparing rates of erosion that occur on croplands and forests, researchers found that human activities have the greatest impact on soil erosion. Farmland accounts for 86 percent of agri-cultural land with erosion issues. Fifty-seven percent of farms on slopes exceeding 10 degrees show some level of erosion. On slopes of 25 degrees or more, 42 percent of the land showed strong or very strong signs of erosion.[69]

As of 2005, farmland and grazing areas with a slope greater than 25 degrees amounted to 15 million acres of land. Much of this acreage is located in the Yangtze River and Yellow River basins. In order to address the extensive soil erosion occurring on this land, the Chinese government created the Grain for Green Program (also known as the Conversion of Cropland to Forests and Grasslands Program). Under the program, the government pays farmers in cash and grain to plant grassland, to establish orchards of commercial tree species such as fruit trees, or to reforest the land for ecological benefits.

As a result of Grain for Green, the Chinese government intends to convert 37 million acres of low-yielding farmland to grassland and to reforest another 42 million acres of barren land and mountains. As well, the Chinese State Council passed laws that prohibit clearing sloped land for agricultural purposes. In 2002, new laws required farmland on slopes greater than 25 degrees to be restored to grassland and forests or to be converted to terraced land.

Researchers estimate that the total loss of grain production from the program will be 2 to 3 percent. Since the change will take place over a 10–30-year period, effects at the national level are expected to be minimal. However, at the local level, some areas will lose as much as 20 percent of agricultural lands, and some counties in western China are projected to experience a 50 percent drop in grain production.[70] Considering that western China has some of the poorest areas in the nation, many challenges still remain to ensure the availability of food for the farmers participating in the project.

Future Issues

Estimates place the population of China at 1.6 billion by 2030. As the population becomes more dense, the impacts of deforestation, land reclamation, road construction, and mining may have significant negative influences on flood control.

Forecasts project up to 600 million people relocating from rural locations to urban areas by 2030, placing 68 percent of China's population in cities and towns.[71] In urban areas, residents have higher dependence on city infrastructure, such as water supply, electricity, and roads. The impact of flooding in these areas, therefore, is higher. Flood-mitigation programs in many areas are still unable to handle 20-year floods. Coastal regions will experience further complications from global warming, which is expected to cause sea levels to rise six or more inches by the middle of the century.

Management of water resources in China reaches farther than the mitigation of immediate droughts or floods. China has the highest agricultural production of all nations and has major industries that are highly dependent on agricultural products. Any disruption due to flooding, drought, or mismanagement of water resources can have major economic implications for the entire country as well as for international trade.

To deal with many of these issues, the Chinese government approved the controversial South-North Water Transfer Project in 2001. The project is designed to divert almost 12 trillion gallons of water, equivalent to half the annual water consumption of California, from the Yangtze and Yellow Rivers and their tributaries into northern China through three man-made channels. At a cost of $62 billion, the project is projected to run through 2050.[72] Of the three channels, "The eastern route, which mostly follows the ancient Grand Canal, is largely done," according to a December 31, 2008, article in the *Wall Street Journal.* The article continues, "The mountainous western route, which is the most controversial and technically challenging, isn't slated for completion until 2050. The central section was supposed to start operation in 2010, but officials now say it will be launched in 2014."[73] Criticisms of the project include concerns about increased water pollution and increased sediment in rivers that are tapped for the transfer as well as negative impacts on river navigation, irrigation, and municipal supplies for residents near the siphoned rivers. Some critics assert that the entire project will prove to be only a "temporary fix."[74]

THE FAMINE IN ETHIOPIA IN 1984–1985

In October 1984, a report by the British Broadcasting Corporation (BBC) aired around the world, showing shocking and tragic images from a deadly famine ongoing in Ethiopia. Film footage of emaciated and starving men, women, children, and infants drew worldwide attention to the fact that hundreds of thousands of people had died of malnutrition and disease in Ethiopia and other Sahel countries and an estimated 150 million more Africans were at risk of death in the coming months if aid did not arrive. In response, individuals, groups, and governments sent millions of dollars of aid to Ethiopia and other famine-stricken countries, including Sudan, Chad, Somalia, Yemen, and others. The food and medicine saved countless lives, but the aid could not halt the catastrophic disaster that had been unfolding for two decades in Ethiopia. By 1986, an estimated 1.5 million Ethiopians had perished of malnutrition and disease, rounding out 20 years of famine—the deadliest famine ever recorded in Africa.

Malnourishment is not uncommon in individuals and communities around the world, but famine occurs on a greater scale than a community or town. Famine can cripple and destroy an entire region of a country or multiple countries. Technically speaking, this disaster occurs "when large numbers of people in a region cannot obtain sufficient food, and widespread, acute malnutrition occurs."[75] Acute malnutrition refers to a condition in which energy and nutrients from food intake are insufficient, and the body begins to break down fat and muscle for bodily functions. The condition slows the body's metabolism, interferes with the maintenance of normal body temperature, and weakens the immune system and kidney function. A person suffering from acute malnutrition becomes less able to work or concentrate and more susceptible to disease. In the most severe cases, fluid under the skin, or edema, occurs in the lower limbs in addition to skin lesions, thinning hair, and an enlarged liver. Without immediate treatment, the condition is usually fatal.

Like most famines, the famine disaster in Ethiopia was rooted in multiple factors, some natural and some of human origin. Natural causes of famine include a prolonged shortage of rainfall, or drought, resulting in the failure of crops and death of livestock and, consequently, severe food scarcity. In addition to natural causes of famine, human actions and decisions can contribute to or exacerbate famine conditions. Such factors include war and other forms of armed conflict, political turmoil, and land use. The 1984–85 famine in Ethiopia resulted from a deadly fusion of these natural and human causes, and media coverage routinely linked the natural and human causes of the catastrophe. In November 1984, the journalists James Wilde and Pico Ayer wrote, "After ten years of drought and civil war, 12 of the country's 14 provinces have been laid waste by a famine of biblical proportions. More than 40 percent of the country's 42 million people are malnourished, and 2.2 million have left their homes to wander in search of food."[76]

Geography and History of Ethiopia

The global community has a long history of famines, including the Irish Potato Famine in 1845–50 and China's Great Leap Famine in 1959–61; however, today, famine is largely limited to African nations. "At the beginning of the twentieth century, famine was still a global tragedy," according to the International Food Policy Research Institute. "At the turn of the twenty-first century, famines are largely confined to Africa."[77] Geographically, Africa has a swath of vulnerability to famine in the Sahel, a zone of transition between the Sahara in the north and the savannahs of the Sudan in the south. This zone stretches like a belt from the Atlantic Ocean to the Red Sea, covering more than 1.1 million square miles (2,849 sq km) in the countries of Senegal, Mauritania, Mali, Burkina Faso, Niger, Nigeria, Chad, Sudan, Somalia, Ethiopia, Algeria, and Eritrea. Of these countries, Ethiopia has most notably drawn the world's attention to famine in the Sahel.

The Federal Democratic Republic of Ethiopia is located in the Horn of Africa, in the eastern end of the Sahel zone. It is one of the most ancient countries in the world, with historical roots reaching back to the 10th century

B.C.E. Today, with a population of 88 million people, it is the second most populous country in Africa, after Nigeria, and the 14th most populous country in the world. In 1990, the World Bank ranked Ethiopia as the poorest country in the world. Thirty-eight percent of the population lives below the poverty line, with 85 percent of the workforce based in agriculture.[78] Ethiopia's extreme poverty and dependence on subsistence farming contribute to its great vulnerability to famine.

While the primary natural cause of famine is prolonged periods of below-average precipitation, a population's vulnerability to famine rises sharply when it depends on subsistence farming and livestock for nourishment or when most of the population is poor or underemployed, as is true in much of Ethiopia. In communities in which farmers are required to produce an entire year's food on small plots, the farmers are almost never able to store food or sell food on the open market for income. Consequently, these communities are nearly always vulnerable to food shortages caused by drought. In Ethiopia, a single year of below-average rainfall can be the tipping point to famine conditions.

Famine, however, is rarely the result of drought alone. Rather, this disaster results from drought and its resulting crop failures combined with one or more human factors such as armed conflict and governmental policies. The famines in Africa during the 1970s and 1980s illustrate this point. Since around 1970, all of northeastern Africa had experienced lower-than-normal rainfall and recurring droughts, causing food scarcity again and again. In 1984, an estimated 150 million Africans were threatened by famine. While farmers in some countries were able to replenish food supplies and reserves between droughts, thereby lessening (though not negating) the impact of famine, Ethiopians experienced a harsher reality. Here, factors of armed conflict, political turmoil, and poverty combined with droughts to produce two decades of unrelenting famine.

In late 1984, at the time the BBC report came out, Ethiopia was mired in civil war, and this was not the first such event in the country's recent history. When war and drought occur simultaneously, food prices increase quickly and dramatically. Armed conflict disrupts the agricultural cycle, displaces farmers from cropland, prevents goods from being carried to market, depletes food stores, and drives prices to unrealistic levels. Fighting in Ethiopia cut off supply routes, preventing people from traveling in search of food, and placed men's labors in warfare instead of in the cultivation of food. At the same time, the Ethiopian government splurged on a $100 million celebration of its 10th year in power, centering the celebration in the capital city of Addis Ababa while rural communities in the north were starving. Not only that, but critics charged the Ethiopian government with withholding food supplies that would have gone to rebel-occupied territory.

Social and economic factors also often influence the progress of a famine. Levels of food consumption define the extent of the disaster, not the amount of food available. In some cases, economic factors have a significant impact on the severity of famine. When crops fail in an extremely poor area, enough food may be available but at prices unaffordable to those who need the food. In this case, demand failure causes food supplies to be moved away from the areas that most need it to areas where merchants can sell it at current prices. Conversely, if the population in need of food has the necessary money, but sufficient food cannot be provided, supply failure occurs.

International Response in 1984–1985

The BBC report galvanized people and governments to action, but it was not the first report of famine and food scarcity in Ethiopia and other African countries. For the previous two years, relief organizations and the Ethiopian government had issued appeals for aid, but the appeals went unheeded. Earlier in 1984, the U.S. government had canceled "an economic-policy initiative that would have provided Africa with $75 million for development next year [1985]. Why? The administration had insisted that the money go only to governments that reject socialism," but Ethiopia's government was Marxist-Leninist.[79] In light of such failures to provide assistance in a timely manner, the international aid community was charged in the media with being slow to respond to the disaster. Perhaps it was the emotional impact of graphic images of near-death human beings that finally triggered the sudden and large-scale response, as opposed to the drier numerical figures in previous warnings. Mohammed Amin, the camera operator whose images helped spark the international outpouring of aid, lamented, "Why did [the famine] have to wait for a ten-minute TV film to awaken public sympathy?"[80]

FIRST RESPONSE

The major causes of death during a famine are malnutrition resulting from lack of food, dehydration brought on by diarrhea, and communicable diseases such as measles. Emergency response measures must address all three areas to deal effectively with the famine. Typically aid organizations distribute general rations as the main component of food relief efforts. A general ration is composed of grains, oils, protein sources, and a mix of vitamins to supplement the nutritional needs of children. The energy content of the ration ranges from 2,250 to 2,500 calories. When workers can maintain distribution without interruption, most people within the famine area can survive on the general ration. Typically, a family receives an allotment of general rations equal to the total number of family members. The content of the ration is based on adult needs, and children may need fewer calories. The result is that families with young children have a small excess of calories. This buffer helps make up for food spoilage or interruptions of relief efforts. Workers distribute rations every four to 10 days. This regular distribution pattern builds confidence in the food supply and discourages hoarding. It also helps to reduce deaths that occur when people devour rations too quickly and then are left starving, or when people sell some of their rations for income.

In the days following the BBC broadcast, the United Nations approved $416 million for food assistance to Ethiopia and other famine-stricken countries in Africa. The United States pledged $45 million in food aid to Ethiopia, and the European Community pledged $42 million to various African countries. Individuals in the United States and other countries donated money to relief organizations including CARE, Oxfam America, the Red Cross, and UNICEF. People raised money by hosting fund-raising lunches and staging hunger fasts, among other efforts. By November, relief workers were arriving in Africa with planeloads of food and medicine. Already in Ethiopia, 300,000 or more people had died, and another 200,000 had died in Mozambique. In Senegal, Chad, Mali, and Sudan, people were dying by the thousands.

Getting food and medicine to Ethiopia and other countries was an accomplishment born of enormous effort and organization, but the work did not stop there. Now the supplies had to be distributed to the people in need as rapidly as possible in the face of challenges including poor or absent road systems and civil war. Half of the population lived a two-day walk from the nearest road, and in all of Ethiopia there existed only about 6,000 trucks, most of which were in use by the military. A few hundred trucks were located to use in the transport of food, medicine, and other supplies. The first shipments of supplies forced workers to iron out numerous kinks in the supply chain, from aging and slow-moving cranes at the docks to government red tape. A month later, the process was moving more smoothly. By then, the Ethiopian government had waived handling charges on relief shipments, and two transport planes were shuttling loads of supplies from Addis Ababa to distribution points around the country.

During a famine, relief programs must address victims who have developed malnutrition or who have other special needs. For instance, a malnutrition rate above 10 percent for children under age five signals the need for supplementary feeding. The most effective means of distributing supplementary foods is to set up feeding centers that prepare meals for immediate consumption. Workers may also distribute extra rations as needed, but since there is no assurance the food will be consumed by the person for whom it is intended, this method does not guarantee the nutrition of the targeted groups.

In Ethiopia, aid workers set up camps for direct feeding and ration distribution, and starving people walked for miles to stand in long lines waiting for a meal. During 1984–85, nearly 300 camps were set up around Ethiopia. In early 1985, the camp at Korem fed more than 10,000 children daily. In the early days of relief work, the camp at Bati was receiving more than 1,000 people a day, with people arriving faster than sufficient supplies to aid them. A survey by the International Food Policy Research Institute showed that 35 percent of households in the country had at least one child admitted to a feeding camp, and "most of these camps admitted children who were under intensive care for up to nine months."[81] Along travel routes and in the camps, people continued to die of malnutrition and disease, but relief workers were making a difference. The death rate began to fall, from hundreds of dead per day to a couple of dozen per day. Still, the famine continued, reaching its peak in 1985, causing a continued need for the food and medicine supplied by relief agencies.

At the time of the famine's peak, popular musicians, led by Bob Geldof, orchestrated the now-famous Live Aid event that raised millions of dollars for Ethiopian famine relief. On July 13, 1985, the concert was held simultaneously in Wembley Stadium in London, England, where 72,000 people attended, and in John F. Kennedy Stadium in Philadelphia, Pennsylvania, where 99,000 people attended. Satellites linked the concerts and broadcast the event by television to millions of viewers (a BBC report states 1.5 billion viewers)[82] around the world. The British concert concluded with a performance of "Do They Know It's Christmas," a song Bob Geldof and his charity group Band-Aid had released the previous December to raise money for Ethiopia. The American concert concluded with "We Are the World," a song performed by a group of 47 artists working together as USA for Africa, a group led by Michael Jackson and Lionel Richie. The Live Aid event raised an estimated $234–$283 million.

Twenty-five years later as the anniversary of Live Aid approached, the BBC reported that millions of dollars of funds raised by Live Aid and other relief organizations had been diverted in Ethiopia by rebel soldiers disguised as grain traders, who used the money to buy weapons.[83] This rebel army, led by Meles Zenawi, went on to overthrow the government, and Zenawi became prime minister of Ethiopia in 1991. The BBC report drew angry denials from Bob Geldof and doubt from some others; nevertheless, the report cites an

American CIA report and testimonials of former rebels as evidence of the theft of aid.

THE RELIEF SUPPLY CHAIN

Aid agencies that carry out famine-relief operations need to move supplies (typically food and medicine) from suppliers to the victims of the disaster. With large-scale operations, this work is extremely complicated. Often supplies must be obtained from international sources that are thousands of miles from the disaster victims. Suppliers ship the goods using bulk transportation such as ships, railways, or cargo planes. A medium transport plane, for example, can carry shipments of 15–20 tons; ships or barges can carry larger quantities. Generally, the supplier of the goods is responsible for transportation of supplies to the port of entry, the location where the supplies enter the famine-stricken country.

As goods arrive at the port of entry, the humanitarian organization becomes responsible for transportation. Workers must transfer goods to short-haul vehicles capable of carrying loads of two or three tons so that goods can be moved from the port to warehouses as soon as possible. This task requires a sophisticated monitoring system to receive, catalog, and dispatch goods to appropriate storage facilities. As a constant stream of supplies enters the port of entry, short-haul transportation must be managed precisely to keep the goods moving toward their final destination. The complexities of this and the previous stage can be greatly reduced or eliminated if supplies can be obtained locally.

During the final stage, relief workers deliver the supplies to operation centers in the field that will distribute the provisions to those in need. At this stage of the delivery process, the conditions of roads and landing strips, or the lack thereof, limit what sorts of transport vehicles can be used. Relief workers commonly use trucks, planes, or helicopters capable of carrying one ton of supplies.

During all three stages of supply transportation, a cold chain must be maintained. The cold chain maintains precise temperature requirements for transporting vaccines. Composed of cold-storage rooms, sealed cold boxes, and self-powered refrigerators, this system must provide an unbroken means of refrigeration from the supplier to the field operations center. Otherwise, vaccines may spoil or lose effectiveness.

Throughout the operation, three- or four-person assessment teams continually monitor the crisis. These teams gather statistics such as daily death rates, types of disease encountered, and nutritional status of victims. The teams also examine the types and amounts of food available, the effectiveness of food distribution, the extent of health programs such as measles immunizations, and the quality of water and sanitation programs.

MEDICAL AID

Centers set up for supplemental feeding also serve several other purposes. As children arrive for prepared meals, workers measure their height and weight to ensure their weight gain is adequate. These measurements also provide a measure of the population's overall health during the crisis. Workers often use supplementary food centers to implement immunization programs. As the second major component of famine response, vaccinations against communicable diseases, mainly measles, must be administered to prevent outbreaks. Health workers also watch for other diseases known to cause increased mortality in malnourished children, such as diarrheal disease, malaria, and acute respiratory infections. Such diseases often spread rapidly in densely populated refugee camps due to overcrowding, inadequate sanitation, and contaminated water supplies.

The third major condition that health workers must control during a famine is diarrhea. Workers take preventative actions against this condition by dealing with possibilities of contamination anywhere in the hygiene loop. The hygiene loop is made up of the water supply, water transport, water storage, food preparation, latrines, and personal hygiene. Contamination in any of these areas often leads to diarrheal disease. Patients suffering from diarrhea die most often from dehydration, so the major focus of treatment is oral rehydration therapy (ORT). ORT uses a solution of sugar, salt, baking soda, potassium chloride, and water to restore the body's electrolyte balance and hydration levels. The dry ingredients of ORT are available in standard medical kits from WHO, UNICEF, and other NGOs. In emergency situations, the solution can be made from table salt, baking soda, potassium chloride, sugar, and water.

EFFECTIVENESS OF RELIEF EFFORTS

Relief efforts that were centered on the provision of food and medical care were carried out in Ethiopia from November 1984 through September 1986, a period of nearly two years. The activities achieved measurable progress. By June 1986, "the proportion of children below 80 percent of the average weight for height had fallen to less than 3 percent—from a high of 35 percent in July 1985."[84]

The emergency relief system was not without complications and complaints. Despite mass vaccinations of children entering the camps, half of the children admitted contracted an illness during their stay. Most of these illnesses were due to poor sanitation facilities at the makeshift camps, and it is worth noting that epidemics of more serious diseases such as cholera were successfully avoided. The main foods supplied at the feeding camps were wheat flour and soy-fortified sorghum grits, *faffa* (a processed supplement), corn-sugar milk, and fish powder. However, the sorghum grits and fish powder were so unpalatable to children and adults that these rations were soon designated for babies. Other problems involved communication problems between camp staff and community members, often as a result of language barriers. Recipients of food aid reported that families were not receiving enough rations for household size. The means used to register and monitor participants in food programs has a direct effect on the success of the program. An adequate registration system ensures that families periodically receive food but cannot receive more than their allotted share, but a registration system that fails to account for all people consuming an allotment of rations can result in a nutrition shortfall in a family or group.

In September 1986, organizers shifted the focus of their activities from emergency relief to rehabilitation activities, and the feeding camps were closed. Unfortunately, drought struck the country again in 1987 and then again in 1988, and epidemics of malaria and meningitis broke out. Emergency feeding measures were resumed in late 1988, but by mid-1989 the rate of severe malnutrition had climbed back to 18 percent. This chain of occurrences helped to support the predictions of experts that the famine problem in Ethiopia and other African countries could not be fixed with one energetic relief effort. The social and economic infrastructure of the countries would have to change.

Current Situation

A quarter-century after the 1984–85 famine in Ethiopia, the deadly cycle of drought and famine continues. A June 2010 report from IRIN, a project of the UN Office for the Coordination of Humanitarian Affairs, stated that "about 13 million Ethiopians—nearly one-sixth of the population—receive some form of foreign aid."[85] With 13 million people in need of food aid, humanitarian organizations lament the stagnating economy that seems to be slowing donations and lowering available funding for relief programs. Half of the children in Ethiopia suffer from malnutrition, and in August 2009 UNICEF said, "There are growing concerns about the impact of relief food shortfalls on already vulnerable children" in Ethiopia.[86] In 2009, donors to international aid organizations like UNICEF gave around $176 million in cash and kind (such as food donations), less than half of what they gave in 2008.[87]

As in the past, Ethiopia's famine problem is not attributable to drought alone but also encompasses factors of political and social policies, both within the country and internationally. Land ownership is one such policy. In accordance with the Ethiopian constitution, since 1975, private ownership of land has been prohibited. Land is owned by the state, which leases it to tenants for agricultural or other purposes. This did not change when the country established a new constitution in 1995, which granted citizens the right to own private property. However, land is still considered common property, and the constitution specifically provides that land cannot be sold or exchanged. Land allotments for family farms are so small as to be insufficient. According to the International Food Policy Research Institute, the system of state ownership of land helped to hobble the country's ability to feed itself. "Individual households, which were responsible for at least 90 percent of national production, were granted access rights to a maximum of 10 hectares [about 25 acres] for private production. . . . In practice, average holdings were between 1 and 2 hectares [between 2.5 and 5 acres] per household."[88] The system makes it impossible to increase the size of a farm to increase profits, to borrow money using personally owned land as collateral, or to perform other business practices that could help a family rise out of poverty. Most of Ethiopia's agricultural land was placed in state farms. "Of an estimated 750,000 hectares [1.85 million acres] of commercial farmland under cultivation before 1974, 67,000 [165,557 acres] were converted to state farms that, beginning in 1979, were operated under the jurisdiction of a new Ministry of State Farms. . . . However, despite large investments and operating costs, these grossly inefficient farms were responsible for only 4.2 percent of main-season cereal production in 1988–89."[89] Because of the absence of a sufficient food safety net, in the 1980s and on into the 21st century, Ethiopians have been continuously at high risk of food insecurity.

Even when humanitarian organizations have food or other aid to distribute, governmental policies in Ethiopia can make it difficult to carry out vital work. In 2009, Ethiopia's parliament passed a law to regulate the activities of charities and foreign humanitarian groups in the country. Per the law, any organization that gets 10 percent or more of its funding from outside the country is considered foreign. This law, according to the journalist Kassahun Addis, "restricts charity work on issues related to gender, ethnicity, children's rights and conflict resolution, and bars advocacy activities. The government says the law is meant to ensure that charities focus on development, but many fear it will deter those working in the field from taking bold actions like advocating for the hungry." Dealing with policies such as this one can slow or halt the administration of lifesaving activities.

Counterstrategies

In the arsenal used against famine and food insecurity in Ethiopia and other countries in the Sahel, famine prediction tools are among those most relied upon. In predicting famine, there are two important models: The food-supply model compiles statistics from national harvest predictions, measurements of food storage, and expected food consumption rates to determine if adequate food is available within a region. The food-demand model compares the market availability and price of food with the buying power of the region's population. Sudden increases in prices or decreases in income signal the beginning of famine. Humanitarian organizations such as the USAID Famine Early Warning System (FEWS) and the Global Information and Early Warning System of FAO use famine prediction data to map the vulnerability of regions, countries, and geographical zones to food insecurity. Famine prediction and vulnerability mapping cannot prevent but can mitigate disaster by providing information necessary to save lives.

Vulnerability mapping produces an assessment of areas with a predisposition to famine by examining leading indicators to famine, including rainfall shortages, heavy debt burden, low food reserves, political instability, economic crises, and income decline. For example, in 1995–96, the mapping of leading indicators revealed that the population in the Sahel region of Africa that was at risk for a lack of sufficient food leapt from 70,000 to 1.7 million.[90] As a famine takes hold, this mapping process provides valuable information concerning the potential geographic boundaries of the disaster and its ability to spread to surrounding areas.

Vulnerability maps are used in famine early warning systems. Internationally, the most significant early warning systems for famine are FEWS and the Global Information and Early Warning System of FAO. Both organizations collect data and provide support within the Sahel zone. Funded by USAID, FEWS collaborates with other U.S. organizations such as NOAA, NASA, and the USGS, as well as international partners, to provide famine vulnerability data for 25 countries in Africa, Central America, and Asia. FEWS publishes a monthly food security update detailing the conditions in areas susceptible to famine. For instance, FEWS determined that during 2010, 5.23 million people in Ethiopia would need emergency food assistance through the end of the year. In addition to people being supported through the emergency relief program, 7.8 million chronically food insecure people in 2010 were receiving cash and food under the Productive Safety Net Program, an Ethiopian government aid program launched in 2005.[91] FEWS also publishes alerts for unfolding situations. For instance, in October 2009, crop yield reductions in Niger, in the Sahel zone, prompted FEWS to forecast sufficient food supplies through the end of 2009. "Thereafter, declining food stocks, rising cereal prices, and falling livestock prices will reduce household purchasing power, causing moderate to high food insecurity, especially among agropastoral and pastoral households, through August 2010."[92]

Traditionally, one of the chief ways in which vulnerability maps and famine predictions have been used has been to organize and distribute food aid, whether on a large scale, as was done in Ethiopia during 1984–85, or on a smaller scale, as has been done in Ethiopia in years since then. But after a quarter-century of food aid flowing into Ethiopia and little evident change in the country's ability to feed itself, some experts have begun to question whether the global system of food aid should be reworked.

In January 2007, the FAO issued a report stating the need for an overhaul of the international food aid system, pointing out that "one third of aid resources never reach beneficiaries" for reasons such as "tying" the aid to specific conditions. For example, according to FAO, "The world's leading food donors spend as much as half of their food aid budgets on domestic processing and shipping by national carriers."[93] Better uses of these funds would include supplying cash or food coupons directly to those in need and repairing roads and infrastructures within the country to increase access to available food. In eight of the nine years from 2000 to 2008, Ethiopia was either the largest or second-largest recipient of food aid worldwide. In countries like Ethiopia where food insecurity is chronic, FAO warns that "donors and recipients can get caught in a 'relief trap' in which longer-term development strategies are neglected."[94] Despite the push for food aid reform, FAO acknowledges that in humanitarian crises there is no substitute for food aid, and it has certainly saved millions of lives.

ERUPTION OF NEVADO DEL RUIZ, COLOMBIA, 1985

On November 13, 1985, the Nevado del Ruiz volcano in western Colombia erupted, ejecting approximately 26 million cubic yards (20 million cubic meters) of hot ash and rocks into the air and across the ice- and snow-covered slopes of the volcano. Pyroclastic flows, consisting of hot volcanic debris, and pyroclastic surges, consisting of hot clouds of gas and ash, raced across the frozen slopes, rapidly melting the snow and ice. The hot water mixed with dirt and volcanic debris to form dense, deadly lahars that rushed down crevices and canyons toward Chinchiná, Armero, and other smaller villages. A few hours later, more than 23,000 people had lost their lives in the lahars, with at least 20,000 of these deaths occurring in Armero. The volume of magma ejected from Nevado del Ruiz was only about 3 percent of that ejected by Mount St. Helens in 1980; however, the high-volume debris flows of the Colombian volcano produced the second deadliest eruption of the century worldwide (after Mt. Pelée in 1902, which killed 30,000). Twice before in Nevado del Ruiz's recorded history, it had produced deadly lahars in the same areas where Armero and other villages had since grown up. In 1985, the deadly Nevado del Ruiz disaster focused the world's attention on the crucial need for understanding a volcano's history as part of the assessment of its hazards.

The Eruption and Lahars

The lahars struck Armero after nightfall on November 13, 1985—a Wednesday—but the deadly sequence of events at Nevado del Ruiz had begun hours earlier. At 3:06 P.M. (some reports say 3:08 P.M.), a cloud of ash and steam exploded from the mountain and rose several miles into the air. Because thick clouds obscured the volcanic activity from view, no one was immediately aware of the eruption. Two hours later, ash and pumice began to fall on Armero. Residents, however, remained calm and went on with their routines after the mayor, via radio, and the local priest, using the church's public address system, assured them that they were in no danger. At around 7:00 P.M., the Red Cross ordered an evacuation of the town, but the shower of ash and pumice stopped shortly after this, and the evacuation order was canceled. No one expected the series of explosions that occurred just after 9:00 P.M., which sent molten rock into the air and triggered the lahars that struck Armero nearly two hours later.

Eyewitness accounts of the disaster were later provided by outsiders who, for one reason or another, were caught up in the events. One group of outsiders was a small team of Colombian scientists who had been sent to monitor Nevado del Ruiz by the Central Hydroelectric Company of Caldas (CHEC), which maintained several local dams that produced electrical power for the state of Caldas. The team included Bernardo Salazar, Néstor García, and Marta Lucía Calvache; they had set up their operations in a small house in the farming community of Arbolito, just three miles northwest of the volcano. A second group of outsiders were two dozen geology students from the University of Caldas who had stopped in Armero for the night, two hours short of their intended destination of Ibagué. The third outside account came from Manuel Cervero, a pilot who happened to be flying overhead when Nevado del Ruiz erupted.

Late in the afternoon on the day of the eruption, the Colombian scientists followed their usual routine of collecting data from the equipment on the mountain. Pulling the drum from the seismograph on the ridge above Arbolito, they noticed an unusual scribble. But clouds obscured the mountain and there had been no discernable sign of activity an hour earlier when the signal registered. A better-trained eye might have realized that the markings indicated that an eruption had already occurred. By 7:30 that evening, ash ejected in the earlier steam explosion had stopped falling, replaced by heavy rain. Around 9:00 P.M., the scientists in Arbolito went to sleep. Meanwhile, in response to the ashfall, the emergency committee of the state of Tolima instructed civil defense stations and the Red Cross to be on alert and to sound the alarm in low-lying areas if reports of mudflows were received.

Colombia had alerted all aircraft to avoid the volcano by 30 miles, but the notice apparently did not reach Manuel Cervero, the pilot of the DC-8 cargo transport flying 7,000 feet above Nevado del Ruiz when it erupted violently just after 9:00 P.M. Cruising at 24,000 feet, Cervero later described a reddish illumination reaching to 26,000 feet and his plane's being covered with ash. With the aircraft's windows obscured by ash, he flew by instruments only. He diverted his flight to the city of Cali, where he landed safely 20 minutes later.

In Arbolito, the blast from the volcano awakened the scientists, and at approximately 9:15 P.M., Bernardo Salazar delivered a frantic message to the capital of Colombia: "The volcano erupted . . . Nevado del Ruiz erupted!" Moments later, another blast came, and then a third explosion lit up the clouds around the mountain's summit. The sky flashed with lightning as the heat of the blasts created a thunderstorm. Pebbles of volcanic rock, or pumice, began to rain down, forcing Salazar and the others to retreat into the house. Minutes later, the ground began to rumble, and the sound grew stronger every second. The scientists stared down into the valley beneath them. They could see nothing in the darkness, but they knew what was happening. Salazar immediately transmitted another message, yelling over the noise. "Manizales, the mudflows are coming. Alert the state of Tolima immediately!"[95] Radio stations soon issued red alerts for the western communities of Caldas. Only minutes later, the national television station broadcast news of the eruption.

In Armero, the ash stopped falling, replaced by torrential rain. Over the church's loudspeaker, the priest entreated the residents of the city to remain calm. Several of the geology students staying there reported that the mayor announced over the radio that there was nothing to worry about. The announcement, made around 9:35 P.M., instructed the population to cover their mouths with a wet handkerchief if ash began to fall again. No word of Salazar's call had yet reached the town.

At the peak of Nevado del Ruiz's eruption, the heat began melting millions of tons of ice on the mountain's glacial cap. The melting ice mixed with pumice, steam, and ash from the eruption, forming a thick wall of hot mud that rushed down the riverbeds. Picking up more debris as it thundered along, the 60-foot wall ripped through the valley at around 50 MPH. At 10:30 P.M., the mudflow hit the city of Chinchiná, 15 miles west of the volcano. Only the residents living closest to the river had been evacuated, under the assumption that only the lowest areas would be affected. But the mud flowed faster and higher than anyone expected. A few villagers fled up the valley's steep slopes, but most had too little time. In less than a minute, 1,000 of the city's residents perished.

The lahars sped down the mountain, funneled by the valleys of the Azufrado and Lagunillas Rivers. In the Azufrado River valley, the level of the mud peaked at 100 feet. In response, officials in Ibagué, the capital of Tolima, decided to set the evacuation of Armero in motion. Before they could transmit their orders, Armero lost power, and the town's telephones, radios, and televisions ceased to work. The earlier assurances from the priest and mayor were the last official information that the 29,000–30,000 residents of the city had received. A few warnings filtered in from ham radio operators. Living high on the valley's walls, they witnessed the mud as it rushed down the valley, and they transmitted warnings to Armero. A few residents who received the information immediately fled, but as the news spread, chaos filled the darkened streets. Authorities had told the people of Armero that they would have two hours to evacuate, but with the mud already halfway to the city, no official evacuation order had been given.

At 10:45 P.M., more than 90 minutes after the eruption began, civil defense finally alerted firefighters in Armero. Having no electrical power to sound the alarm and no plan for evacuation, they went from street to street blowing whistles, knocking on doors, and calling for people to leave for higher ground. Even then, many refused to believe the danger. Even the mayor, when alerted by civil defense, discounted the threat and returned home to his family. His previous statement concerning the people's distrust of the authorities rang true. Many residents trusted only the words of the mayor and the priest, ignoring warnings from other sources. At 10:50 P.M., the mayor made his last transmission over the rumble of the mud slamming into the town. "Wait a minute. I think Armero is being flooded."[96]

Across town, the geology students climbed to the third floor of their hotel. From there, they witnessed a chaos of cars and people trying to flee. They thought it was an earthquake. Then, the wall of mud hit the hotel, fracturing the cement walls. Seconds later, the mud burst through the walls. José Restrepo, one of the students, held tightly to a slab of concrete as the mud flowed by. A large water tank, carried by the mudflow, stopped just short of crushing him. The mud lifted the university's bus level with him, and seconds later it exploded. After several minutes, the mud ceased to flow. Cries for help rang out all around. Restrepo tried to help a young girl, but she was already dead. Heavy rain continued to fall as people attempted to dig themselves out of the thick mud. Then, the ground rumbled again as a second wave of mud hit the city. A concrete wall protected Restrepo, but many people around him disappeared into the mud.

Disaster Assessment

As the sun rose the next morning, the nightmare became fully evident. The mudflows had completely destroyed Armero. Mud covered everything, including the living and the dead. Some people called for help, and others struggled against the mire, but most were motionless. Dead livestock littered the area as human survivors crawled to higher ground, hoping for rescue. Survivors heard a crop duster flying over the city just after 8:00 A.M., but the plane did not turn, and no other aircraft were heard for over an hour. Fernando Rivera, the pilot of the plane, surveyed the destruction and reported that the city of Armero no longer existed. He saw no survivors in the sea of mud, likely because the mud completely covered the people on the ground. Red Cross and civil defense immediately sent emergency workers to the city, but the mudflows had destroyed the bridges and roads, blocking access.

Survivors did what they could to help one another, wrapping wounds with whatever they could find and scrounging food scattered in the mud. Late in the afternoon, a helicopter arrived and transported a handful of people away from the devastation. The day turned to night, and no other help arrived. The next morning, the victims continued to move to higher ground, scavenging food and clothing from the few houses that remained standing. More helicopters arrived during the second day, but rescuing the thousands of stranded people progressed at an agonizingly slow pace.

On the third day, the mud had dried enough to walk on, enabling people to move more easily to higher ground. Ironically, one of the mostly undamaged parts of town was a cemetery perched high on a hill, and helicopters dropped food and water for the starving people gathered there. Many of Armero's survivors forced themselves to walk to Guayabal, a village about two miles away. They crowded into the small town, which by now was lined with hospital tents. Rescue workers passed out water and food. Victims searched for lost relatives in the crowds, but most never found their loved ones. Of the 29,000–30,000 residents of Armero, more than 20,000 perished in the disaster, and 5,000 others were injured. An additional 3,000 residents of other villages perished as well. The lahars killed thousands of livestock, completely covered 8,500 acres of farmland, and destroyed 4,650 houses.[97]

Geography and History of Nevado del Ruiz

Nevado del Ruiz is a stratovolcano in the Andean Volcanic Belt that runs north-south in western South America. The volcano is located approximately 80 miles west of Colombia's capital city, Bogotá. With a summit reaching more than 17,500 feet, Nevado del Ruiz is Colombia's tallest volcano. Blanketing the summit are nearly 10 square miles of snow and ice, which, when combined with the heat of a volcanic eruption, can create lahars such as the ones that have occurred three times in recorded history.

The volcano's deadly history goes back to at least 1595. That year, an eruption sent mudflows into the same areas as in 1985, killing hundreds of people. In a text written in 1625, the Spanish priest Pedro Simón described the 1595 event as four extraordinary thunderclaps followed by pumice the size of ostrich eggs falling to the ground. After that, according to the account, the Río Gualí and the Río Lagunillas, rivers fed by snow melting on the mountain, "ran so full of ash that it looked more like a thick soup of cinders than water. Both overflowed their channels leaving the land over which they flowed so devastated that for many years afterward it produced nothing but weeds."[98]

In 1845, another eruption generated lahars that killed hundreds more people. The following year, a Colombian naturalist named Joaquín Acosta described the eruption as first "a subterranean noise" and "trembling of the ground" followed by "an immense flood of thick mud which rapidly filled the bed of the river, covered or swept away the trees and houses, burying men and animals."[99] According to Acosta's account, the mudflow wiped out the entire population of the upper Lagunillas Valley. He estimated that 300 tons of mud and debris washed down into the valley.

With written records of Nevado del Ruiz's deadly eruptions as recently as 140 years previous, why was the 1985 eruption so unexpected and, consequently, so deadly? In the intervening years, people forgot about the eruption and built towns on top of the deposits of hardened mud, volcanic ash, and debris of previous lahars. The first warning signs that the volcano was awakening were observed in December 1984, when mountain climbers reported feeling earthquakes and seeing clouds of gases rising from the summit of Nevado del Ruiz. Guests at the *refugio*, a rustic two-story hotel just 2,000 feet below the volcano's summit, felt tremors. Climbers staying at the *refugio* reported that the ice that normally obscured the crater had melted, and there was a strong smell of sulfur. Two seasoned climbers, John Jairo Gutierrez and Luis Fernando Toro, set out to examine the summit and found the formerly smooth summit had become a 500-foot-wide pool of mud with multiple columns of steam billowing upward. They immediately notified authorities, who in turn notified the civil defense for the state of Caldas. Nevado del Ruiz had awakened.

Local Preparedness

THE SCIENCE TEAM

The first local response to the unfolding situation came from CHEC out of concern for the damage an eruption could cause to several of its dams that produced electrical power for the state of Caldas. CHEC assigned three of its scientists the task of recording activity at the volcano. The team consisted of Bernardo Salazar, a civil engineer, Nestor Garcia, a chemical engineer assigned to test the acidity and temperature of steam venting from the mountain, and Marta Lucía Calvache. Only 26 years old, Calvache had no experience with active volcanoes. Ironically, her three months' experience in New Zealand studying geothermal energy made her Colombia's leading expert on volcanoes. The team set up its operation in a small house in the farming community of Arbolito, a few miles northwest of the volcano.

Pablo Medina was vice president of the financial corporation of Caldas, a division of the Colombian government. He convinced local coffee growers to assist in the task of monitoring the volcano, and an entire floor of a local bank building, which became known as Piso Once ("eleventh floor"), was lent to the cause, becoming a sort of headquarters. Finding that Colombia had no native experts on volcanoes, Medina contacted the United Nations for assistance. In response to Medina's request, help soon arrived from Ecuador, Switzerland, the United States, and Italy. However, conflicting opinions among the experts confused the local team and Colombian officials. In May 1985, an American volcanologist named Pete Hall, who ran the volcano observatory in Quito, Ecuador, arrived as a representative of the United Nations. He made several recommendations concerning the monitoring system on the mountain and evacuation procedures.

EQUIPMENT

The equipment necessary to monitor volcanic activity within Nevado del Ruiz consisted of two key pieces. First was a seismograph, a rotating drum eight inches in diameter and 18 inches long wrapped with a piece of paper on which data would be recorded. The second piece of equipment was a seismometer, a sensor buried in the ground to detect activity within the mountain. A seismometer can detect the movement of gases, the fracturing of rocks, and the vibrations created by flowing molten rock. As the seismometer sends data to the seismograph, a graphite needle (similar to a pencil) set against the paper-covered drum records the data, much like an electrocardiogram records rhythms of the heart.

The problem at Nevado Ruiz was that, by late June, the scientists had collected only three seismographs from the electric company, but they had no seismometers. At the time that Nevado del Ruiz began to rumble, only six seismometers existed in Colombia. Spread across the country, the instruments were used by the Los Andes Geophysics Institute to provide data for earthquakes, and the seismometers were too distant to record activity at Nevado del Ruiz. An appeal to the United Nations Disaster Relief Office went through to the USGS. In the aftermath of the eruption of Mount St. Helens five years earlier, the USGS had become one of the world's most seasoned veterans of volcanic emergencies. The USGS agreed to send three seismometers to the scientists in Colombia but declined to send personnel with the equipment, citing the urgency of volcanic activity at Mount St. Helens and in Hawaii.

The scientists at Arbolito installed the seismometers as best they could and managed the seismic monitoring stations around the mountain to measure its activity. Each seismograph contained a drum that slowly revolved during the day as a pen etched the activity on a piece of paper. As the day progressed, the pen shifted, leaving a spiral of seismic data for a 24-hour period. Each day started before sunrise, because the task of changing out these drums in the equipment took much of the day. The poor quality of the roads slowed travel, and some of the machines required access by foot. In addition, in order to gain the best measurements possible, the team scattered the seismographs around the massive mountain, which increased travel times significantly.

The first expert technical assistance did not come until August, when the Swiss volcanologist Bruno Martinelli arrived with three additional sets of seismographs. Together with Bernardo Salazar and Juan Duarte, one of the technicians who maintained the equipment, he set about installing the new seismographs on the mountain and repositioning the existing equipment for better coverage. Working out of the *refugio*, the three braved harsh conditions. They carried some of the equipment by hand up to the crater, and by late August they had completed the task. After briefly training the scientists in how to collect data from the equipment, Bernardo Salazar and Bruno Martinelli returned to Piso Once on September 10. Martinelli was to depart the country the next day. Instead, reports of ash falling at the *refugio* prompted them to return the next day to retrieve the two remaining occupants at the hotel. As they drove, lightning lit up the night and rocks the size of golf balls, ejected from the volcano, pelted their car.

Following the September eruption, another civil engineer, Fernando Gil, became part of the scientific team in Arbolito. Gil's task was to assist in gathering and analyzing data from the seismographs. He and Bernardo Sala-zar, the other civil engineer, meticulously gathered and cataloged each day's data and delivered it to Piso Once.

INTERPRETING THE DATA

At Piso Once, Pablo Medina received the biweekly deliveries of seismo-graphic data from the scientists monitoring Nevado del Ruiz. The Los Andes Geophysics Institute at the Jesuit University in Bogotá tried to take control of the data coming from Nevado del Ruiz, but since their expertise was limited to earthquakes and their funding was limited, the responsibility of interpret-ing the data fell to the National Institute of Geology and Mines, a government agency in Bogotá. Unfortunately, this organization possessed no expertise in volcanoes either, had scarce funding, and had no equipment. Critics often target this organization for its lack of action, but as early as August 1985, Alfonso Lopez issued a warning in the El Tiempo newspaper. He described the volcano as unstable and gave it a one in four chance of eruption.

The eruption on September 11 brought new attention to the situation at Nevado del Ruiz. Soon, Piso Once bustled with scientists, press, and occa-sional visits by government officials. The eruption also prompted a new appeal for help, which brought Pete Hall back on a second trip. He found that very little had been done with the recommendations he made during his first visit. The monitoring system could barely accomplish its task, and no evacu-ation plans had been put in place. He also learned that the scientists at Piso Once still delivered all of the seismic data to Bogotá for analysis, but Bogotá had not returned a single report in over a month and a half. Seeing the situ-ation, Pete Hall recruited Fernando Muñoz, a professor at the University of Caldas, and paired him with Bernardo Salazar, the scientist working for the electric company who had been monitoring the volcano for months. Hall gave them a half-hour crash course in reading the seismographs. Later Muñoz reflected, "What Pete Hall was showing us was how to read earth-quake signals, not volcanic signals. At the time, I had no idea there was a difference."[100]

DISASTER PREPAREDNESS

In September, Father Raphael Goberna, director of the Geophysics Institute, published the following statement in Magazin 8 Dias: "Today, because of our monitoring, we can tell the entire population that nothing is happening on the volcano that threatens the inhabitants of the region. If and when it is nec-essary to declare an emergency, the Instituto Geofísico de los Andes will do it. Before would only alarm the population without reason."[101] Geographic boundaries complicated the task of preparing potential victims for the possibility of disaster. The more prosperous Cafetera region to the west took measures to prepare for a possible disaster, but the Tolima region to the east moved much more slowly, devoting serious attention to the threat only after dire warnings from scientists in Bogotá. On October 7, 1985, the National Institute of Geology and Mines issued a report calling for the evacuation of towns at the base of the volcano. That same day *El Espectador* reported that an eruption would likely give the residents of Tolima's low-lying city of Armero only two hours to evacuate before a mudflow obliterated the city.

The activity on Nevado del Ruiz had already caused problems for the surrounding area. Snowmelt contaminated with sulfur drained down into the valleys, causing several aqueducts to be shut down. Ash from the eruption sickened at least 4,000 cattle; some had died already. A landslide created a natural dam on the Lagunillas River. A large lake had formed behind it, threatening to flood the cities below. The mayor of Armero filed a request for government funds to drain the gathering water. As studies to determine how to demolish the dam dragged on, he expressed his frustration in an October 8 interview in *Consigna* magazine, saying, "The new Emergency Committee does not have the necessary information or the financial resources to do anything in the event of a catastrophe. For this reason, the people have lost confidence in the veracity of the information and have commended their fate to God."[102]

Just weeks before the Nevado del Ruiz eruptions, scientists at the University of Manizales created hazard maps with colored pencils. They showed pyroclastic flows in red, ash falls in yellow, and mudflows in brown. The maps again confirmed that an eruption would completely bury the city of Armero.

International Response in 1985

Following the eruption of November 13, 1985, international aid arrived quickly. The United States dispatched helicopters from Panama to assist with the rescue operations, and public and private contributions from the United States neared a million dollars. The League of Red Cross and Red Crescent Societies in Geneva managed the distribution of $1,250,000 worth of equipment and food donated by 12 other countries.

Within hours of the eruption, images of the disaster filled the world's airwaves. The world awoke to images of a city destroyed by mud and news of the rising death toll. The photographer Frank Fournier snapped a photograph of 13-year-old Omayra Sanchez. Pinned by concrete and debris, Omayra sat neck-deep in water for three days before dying from hypothermia. She became the face of the disaster, and Fournier's photograph won the 1985 World Press Photo award. Twenty years later, a British news article stated,

"When the photo was published, many were appalled at witnessing so intimately what transpired to be the last few hours of Omayra's life. They pointed out that technology had been able to capture her image for all time and transmit it around the globe, but was unable to save her life."[103]

John Tomblin from the United Nations Disaster Relief Office and Bruno Martinelli, the Swiss volcanologist, returned immediately to Piso Once. A borrowed telemetered seismograph also arrived. After five days, scientists from around the world had overrun Piso Once. USGS scientists arrived with six more telemetered seismographs, which were promptly installed on Nevado del Ruiz. Norm Banks of the USGS also received the okay to bring his Volcano Crisis Assistance Team to Colombia, and he arrived and took charge of the USGS scientists. Soon the equipment on the mountain began transmitting information to computers at Piso Once, allowing scientists to analyze the data significantly faster than before. But by a month later, the mountain had begun to calm.

Current Situation and Counterstrategies

At the time of the disaster, businesses in Armero were the source of a significant portion of the region's economic activities, especially of the coffee industry. Just two weeks after the tragedy, the Colombian government created the *Resurgir* program to direct the reconstruction of the area. (*Resurgir* means "to arise.") Created under the economic emergency article of the Colombian Constitution, the program was granted special powers to accomplish its task, including the abilities to pardon mortgages, issue death certificates quickly, reduce taxes, grant tax benefits to encourage donations, and import equipment tax-free. The total estimates for reconstruction came to $305 million in U.S. dollars.[104]

When Nevado del Ruiz became active in 1984, there existed no emergency team of volcanologists, in Colombia or elsewhere, who could deploy to the scene of an awakening volcano to assess and mitigate the hazard and issue authoritative warnings to local officials and the public. The tragedy at Armero helped to convince experts of the need for such a contingency. Following the tragedy at Armero, "the U.S. Agency for International Development's Office of Foreign Disaster Assistance (OFDA) asked the USGS to help create a program to reduce fatalities and economic losses in countries experiencing a volcano crisis. Toward this goal," according to the USGS, "the two agencies jointly established the Volcano Disaster Assistance Program (VDAP). VDAP consists of a small core group of scientists at CVO [Cascades Volcano Observatory], a larger group of other contributing USGS scientists, and portable volcano-monitoring equipment ready for rapid deployment.

The VDAP crisis-response team is mobilized and sent overseas only when the U.S. State Department receives an official request from a country with a restless volcano. Once on site, the VDAP team works with local scientists and technicians to help them provide timely information and analysis to emergency managers and public officials. VDAP also conducts training exercises and workshops in volcano-hazards response with foreign scientists and emergency-management officials."[105]

Nearly six years after the Armero tragedy, the new VDAP demonstrated how volcano crisis-response could save tens of thousands of lives. On April 2, 1991, the Mount Pinatubo volcano in the Philippines awoke with a series of steam explosions and earthquakes. The USGS scientists with VDAP deployed to the scene, where they worked with local scientists to evaluate the volcano's threat. They issued evacuation warnings to towns and cities in the hazard zone as well as to residents of the U.S. Clark Air Force Base and the U.S. Naval Station at Subic Bay. More than 75,000 people, including 18,000 American servicemen and dependents, safely evacuated, and millions of dollars' worth of military equipment was moved to safety.[106] On June 15, Mount Pinatubo erupted, sending pyroclastic flows and lahars surging through 30 miles of valleys. Falling ash destroyed towns and cities and damaged and destroyed buildings on the military bases.

Since its first deployment in 1991, VDAP has responded to more than a dozen volcano crises in Central and South America, the Caribbean, Africa, Asia, and the South Pacific, saving countless lives and learning crucial lessons regarding how best to reduce risks from volcano hazards in the United States and abroad.[107]

The 1985 tragedy at Armero, Colombia, is the largest volcano-glacier disaster in recorded history and is among the world's greatest volcano disasters. Nevado del Ruiz, along with sister volcanoes in Los Nevados National Natural Park, continues to pose significant hazards to surrounding communities. In 1989, the Colombian government established the National Disaster Prevention and Attention System to coordinate local and national agencies for disaster risk prevention, mitigation, and emergency preparedness for volcanic eruptions and other natural and human-caused disasters.

In the quarter-century since Nevado del Ruiz's deadly eruption, populations in communities downstream from the volcano's glaciers have continuously grown. According to a 2007 report in *Annals of Glaciology*, "Since 1985, fumarolic activity has been continuously observed at the Arenas crater," which was the vent for the 1985 eruption.[108] Hazards today include the risk of avalanche resulting from unstable slopes or fractured glaciers as well as the risk of an eruption that triggers glacier melt. Scientists have stated that a new eruption of Nevado del Ruiz could be equally destructive, if not more destructive, than the eruption in 1985.[109]

[1] EM-DAT. "The International Disaster Database." Available online. URL: http://www.emdat. be/advanced-search. Accessed May 25, 2010. These figures are for the decade 2000–2009.

[2] United States Geological Survey. "Magnitude 9.0 Sumatra-Andaman Islands Earthquake FAQ." Available online. URL: http://earthquake.usgs.gov/eqcenter/eqinthenews/2004/ usslav/faq.php. Accessed November 19, 2009.

[3] USGS Earthquake Hazards Program. "Tectonic Summary." Available online. URL: http:// neic.usgs.gov/neis/eq_depot/2004/eq_041226/neic_slav_ts.html. Accessed November 19, 2009.

[4] National Geophysical Data Center. "Tsunami Events Full Search, 2004." Available online. URL: http://www.ngdc.noaa.gov/nndc/struts/results?bt_0=2004&st_0=2004 &type_ 8=EXACT&query_8=None+Selected&op_14=eq&v_14=&st_1=&bt_2=&st_2=&bt_1= &bt_10=&st_10=&ge_9=&le_9=&bt_3=&st_3=&type_19=EXACT&query_19=None+ Selected&op_17=eq&v_17=&bt_20=&st_20=&bt_13=&st_13=&bt_16=&st_16=&bt_6=&st _6=&bt_11=&st_11=&d=7&t=101 650&s=70. Accessed November 19, 2009.

[5] RMS. "Managing Tsunami Risk in the Aftermath of the 2004 Indian Ocean Earthquake and Tsunami." Available online. URL: http://www.rms.com/Publications/IndianOcean TsunamiReport.pdf. Accessed November 19, 2009.

[6] ———. "Managing Tsunami Risk."

[7] UN Office of the Special Envoy for Tsunami Recovery. "The Human Toll." Available online. URL: http://west.wwu.edu/atus/graphics/museum/images/room4.pdf. Accessed November 19, 2009.

[8] ———. "The Human Toll." Available online. URL: http://west.wwu.edu/atus/graphics/ museum/images/room4.pdf. Accessed May 25, 2010.

[9] RMS. "Managing Tsunami Risk."

[10] ———. "Managing Tsunami Risk."

[11] *Indonesian Health Profile 2004.* Jakarta: Ministry of Health Republic of Indonesia, 2006, p. 92.

[12] *Indonesian Health Profile 2004.*

[13] George Pararas-Carayannis. "The Great Earthquake and Tsunami of 26 December 2004 in Southeast Asia and the Indian Ocean." Available online. URL: http://www.drgeorgepc.com/ Tsunami2004Indonesia.html. Accessed November 19, 2009.

[14] RMS. "Managing Tsunami Risk."

[15] Asian Disaster Preparedness Center. "Thailand-Post Rapid Assessment Report: Dec 26th 2004 Tsunami." Available online. URL: http://www.adpc.net/tudmp/Tsunami%20Rapid%20 Assessment%20Report_15%20Feb.pdf. Accessed November 19, 2009.

[16] Phi Phi Islands. "How Ko Phi Phi Was Affected by the Asian Tsunami." Available online. URL: http://www.phi-phi-islands.com/phi-phi-tsunami.php. Accessed November 19, 2009.

[17] UN Office for the Coordination of Humanitarian Affairs. "Earthquake and Tsunami: Indonesia, Maldives, Sri Lanka." Available online. URL: ftp://ftp.fao.org/FI/DOCUMENT/tsunamis_05/OCHA/ReportNo18.pdf. Accessed November 19, 2009.

[18] Indian Institute of Technology Kanpur. "Effects of the December 2004 Indian Ocean Tsunami on the Indian Mainland." Available online. URL: http://www.iitk.ac.in/nicee/RP/2006_Effect_EQSpectra.pdf. Accessed November 19, 2009.

[19] ———. "Effects of the December 2004 Indian Ocean Tsunami."

[20] Earthquake Engineering Research Institute. "The Great Sumatra Earthquake and Indian Ocean Tsunami of December 26, 2004." Available online. URL: http://www.eeri.org/lfe/pdf/sri_lanka_tsunami_eeri_survey.pdf. Accessed November 19, 2009.

[21] Department of Census and Statistics Sri Lanka. "Summary of the Number of Housing Units Damaged Due to Tsunami Waves by Affected Districts." Available online. URL: http://www.statistics.gov.lk/Tsunami/census/Summarynew.pdf. Accessed November 19, 2009.

[22] Republic of the Maldives. "Maldives Tsunami: Impact and Recovery." Available online. URL: http://www.presidencymaldives.gov.mv/publications/TsunamiImpactandRecovery.pdf. Accessed November 19, 2009.

[23] International Federation of Red Cross and Red Crescent Societies. "Asia: Earthquakes & Tsunami." Available online. URL: http://www.reliefweb.int/rw/RWFiles2009.nsf/FilesByRWDocUnidFilename/ EDIS-7X2KXG-full_report.pdf/$File/full_report.pdf. Accessed November 21, 2009.

[24] United Nations Environment Programme. "After the Tsunami: Rapid Environmental Assessment," p. 128. Available online. URL: http://www.unep.org/tsunami/reports/TSUNAMI_report_complete.pdf. Accessed November 19, 2009.

[25] Georgia Tech Savannah Campus. "Tsunami Research: Madagascar." Available online. URL: http://www.gtsav.gatech.edu/cee/groups/tsunami/madagascar.html. Accessed November 19, 2009.

[26] Vanessa Gezari. "A Reporter's Account of Tsunami Crisis." Available online. URL: http://www.poynter.org/column.asp?id=60&aid=77249. Accessed May 25, 2010.

[27] UNESCO. "Indian Ocean Tsunami Warning System Up and Running." Available online. URL: http://portal.unesco.org/en/ev.php-URL_ID=33442&URL_DO=DO_TOPIC&URL_SECTION=201.html. Accessed May 25, 2010.

[28] Council on Foreign Relations. "Tsunami Disaster: Relief Effort." Available online. URL: http://www.cfr.org/publication/7792/tsunami_disaster.html. Accessed November 19, 2009.

[29] Tsunami Evaluation Coalition. "The Tsunami in Numbers—Stats and Facts." Available online. URL: http://www.alnap.org/initiatives/tec/resources.aspx. Accessed May 27, 2010.

[30] ———. "Funding the Tsunami Response," p. 8. Available online. URL: http://www.oecd.org/dataoecd/43/31/39412447.pdf. Accessed May 27, 2010.

[31] Council on Foreign Relations. "Tsunami Disaster: Relief Effort."

[32] ———. "Tsunami Disaster: Relief Effort."

[33] World Health Organization. "South Asia Earthquake and Tsunamis." Available online. URL: http://www.who.int/hac/crises/international/asia_tsunami/en. Accessed May 26, 2010.

[34] Council on Foreign Relations. "Tsunami Disaster: Relief Effort."

[35] Human Rights Watch. "After the Deluge." Available online. URL: http://www.hrw.org/reports/2005/india0505. Accessed May 26, 2010.

[36] ———. "After the Deluge."

[37] United Nations Development Programme, World Health Organization, and International Federation of Red Cross and Red Crescent Societies. "3rd TRIAMS Workshop Report." February 2009. Available online. URL: http://www.who.int/hac/crises/international/asia_tsunami/triams/en/index.html. Accessed May 26, 2010.

[38] Tsunami Evaluation Coalition. "Funding the Tsunami Response," p. 16.

[39] ———. "Funding the Tsunami Response," p. 21.

[40] ———. "Funding the Tsunami Response," p. 25.

[41] RMS. "Managing Tsunami Risk," p. 5.

[42] Pacific Tsunami Warning Center. "Tsunami Information Bulletin Numbers 001–002." ITIC Tsunami Bulletin Board, December 26, 2004. Available online. URL: http://ioc3.unesco.org/itic/files/PTWCmsg_26dec04.doc. Accessed November 19, 2009.

[43] *Rebuilding Livelihoods: Tsunami Response, Indonesia.* Food and Agriculture Organization of the United Nations, 2006, pp. 1–2.

[44] Food and Agriculture Organization of the United Nations. "Overall FAO Tsunami Emergency and Rehabilitation Response." Available online. URL: ftp://ftp.fao.org/FI/DOCUMENT/tsunamis_05/FAO_guiding/overview.pdf. Accessed May 26, 2010.

[45] USAID: Asia. "Highlights of Completed Projects." Available online. URL: http://www.usaid.gov/locations/asia/tsunami/photo_gallery.html#Indonesia. Accessed June 4, 2010.

[46] ———. "Aceh West Coast Highway Project: 2010." Available online. URL: http://indonesia.usaid.gov/en/USAID/Activity/210/Banda_Aceh_to_Calang_Road_ Project. Accessed May 25, 2010.

[47] English spellings of Chinese names for rivers and provinces vary slightly among sources, as do the measurements of river basins. The figures used here are taken from Peter H. Gleick. "China and Water" in *The World's Water: 2008–2009: The Biennial Report of Freshwater Resources.* Peter H. Gleick, et al. Washington, D.C.: Island Press, 2009, pp. 79–100.

[48] Ministry of Water Resources: The People's Republic of China. "Flood Control and Drought Relief in China." Available online. URL: http://www.mwr.gov.cn/english/fcdrc.html. Accessed June 2, 2010.

[49] Central Intelligence Agency. "The World Factbook: East and Southeast Asia: China." Available online. URL: https://www.cia.gov/library/publications/the-world-factbook/geos/ch.html. Accessed June 1, 2010.

[50] Bryan Lohmar, et al. "China's Agricultural Water Policy Reform." U.S. Department of Agriculture, Bulletin Number 782, p. 3.

[51] Richard Cowen. "Huanghe, the Yellow River." Available online. URL: http://mygeologypage.ucdavis.edu/cowen/~gel115/115CHXXYellow.html. Accessed November 19, 2009.

[52] Ministry of Water Resources: The People's Republic of China. "Flood Control and Drought Relief in China."

[53] Ministry of Culture, People's Republic of China. "Canals in Ancient China." Available online. URL: http://211.147.20.24/library/2008-02/01/content_26449.htm. Accessed June 2, 2010.

[54] Thayer Watkins. "The San-men (Three Gate) Gorge Dam on the Yellow River." Available online. URL: http://www.sjsu.edu/faculty/watkins/sanmen.htm. Accessed June 3, 2010.

[55] *Sydney Morning Herald.* "One Dam Mistake after Another Leaves $4.4bn Bill." Available online. URL: http://www.smh.com.au/articles/2004/05/21/1085120121829.html. Accessed November 19, 2009.

[56] Ministry of Water Resources: The People's Republic of China. "Flood Control and Drought Relief in China."

[57] Zhang Hai-Lum, et al. "Flood Control and Management for Large Rivers in China." *Workshop on Strengthening Capacity in Participatory Planning and Management for Flood Mitigation and Preparedness in Large River Basins.* Nanjing, China: Nanjing Institute of Hydrology and Water Resources, 2001.

[58] *China Daily.* "Floods Ravage South and East Provinces." Available online. URL: http://www.chinadaily.com.cn/english/doc/2005-06/23/content_453729.htm. Accessed November 19, 2009.

[59] Associated Press. "Flood Kills 115 in Southern China." Available online. URL: http://www.msnbc.msn.com/id/37328252/ns/weather. Accessed June 3, 2010.

[60] Ministry of Water Resources: People's Republic of China. "Flood Control and Drought Relief in China."

[61] ———. "Flood Control and Drought Relief in China."

[62] Lohmar, et al. "China's Agricultural Water Policy Reform," p. 4.

[63] ———. "China's Agricultural Water Policy Reform," p. iv.

[64] China Three Gorges Project. "Biggest Flood Control Benefit in the World." Available online. URL: http://www.ctgpc.com/benefifs/benefifs_a.php. Accessed November 19, 2009.

[65] International Rivers. "China's Three Gorges Dam." Available online. URL: http://www.internationalrivers.org/files/3Gorges_FINAL.pdf. Accessed November 19, 2009.

[66] Zhang Hai-Lum, et al. "Flood Control and Management for Large Rivers in China." *Workshop on Strengthening Capacity in Participatory Planning and Management for Flood Mitigation and Preparedness in Large River Basins.* Nanjing, China: Nanjing Institute of Hydrology and Water Resources, 2001, p. 58.

[67] International Rivers. "China's Three Gorges Dam."

[68] ———. "A Faultline Runs Through It." Available online. URL: http://www.internationalrivers.org/en/node/3980. Accessed November 19, 2009.

[69] H. L. Long, et al. "Land Use and Soil Erosion in the Upper Reaches of the Yangtze River: Some Socio-economic Considerations on China's Grain-for-Green Programme." *Land Degradation & Development,* vol. 17, pp. 593–594.

[70] Zhiming Feng, et al. "Grain-for-Green Policy and Its Impacts on Grain Supply in West China." *Land Use Policy,* vol. 22, 2005, pp. 301–302.

[71] Zhang Hai-Lum, et al. "Flood Control and Management," pp. 62–67.

[72] International Rivers. "South-North Water Transfer Project." Available online. URL: http://www.internationalrivers.org/en/china/south-north-water-transfer-project. Accessed November 19, 2009.

[73] Shai Oster. "China Slows Water Project." Available online. URL: http://www.international rivers.org/en/node/3705. Accessed June 4, 2010.

[74] ———. "China Slows Water Project."

[75] Frederick C. Cuny, et al. *Famine, Conflict, and Response.* West Hartford, Conn.: Kumarian Press, 1999, p. 1.

[76] James Wilde and Pico Iyer. "Ethiopia: The Land of the Dead." *TIME* (November 26, 1984). Available online. URL: http://www.time.com/time/magazine/article/0,9171,926987,00.html. Accessed June 10, 2010.

[77] Joachim von Braun, Tesfaye Teklu, and Patrick Webb. *Famine in Africa: Causes, Responses, and Prevention.* Baltimore, Md.: Johns Hopkins University Press, 1998, p. 1.

[78] Central Intelligence Agency. "The World Factbook: Ethiopia." Available online. URL: https://www.cia.gov/library/publications/the-world-factbook/geos/et.html. Accessed June 8, 2010.

[79] Wilde and Iyer. "Ethiopia: The Land of the Dead."

[80] ———. "Ethiopia: The Land of the Dead."

[81] Von Braun, Teklu, and Webb. *Famine in Africa*, p. 133.

[82] BBC. "1985: Live Aid Makes Millions for Africa." Available online. URL: http://news.bbc.co.uk/onthisday/hi/dates/stories/july/13/newsid_2502000/ 2502735.stm. Accessed June 11, 2010.

[83] Martin Plaut. "Ethiopia Famine Aid 'Spent on Weapons.'" Available online. URL: http://news.bbc.co.uk/2/hi/8535189.stm. Accessed June 11, 2010.

[84] Von Braun, Teklu, and Webb. *Famine in Africa*, p. 134.

[85] IRIN. "Ethiopia: Government Denies Food Aid 'Manipulated' for Political Gain." Available online. URL: http://www.irinnews.org/Report.aspx?ReportId=89382. Accessed June 11, 2010.

[86] Kassahun Addis. "Drought and Famine: Ethiopia's Cycle Continues." *TIME* (August 15, 2009). Available online. URL: http://www.time.com/time/world/article/0,8599,1915544,00.html. Accessed June 10, 2010.

[87] ———. "Drought and Famine: Ethiopia's Cycle Continues."

[88] Von Braun, Teklu, and Webb. *Famine in Africa*, pp. 27–28.

[89] ———. *Famine in Africa*, p. 28.

[90] Cuny, et al. *Famine, Conflict, and Response*, p. 35.

[91] Famine Early Warning System Network. "Ethiopia Food Security Update: May 2010." Available online. URL: http://www.fews.net/docs/Publications/ethiopia_FSOU_05_2010_final.pdf. Accessed June 15, 2010.

[92] ———. "Niger Food Security Alert." Available online. URL: http://www.fews.net/docs/Publications/Niger_alert_10_2009_final.pdf. Accessed November 19, 2009.

[93] Food and Agriculture Organization. "FAO Urges Food Aid Reform." Available online. URL: http://www.fao.org/newsroom/en/news/2007/1000482/index.html. Accessed June 11, 2010.

[94] ———. "FAO Urges Food Aid Reform."

[95] Victoria Bruce. *No Apparent Danger.* New York: HarperCollins, 2001, pp. 19–20.

[96] ———. *No Apparent Danger,* p. 64.

[97] Mauricio Ramirez. *MIS for Disaster Handling.* Cambridge: Massachusetts Institute of Technology, 1987, pp. 45–47.

[98] Bruce. *No Apparent Danger,* p. 51.

[99] ———. *No Apparent Danger,* p. 49.

[100] ———. *No Apparent Danger,* p. 56.

[101] ———. *No Apparent Danger,* p. 58.

[102] ———. *No Apparent Danger,* p. 58.

[103] BBC News. "Picture Power: Tragedy of Omayra Sanchez." Available online. URL: http://news.bbc.co.uk/2/hi/4231020.stm. Accessed November 18, 2009.

[104] Ramirez. *MIS for Disaster Handling,* p. 116.

[105] United States Geological Survey. "Mobile Response Team Saves Lives in Volcano Crises." USGS Fact Sheet 064-97. Available online. URL: http://pubs.usgs.gov/fs/1997/fs064-97. Accessed June 16, 2010

[106] ———. "Mobile Response Team Saves Lives in Volcano Crises."

[107] ———. "Mobile Response Team Saves Lives in Volcano Crises."

[108] Christian Huggel, et al. "Review and Reassessment of Hazards Owing to Volcano-Glacier Interactions in Colombia." *Annals of Glaciology,* vol. 45, 2007, p. 129. Available online. URL: http://www.igsoc.org/annals/45/a45A019.pdf. Accessed June 21, 2010.

[109] ———. "Review and Reassessment of Hazards," p. 130.

PART II

Primary Sources

4

United States Documents

This chapter contains primary sources with information about natural disasters that have occurred in the United States. The documents are divided into three sections—historical events, recent events, and disaster management information—and each section is presented in chronological order. Each section is an excerpt from a longer document. The complete document or article can be accessed by looking at the source information that follows each passage.

HISTORICAL EVENTS

1811 Missouri Earthquakes: Myron L. Fuller, "The New Madrid Earthquake" (1912) (excerpt)

Beginning on December 16, 1811, a series of earthquakes lasting nearly two months struck the small town of New Madrid, Missouri. The worst natural disaster to date in the United States, the tremors caused six confirmed deaths and completely destroyed New Madrid. The U.S. Department of the Interior prepared the following report concerning the disaster.

Introduction.
General Statement.
The succession of shocks designated collectively the New Madrid earthquake occurred in an area of the central Mississippi Valley including southeastern Missouri, northeastern Arkansas, and western Kentucky and Tennessee. . . . Beginning December 16, 1811, and lasting more than a year, these shocks have not been surpassed or even equaled for number, continuance of disturbance, area affected, and severity by the more recent and better-known shocks at Charleston and San Francisco. As the region was almost unsettled at that time relatively little attention was paid to the

phenomenon, the published accounts being few in number and incomplete in details. For these reasons, although scientific literature in this country and in Europe has given it a place among the great earthquakes of the world, the memory of it has lapsed from the public mind.

. . .

Field Work and Acknowledgments.

The writer's attention was first called to the region by Prof. E. M. Shepard, who had become interested in it because of the relation between artesian conditions and certain effects of the earthquake, which had become apparent during an investigation of underground waters. In the fall of 1904 Prof. Shepard and the writer traversed in a dugout the sunk lands along Varney River, near Kennett, Mo., and later made a trip on horseback up the old De Soto trail and along St. Francis River. The second trip was made under the guidance of Mr. C. B. Baily, city engineer of Wynne, Ark., who from timber explorations had become familiar with the earthquake features in the still almost untouched forests north of the St. Louis, Iron Mountain & Southern Railway between Memphis and Wynne. In the following year the writer, in connection with studies of underground water, covered most of the region reached by railroads in Arkansas and Missouri and made a wagon trip, again in company with Prof. Shepard, around Reelfoot Lake, in Tennessee. A short account embodying Prof. Shepard's observations appeared in 1905,[1] and a number of preliminary notes and popular papers by the writer have been published in magazines.[2]

In the present report an attempt is made to present a systematic record of some of the phenomena of this great earthquake, including both the contemporaneous and the present aspect of the resulting features.

The Story of the Earthquake.
Sources of Information.

The story of the earthquake is told in two ways—in the quaint, picturesque, and graphic accounts of contemporaries, and in the equally striking geographic and geologic records, which even now may be clearly read at many

---

[1] The New Madrid earthquake: Jour. Geology, vol. 13. pp. 45–62.

[2] Causes and periods of earthquakes in the New Madrid area, Missouri and Arkansas: Science, new ser. vol. 21, 1905, pp. 340–350; Comparative intensities of the New Madrid, Charleston, and San Francisco earth; quakes: Idem, vol. 23, 1905, pp. 917–918; Our greatest earthquakes: Pop. Sci. Monthly, July, 1906, pp. 76–86. Earthquakes and the forest: Forestry and Irrigation, vol. 12, 1906, pp. 261–267.

points in the region. For the sake of brevity only a single general account compiled from the early descriptions will be presented, but in the bibliography (pp. 111–115) references to original publications containing detailed narratives are given. In the discussion of the physiographic phenomena many references to reports of the old writers will also be found.

The contemporary accounts are doubtless exaggerated, for calm observation and accurate recording of an earthquake is impossible if the shocks are severe and dangerous. It is interesting to note, however, that, except a few features, such as the flashes of light (doubtless resulting from the general belief in the volcanic origin of the disturbance), most of the reported phenomena have been verified by the recent investigations. Fortunately a number of scientists or men of education were in or near the region during the period of disturbance and have given vivid pictures of their experiences.

. . .

Summary of the Disaster.

The evening of December 15, 1811, in the New Madrid area was clear and quiet, with no unusual conditions which could be regarded as portending the catastrophe soon to take place. A little after 2 o'clock on the morning of December 16, the inhabitants of the region were suddenly awakened by the groaning, creaking, and cracking of the timbers of the houses or cabins in which they were sleeping, by the rattle of furniture thrown down, and by the crash of falling chimneys. In fear and trembling they hurriedly groped their way from their houses to escape the falling débris, and remained shivering in the winter air until morning, the repeated shocks at intervals during the night keeping them from returning to their weakened or tottering dwellings. Daylight brought little improvement to their situation, for early in the morning another shock, preceded by a low rumbling and fully as severe as the first, was experienced. The ground rose and fell as earth waves, like the long, low swell of the sea, passed across its surface, tilting the trees until their branches interlocked and opening the soil in deep cracks as the surface was bent. Landslides swept down the steeper bluffs and hillsides; considerable areas were uplifted, and still larger areas sunk and became covered with water emerging from below through fissures or little "craterlets" or accumulating from the obstruction of the surface drainage. On the Mississippi great waves were created, which overwhelmed many boats and washed others high upon the shore, the return current breaking off thousands of trees and carrying them out into the river. High banks caved and were precipitated into the river, sand bars and points of islands gave way, and whole islands disappeared.

During December 16 and 17 shocks continued at short intervals but gradually diminished in intensity. They occurred at longer intervals until January 23, when there was another shock, similar in intensity and destructiveness to the first. This shock was followed by about two weeks of quiescence, but on February 7 there were several alarming and destructive shocks, the last equaling or surpassing any previous disturbance, and for several days the earth was in a nearly constant tremor.

For fully a year from this date small shocks occurred at intervals of a few days, but as there were no other destructive shocks the people gradually became accustomed to the vibrations and gave little or no further attention to them.

Source: Myron L. Fuller. "The New Madrid Earthquake." *United States Geological Survey*, Bulletin 494 (1912). Washington, D.C.: Government Printing Office, pp. 7–11.

1816 Cold Summer: John H. Dillingham, ed., "Science and Industry" (1904) (excerpt)

The massive volcanic eruption of Tamboro on the island of Sumbawa in Java caused "The cold summer of 1816," a period of unusually cold weather in the United States and in Europe. The debris from the explosion affected weather around the world. The following excerpt provides a monthly summary of the weather and its effects on crops for the northeastern United States in 1816.

Science and Industry.

A YEAR WITHOUT A SUMMER.—Under the above caption, a correspondent of the Boston *Globe* has detailed the very remarkable weather record of the year 1816. The vagaries of our climate are too well known to require comment; in fact, one of the early settlers of Pennsylvania was so struck with this feature of his new climatic environment that when writing to friends in England, be disposed of this ever at hand topic for conversation by saying: "We do not seem to have a climate here; only samples of weather." The summer of 1816 was the coldest ever known through Europe and America.

The following is a brief abstract of the weather during each month of the year.

First Month was so mild as to render fires almost needless in parlors. Twelfth Month previous was very cold.

Second Month was not very cold, being, with the exception of a few days, mild like its predecessor.

Third Month was cold and boisterous during the first part of it, but the remainder of the month was mild. A great freshet on the Ohio and Kentucky rivers caused a great loss of property.

Fourth Month began warm, but grew colder as the month advanced, and ended with snow and ice, and a temperature more like winter than spring.

Fifth Month was more remarkable for frowns than smiles. Buds and fruits were frozen; ice formed half an inch thick; corn was killed and replanted again and again, until deemed too late.

Sixth Month was the coldest ever known in this latitude. Frost, ice and snow were common. Almost every green thing was killed. Fruit was nearly all destroyed. Snow fell to the depth of ten inches in Vermont.

Seventh Month was accompanied by frost and ice. On the 5th ice formed of the thickness of common window glass throughout New England, New York, and some parts of Pennsylvania. Indian corn was nearly all destroyed, though some favorably situated fields escaped; this was true of some of the hill farms of Massachusetts.

Eighth Month was more cheerless, if possible; than the summer months already passed. Ice was formed half an inch thick. Indian corn was so frozen that the best part it was cut down and dried for fodder. Almost every green thing was destroyed, both in this country and in Europe. Very little corn ripened in the New England and Middle States. Farmers supplied themselves with con produced in 1815 for the needs of the spring of 1817. It sold at from $4 to $5 per bushel.

Ninth Month furnished about two weeks of the mildest weather of the season. Soon after the middle it became very cold and frosty; ice formed a quarter of an inch thick.

Tenth Month produced more than its share of cold weather; frost and ice were common.

Eleventh Month was cold and blustery. Snow fell so as to make good sleighing. Twelfth Month was mild and comfortable.

The above is a brief summary of "the cold summer of 1816," as it was called, the year showing the remarkable record of frost and ice every month therein. The sun's rays seemed to be destitute of beat throughout the year, and all nature was clad in a sable hue. The average wholesale price of flour during the year in Philadelphia was $18 per barrel.

Source: "Science and Industry." *The Friend,* vol. 77, no. 38 (1904). Philadelphia: Edwin P. Sellew, pp. 303–304.

1871 Peshtigo Fire, D. Appleton and Company, "Wisconsin" (1872) (excerpt)

On October 8, 1871, wildfires fanned by high winds cut off the small town of Peshtigo, Wisconsin, on all sides. Often called the "forgotten fire" because it occurred on the same day as the Great Chicago Fire, it caused between 1,200 and 2,400 deaths. This excerpt includes a description of the disaster prepared by the Milwaukee Relief Committee and the Wisconsin governor's appeal for help with the relief efforts.

WISCONSIN. A most calamitous event has rendered the year 1871 forever memorable in this State. In the last days of September and the first days of October, extensive fires overran the northeastern part, destroying much property and causing great distress. These were but the forerunner of an incomparably more disastrous conflagration on the 8th and 9th of October, wonderful alike for the vastness of its dimensions ten or more miles in width and of indefinite length; for the extraordinary, almost preternatural, manner of its progress; and for the incalculable amount of its destruction in property and in human life. This deluge of fire swept over four counties, and portions of two more, immersing them, as it were, in a tempestuous sea of flame, accompanied by a most violent hurricane, which multiplied the force of the destructive element. Forests, farm improvements, and entire villages, were consumed, and a very large number of men, women, anti children, perished. Even those who fled before the fire and sought refuge in cleared fields, swamps, lakes, and rivers, found no safety there, multitudes of them having been either burned, or died by suffocation or drowning.

One of the towns swept out of existence was Peshtigo, and its destruction is thus described by the Milwaukee Belief Committee:

The fire which destroyed Peshtigo occurred on the evening of the 8th of October, and history has never furnished a parallel to its terrible destructiveness. Shortly after the church-going people had returned from the evening service, an ominous sound was heard, like the distant roar of the sea, or of a coming storm. This increased in intensity, and soon the inhabitants became apprehensive of coming danger. Balls of fire were observed to fall like meteors in different parts of the town, igniting whatever they came in contact with. By this time the whole population were thoroughly aroused and alarmed, and caught up their children and what valuables they could hastily seize, and began to flee for a place of safety. Now a bright light appeared in the southwest horizon, gradually increasing till the heavens ware aglow with light. But a few moments elapsed after this before the horrible tornado of fire came upon the people, and enveloped them in flame, smoke, burning sand, and cinders. Those who had now reached the river or some other place of safety were suffocated and burnt to a cinder before they could advance a half-dozen steps farther. No pen can describe, no brush can depict, the realities of that night. Exaggeration would be utterly impossible. It defies human ingenuity.

The character of this fire was unlike any we have ever seen described before. It was a flame fanned by a hurricane, and accompanied with various electrical phenomena. Those that survived the terrible ordeal testify that they received electrical shocks, while they saw electrical flames flash in the air and dance over the surface of the earth around them, but the flash was past in ball an hour, though the fire continued to burn during the whole night. The tornado came from the southwest and swept over a tract of country eight or ten miles in width, and of indefinite length. The timber in its course was felled by the wind and burned by the fire, and every vestige of fence and building was swept away, with two or three exceptions. Sometimes the wind struck the earth with such force that the small undergrowth was torn up and heaped in rows, while at other times it would skip away from the earth. The whole population of Peshtigo village and of the farmlands in its vicinity was 2,000, and fully one-third of those perished on that fearful night. On the east shore of the bay, reports place the loss of life fully as high as at Peshtigo, making the entire loss of life reach the fearfully large number of 1,200. Some of the bodies were so thoroughly burned and consumed that they could be scooped up and held in the double hands. But the details and incidents are too harrowing to relate. We saw many children, some only one month old, which had been kept in the water the whole night, and yet survived. Some who were too ill to walk were taken from their beds and thrown into the water. About fifteen per cent. of those injured are so badly burned that it is impossible for them to recover. The others will be able to return to business in a month's time or less. The burns occurred most frequently upon the feet, hands, and face, and nearly all suffer from the inhalation of hot sand and cinders, and from the usual pulmonary complications of burns. The people have been literally stripped of every thing. Not a vestige of house or fences, or any thing of a combustible nature remains. A more desolate spectacle cannot be imagined or described.

At the tidings of so great a calamity, Governor Fairchild hastened to the place of suffering to provide for the relief of as much of it as it was in his power to do. He also appealed to the humanity of the citizens of the State by the following proclamation:

GREEN BAY, *October* 13, 1871.

To the People of Wisconsin:
The accounts of the appalling calamity which has fallen upon the east and west shores of Green Bay have not been exaggerated. The

burned district comprises the counties of Oconto, Brown, Door, and Kewaunee and parts of Manitowoc and Outagamie. The great loss of life and property has resulted from the whirlwind of fire which swept over the country, making the roads and avenues of escape impassable with fallen timber and burned bridges. The long drought had prepared every thing for the flames.

The loss of life has been very great. The first estimates were entirely inadequate, and even now it is feared that it is much greater than present accounts place it. It is known that at least 1,000 persons have been either burned, drowned, or smothered. Of these deaths 600 or more were at Peshtigo and adjacent places, and the others in Door, Kewaunee, and Brown Counties. Men are penetrating that almost inaccessible region for the purpose of affording relief, and I fear that their report will increase this estimate.

From the moat reliable sources of information I learn that not less than 3,000 men, women, and children, have been rendered entirely destitute. Mothers are left with fatherless children; fathers with motherless children. Children are left homeless orphans. Distress and intense suffering are on every hand, where but a few days ago were comfort and happiness. Scores of men, women, and little children, now lie helplessly burned and maimed, in temporary hospitals, cared for by their more fortunate neighbors.

These suffering people must be supplied with food, bedding, clothing, feed for their cattle, and the means of providing shelter during the winter. The response by the good people of Wisconsin has already been prompt and generous in meeting the immediate need, and is being faithfully and energetically distributed through the relief organizations at Green Bay, but provision must be made for many months in the future.

There are wanted flour, salt and cured meats, not cooked, blankets, bedding, stoves, baled hay, building materials, lights, salt, farming implements and tools, boots, shoes and clothing for men, women, and children, log-chains, axes with handles, nails, glass, and house-trimmings, and indeed every thing needed by a farming community which has lost every thing.

To expedite the transfers at Green Bay, all boxes should have cards attached to them, stating their contents. All supplies should be sent to Belief Committee of Green Bay. Money contributed should not be converted into supplies, but should be forwarded to the committee.

Depots have been established at Green Bay, under the management of a committee of public-spirited men, who have the confidence of all for the receiving and dispatching of supplies. They have organized a system of sub-depots contiguous to the burned regions, and steamboats and wagons are being sent out with supplies. Let us uphold their hands in the good work, and see that their depots be kept filled to overflowing. It is fortunate that we live in a wealthy and prosperous State, blessed with prosperity in business and overflowing harvests, and that thus we are by a wise Providence endowed with means to help our less fortunate neighbors.

I am urged by public-spirited citizens of the State to call an immediate extra session of the Legislature, to provide for this calamity. I have given serious attention to this suggestion, and have concluded not to do so, for the reason that the expense of such a session would be likely to equal the amount which the State would be asked to contribute. Believing, therefore, that the people and the Legislature will indorse my action in this emergency, I have, in conjunction with the State Treasurer, decided to advance such a moderate sum of money as seems to be appropriate, in addition to that contributed.

LUCIUS FAIRCHILD,
Governor of State of Wisconsin.

His appeal to the citizens was promptly responded to from all parts of the State, and liberal contributions in money, clothing, and provisions for the sufferers, were sent. The people also of some of the other States, and of British America and Europe, contributed large amounts of money, clothes, and supplies for the sufferers through the Governor, and through the Milwaukee and Green Bay Relief Committees. The money received for that purpose at the Executive office alone, until the end of the year 1871, amounted to $166,789.96; of which sum $111,397.23 still remained unexpended at that date.

As to the loss of property, especially in timber, saw-mills, and farm-products, consumed by this conflagration, it was reported as follows:

A medium estimate of damage to pine-lands in the Green Bay region is $400,000. The damage on the Wolf is figured at $300,000. There is abundance of hard wood left in places; the damage to individuals may amount to $300,000. The loss of the fifteen saw-mills burned is put at $225,000. The loss of cord-wood ties, hemlock-bark, etc., is set at $200,000. The losses of fences, buildings, wagons, cattle, crops, among the six hundred farmers, cannot be less than $600,000—making a total aggregate of more than $3,000,000, aside from those at Peshtigo.

The country through from Brown County north to Big Sturgeon Bay, for 400 square miles, is utterly devastated. At least 400 farms in

this tornado section alone are left desolate, stripped of every improvement. Fences, bans, dwellings, implements, furniture, wagons, har-
ness, and crops, all went up in a "whirlwind of fire." It will take thirty
years in that cold, hard soil for their timber to grow again. In the
aggregate, their losses must foot up to $1,000 a family. Farmers here
have saved half of their teams that were let loose in the woods, and a
third of their stock. But they have no hay, straw, grain, or feed of any
sort—not even the poor chance to browse in the woods. Nearly all,
with large families, have lost their last cow and pig.

Source: "Wisconsin." American Annual Cyclopedia and Register of Important Events of the Year 1871, vol. 11 (1872).
New York: D. Appleton and Company, pp. 773–775.

1900 Galveston Hurricane: Isaac M. Cline, "Special Report on the Galveston Hurricane of September 8, 1900" (1900) (excerpt)

*On September 8, 1900, a hurricane hit the city of Galveston, Texas, and
destroyed much of the city. Below is an excerpt from the official report filed by
the United States Weather Bureau.*

Storm warnings were timely and received a wide distribution not only
in Galveston, but throughout the coast region. Warning messages were
received from the Central Office at Washington on September 4th, 5th, 6th,
7th, and 8th. The high tide on the morning of the 8th, with storm warnings
flying, made it necessary to keep one man constantly at the telephone giving
out information. Hundreds of people who could not reach us by telephone
came to the Weather Bureau office seeking advice. I went down on Strand
Street and advised some wholesale commission merchants who had perishable goods on their floors to place them three feet above the floor. One gen-
tlemen has informed me that he carried out my instructions, but the wind
blew his goods down. The public was warned, over telephone and verbally,
that the wind would go by the east to the south and that the worst was yet
to come. People were advised to seek secure places for the night. As a result
thousands of people who lived near the beach or in small houses moved
their families into the center of the city and were thus saved. Those who
lived in large strong buildings, a few blocks from the beach, one of whom
was the writer of this report, thought that they could weather the wind and
tide. Soon after 3 p.m. of the 8th conditions became so threatening that it
was deemed essential that a special report be sent at once to Washington.

Mr. J. L. Cline, observer, took the instrumental readings while I drove first to the bay and then to the gulf, and finding that half the streets of the city were under water added the following to the special observation at 3:30 P.M.: "Gulf rising, water covers streets of about half city." Having been on duty since 5 A.M., after giving this message to the observer, I went home to lunch. Mr. J. L. Cline went to the telegraph offices through water from two to four feet deep, and found that the telegraph wires had all gone down; he then returned to the office, and by inquiry learned that the long distance telephone had one wire still working to Houston, over which he gave the message to the Western Union Telegraph office at Houston to be forwarded to the Central Office at Washington.

I reached home and found the water around my residence waist-deep. I at once went to work assisting people, who were not securely located, into my residence, until forty or fifty persons were housed therein. About 6:30 P.M. Mr. J. L. Cline, who had left Mr. Blagden at the office to look after the instruments, reached my residence, where he found the water neck-deep. He informed me that the barometer had fallen below 29.00 inches; that no further messages could be gotten off on account of all wires being down, and that he had advised everyone he could see to go to the center of the city; also, that he thought we had better make an attempt in that direction. At this time, however, the roofs of houses and timbers were flying through the streets as though they were paper, and it appeared suicidal to attempt a journey through the flying timbers. Many people were killed by flying timbers about this time while endeavoring to escape from the town.

The water rose at a steady rate from 3 P.M. until about 7:30 P.M., when there was a sudden rise of about four feet in as many seconds. I was standing at my front door, which was partly open, watching the water, which was, flowing with great rapidity from east to west. The water at this time was about eight inches deep in my residence, and the sudden rise of four feet brought it above my waist before I could change my position. The water had now reached a stage ten feet above the ground at Rosenberg Avenue (Twenty-fifth Street) and Q Street, where my residence stood. The ground was 5.2 feet elevation, which made the tide 15.2 feet. The tide rose the next hour, between 7:30 and 8:30 P.M., nearly five feet additional, making a total tide in that locality of about twenty feet. These observations were carefully taken and represent to within a few tenths of a foot the true conditions. Other personal observations in my vicinity confirm these estimates. The tide, however, on the bay or north side of the city did not obtain a height of more than fifteen feet. It is possible that there was five feet of backwater on the gulf side as a result of debris accumulating four to six blocks inland. The debris is piled eight to fifteen feet in height. By 8 P.M. a number of houses had drifted up and lodged to the east and southeast of my residence, and these with the force of the waves acted as a battering ram against which it was impossible for any building to stand for any length of time, and at 8:30 P.M. my residence went down with about fifty persons who had sought it for safety, and all but eighteen were hurled into eternity. Among the lost was my wife, who never rose above the water after the wreck of the building. I was nearly drowned and became unconscious, but recovered, though being crushed by timbers and found myself clinging to my youngest child, who had gone down with myself and wife. Mr. J. L. Cline joined me five minutes later with my other two children, and with them and a woman and child we picked up from the raging waters, we drifted for three hours, landing three hundred yards from where we started. There were two hours that we did not see a house nor any person, and from the swell we inferred that we were drifting to sea, which, in view of the northeast wind then blowing, was more than probable. During the last hour that we were drifting, which was with southeast and south winds, the wreckage on which we were floating knocked several residences to pieces. When we landed about 11:30 P.M., by climbing over floating debris to a residence on Twenty-eighth Street and Avenue P, the water had fallen four feet. It continued falling, and on the following morning the gulf was nearly normal. While we were drifting we had to protect ourselves from the flying timbers by holding planks between us and the wind, and with this protection we were frequently knocked great distances. Many persons were killed on top of the drifting debris by flying timbers after they had escaped from their wrecked homes. In order to keep on the top of the floating masses of wrecked buildings one had to be constantly on the lookout and continually climbing from drift to drift. Hundreds of people had similar experiences.

Sunday, September 9, 1900, revealed one of the most horrible sights that ever a civilized people looked upon. About three thousand homes, nearly half the residence portion of Galveston, had been completely swept out of existence, and probably more than six thousand persons had passed from life to death during that dreadful night. The correct number of those who perished will probably never be known, for many entire families are missing. Where twenty thousand people lived on the 8th not a house remained on the 9th, and who occupied the houses may, in many instances, never be known. On account of the pleasant gulf breezes many strangers were residing temporarily near the beach, and the number of these that were lost cannot yet be estimated.

Source: Isaac M. Cline. "Special Report on the Galveston Hurricane of September 8, 1900." *Galveston in Nineteen Hundred,* edited by Clarence Ousley. Atlanta, Ga.: William C. Chase, 1900, pp. 42–45.

1906 San Francisco Earthquake: Mary Austin, "The Temblor: A Personal Narration" (1907)

Mary Hunter Austin (1868–1934), an American writer who lived in San Francisco at the time of the 1906 earthquake, published her experiences in the form of a short story the following year.

The Temblor

THERE are some fortunes harder to bear once they are done with than while they are doing, and there are three things that I shall never be able to abide in quietness again—the smell of burning, the creaking of house-beams in the night, and the roar of a great city going past me in the street.

Ours was a quiet neighborhood in the best times; undisturbed except by the hawker's cry or the seldom whistling hum of the wire, and in the two days following April eighteenth, it became a little lane out of Destruction. The first thing I was aware of was being wakened sharply to see my bureau lunging solemnly at me across the width of the room. It got up first on one castor and then on another, like the table at a séance, and wagged its top portentously. It was an antique pattern, tall and marble-topped, and quite heavy enough to seem for the moment sufficient cause for all the uproar. Then I remember standing in the doorway to see the great barred leaves of the entrance on the second floor part quietly as under an unseen hand, and beyond them, in the morning grayness, the rose tree and the palms replacing one another, as in a moving picture, and suddenly an eruption of nightgowned figures crying out that it was only an earthquake, but I had already made this discovery for myself as I recall trying to explain. Nobody having suffered much in our immediate vicinity, we were left free to perceive that the very instant after the quake was tempered by the half-humorous, wholly American appreciation of a thoroughly good job. Half an hour after the temblor people sitting on their doorsteps, in bathrobes and kimonos, were admitting to each other with a half twist of laughter between tremblings that it was a really creditable shake.

The appreciation of calamity widened slowly as water rays on a mantling pond. Mercifully the temblor came at an hour when families had not divided for the day, but live wires sagging across housetops were to outdo the damage of falling walls. Almost before the dust of ruined walls had ceased rising, smoke began to go up against the sun, which, by nine of the clock, showed bloodshot through it as the eye of Disaster.

It is perfectly safe to believe anything any one tells you of personal adventure; the inventive faculty does not exist which could outdo the actuality; little things prick themselves on the attention as the index of the greater horror.

I remember distinctly that in the first considered interval after the temblor, I went about and took all the flowers out of the vases to save the water that was left; and that I went longer without washing my face than I ever expect to again.

I recall the red flare of a potted geranium undisturbed on a window ledge in a wall of which the brickwork dropped outward, while the roof had gone through the flooring; and the cross-section of a lodging house parted cleanly with all the little rooms unaltered, and the halls like burrows, as if it were the home of some superior sort of insect laid open to the microscope.

South of Market, in the district known as the Mission, there were cheap man-traps folded in like pasteboard, and from these, before the rip of the flames blotted out the sound, arose the thin, long scream of mortal agony.

Down on Market Street Wednesday morning, when the smoke from the burning blocks behind began to pour through the windows we saw an Italian woman kneeling on the street corner praying quietly. Her cheap belongings were scattered beside her on the ground and the crowd trampled them; a child lay on a heap of clothes and bedding beside her, covered and very quiet. The woman opened her eyes now and then, looked at the reddening smoke and addressed herself to prayer as one sure of the stroke of fate. It was not until several days later that it occurred to me why the baby lay so quiet, and why the woman prayed instead of flying.

Not far from there, a day-old bride waited while her husband went back to the ruined hotel for some papers he had left, and the cornice fell on him; then a man who had known him, but not that he was married, came by and carried away the body and shipped it out of the city, so that for four days the bride knew not what had become of him.

There was a young man who, seeing a broken and dismantled grocery, meant no more than to save some food, for already the certainty of famine was upon the city—and was shot for looting. Then his women came and carried the body away, mother and betrothed, and laid it on the grass until space could be found for burial. They drew a handkerchief over its face, and sat quietly beside it without bitterness or weeping. It was all like this, broken bits of human tragedy, curiously unrelated, inconsequential, disrupted by the temblor, impossible to this day to gather up and compose into a proper picture.

The largeness of the event had the effect of reducing private sorrow to a mere pin prick and a point of time. Everybody tells you tales like this with more or less detail. It was reported that two blocks from us a man lay all day with a placard on his breast that he was shot for looting, and no one denied the aptness of the warning. The will of the people was toward authority, and everywhere the tread of soldiery brought a relieved sense of things orderly and secure. It was not as if the city had waited for martial law to be declared, but as if it precipitated itself into that state by instinct as its best refuge.

In the parks were the refugees huddled on the damp sod with insufficient bedding and less food and no water. They laughed. They had come out of their homes with scant possessions, often the least serviceable. They had lost business and clientage and tools, and they did not know if their friends had fared worse. Hot, stifling smoke billowed down upon them, cinders pattered like hail—and they laughed—not hysteria, but the laughter of unbroken courage.

Source: Mary Austin. "The Temblor: A Personal Narration." The California Earthquake of 1906, edited by David Starr Jordan. San Francisco: A. M. Robertson, 1907, pp. 341–349.

1910 Avalanche at Wellington: "Topping v. Great Northern RY. CO." (October 12, 1914) (excerpt)

Shortly after 1:30 A.M. on March 1, 1910, an avalanche near the town of Wellington, Washington, knocked a passenger train into the Tye River. A total of 96 people perished, making it the worst avalanche in U.S. history in terms of lives lost. The details of the event are preserved in a lawsuit filed against the railroad company that operated the train.

Defendant railroad company operated a train on which decedent was a passenger westward between Spokane and Seattle leaving Spokane on the evening of February 22, 1910. The train reached Cascade tunnel at 5:30 on the morning of the 23d, at which time a snow storm, that had been raging in the mountains for two days, had not abated, and the line at points between W., the west end of the tunnel, and Seattle was blocked with snow. The train remained at the east end of the tunnel until the evening of February 24th, when it was taken through the tunnel and placed on a passing track on the side of the mountain near a hotel where the passengers could obtain food. The train remained here until February 28th, or the morning of March 1st, at about 1:30 o'clock, when an avalanche of snow swept down the mountain side and carried the train with it a distance of about 100 feet below the track, where the train was destroyed and decedent and other passengers killed. The tracks at W. consisted of a main line, two or three switches, and the passing track. The avalanche broke 400 or 500 feet above the track and extended from 1,500 feet to 2,000 feet along the track. The storm was unprecedented, accompanied by wind, thunder, lightning, and excessive snow, and the avalanche came without warning. The railroad had been operated at that point for 17 years, and though similar slides bad occurred, they had always before happened in gullies, and one had never been known to occur at the place where the train was parked. *Held,* that the accident was the result of vis major or act of God, and was not of such a character as to raise a presumption of negligence under the doctrine "res ipsa loquitur."

. . .

MOUNT, J. This action was brought by the plaintiff, through his guardian ad litem, to recover damages on account of the death of his father. The father, Edward W. Topping, was a young man 30 years of age. He was killed while a passenger on one of the trains of the Great Northern Railway Company, which was wrecked by an avalanche at Wellington, in this state, on March 1, 1910.

The cause of action was based upon the following paragraph of the amended complaint:

"That on, to wit, the 1st day of March, 1910, Edward W. Topping was a passenger for hire on one of defendant's trains, between the cities of Spokane, Wash., and Seattle, Wash., and en route to the latter, and while such passenger said train was, on said last-named date, through the negligence and carelessness of defendant, derailed and wrecked at or near the station called Wellington, on said road, the exact nature and extent of said acts of negligence on the part of defendant not being fully known to plaintiff, but well known and understood by defendant, and the said Edward W. Topping was then and there killed in and by said wreck and derailment of said train."

The answer of the defendant was a general denial and an affirmative defense of vis major, or act of God. The cause was tried to the court with a jury. At the close of the plaintiff's evidence, a motion for nonsuit was denied. At the close of all the evidence, a motion for a directed verdict was also denied. After the cause was submitted to the jury, a verdict was returned in favor of the plaintiff for $20,000. This appeal followed.

Numerous errors are assigned. But we are satisfied that the motion for nonsuit and motion for a directed verdict should have been sustained by the trial court, and for that reason we shall notice only these assignments.

The following facts are practically undisputed: Edward W. Topping took passage from Spokane on the defendant's train bound for Seattle on the evening of February 22, 1910. This train was a regular passenger train running from Spokane to Seattle across the Cascade mountains. The distance between Seattle and Spokane is about 375 miles. The running time was about 12 hours. This train did not carry a dining car, but carried sleeping cars. It is conceded that Mr. Topping was a passenger for hire on this train. At the time the train left Spokane there was known to be a storm in the Cascade mountains, which had been raging for a period of one or two days. But at that time trains were running regularly on about schedule time. This train was No. 25, and will hereafter be referred to by that number.

In the Cascade mountains is a tunnel about three miles long through the summit. The station at the east portal is known as the "Cascade Tunnel." The station at the western end of this tunnel is known as Wellington, at which place there is a hotel and a few other buildings located on the mountain side near the railway. Train No. 25 reached Cascade Tunnel at the eastern end of the tunnel at about 5:30 in the morning of February 23d. At that time the storm had not abated in the mountains, and the line at points between Wellington and Seattle was blocked with snow. Train No. 25 remained at Cascade Tunnel until the evening of February 24th, when it was taken through the tunnel and placed on a passing track at Wellington, where it remained until February 28th, or the morning of March 1st at about 1:30 o'clock, when an avalanche of snow swept down the mountain side and carried this train with it a distance of about 100 feet below the track, where the train was destroyed. Edward W. Topping and other passengers were killed. Others were injured.

When train No. 25 arrived at Cascade Tunnel on the morning of February 23d, there was a heavy snowstorm. Snow had been falling about 24 hours. It had been falling at the rate of about three feet per day. Some of the witnesses testified that, when the train went through the tunnel to Wellington, the snow there was between eight and nine feet deep. It was not shown or claimed that this was all fresh snow. A part thereof had accumulated during the winter.

The railroad track at Wellington was built upon the mountain side. The railway tracks at this place consisted of a main line, two or three switches, and what is known as a "passing track." The passing track connected with the main line at each end. During the time train No. 25 stood upon this passing track, the railroad between Wellington and Seattle was blocked with snow and slides, so that trains could not run through. The track was also blocked with snow and slides from the Cascade tunnel east toward Spokane.

From the time train No. 25 was placed on the passing track at Wellington on the 24th of February until the 1st of March, every effort was being made by the railroad company to open the track so that trains could be run through. On the night of the 28th, or the morning of March 1st, an avalanche of snow broke upon the hillside to the north of the train, and slid down, carrying the train with it. This avalanche broke off 400 or 500 feet above the track, and extended from 1,500 feet to 2,000 feet along the track. At that time there was an unusual storm raging—wind, thunder, lightning, and excessive snow. It was an unprecedented storm. The avalanche came without warning and swept the train down the hillside.

Source: "Topping v. Great Northern RY. CO." *Pacific Reporter,* vol. 142, August 31–October 12, 1914. St. Paul, Minn.: West Publishing, pp. 425–426.

1930s Dust Bowl: "Comparison of Dust Bowl Regional Precipitation in 1930s v. 2000s" (May 2009)

The drought of the 1930s is commonly known as the dust bowl. It affected a large portion of Oklahoma, as well as portions of Colorado, Kansas, New Mexico, and Texas. The following report prepared by the National Weather Service compares the rainfall during the dust bowl to the decade beginning in 2000.

Comparison of Dust Bowl Regional Precipitation in 1930s vs. 2000s

1.0 Introduction

With the early 2008 drought in the Western Oklahoma Panhandle and Northwestern Texas Panhandle, there have been many comparisons of the recent dry period with the infamous Dust Bowl of the 1930s. Although the year 2008 began extremely dry, with less than 2.5 inches of rain falling in the first half of the year in some places, the overall dryness has been much less intense and much less widespread than the Dust Bowl of the 1930s.

The High Plains has a semi-arid climate which relies primarily on warm season thunderstorms to bring much-cherished rainfall to the land. Because thunderstorms are most often hit and miss, with one area getting heavy rain while areas nearby get nothing at all, one farmer's field may be flooded while his neighbors a few miles away are left praying for rain for their parched crops and livestock. On the whole, though, most areas in the Southern High Plains average near or less than 20 inches of precipitation per year, with much of this falling in the spring and summer. Figure 1 is a map of average annual precipitation across the Texas and Oklahoma Panhandles.

2.0 Precipitation in the Dust Bowl Era (1930–1940)

The 1930s was an exceptional time to be in the High Plains. The entire region, already a semi-arid climate to begin with, endured extreme drought for almost a decade.

2.1 Extent

Over the 11-year span from 1930–1940, a large part of the region saw 15% to 25% less precipitation than normal. This is very significant to see such a large deficit over such a long period of time. This translates to 50 to 60 inches of much needed moisture which never arrived that decade. For an area which only averages less than 20 inches of precipitation a year, deficits like this can make the region resemble a desert. Deficits like this are the equivalent of missing three entire years of expected precipitation in one decade. Figure 2 is a map of the precipitation departures from normal in terms of a percentage of normal (total precipitation divided by normal precipitation) for the Dust Bowl region for 1930 to 1940.

2.2 Intensity

One of the most impressive things about the Dust Bowl of the 1930s is not only how long the drought lasted, but how intensely dry some of the years were. Some years only had less than half of the normal annual precipitation. Figure 3 is a map of the departure from normal in terms of a percentage of normal (total precipitation divided by normal precipitation) for the single driest year in the 1930 to 1940 time period (for example, the driest year at Boise City was 1934 which was 53% below normal).

2.3 Desertification

Another measure of the intense dryness of this period is how many years were not only dry, but extremely dry. The classic definition of a desert is an area which receives less than 10 inches of rain per year. During the 1930s there were large parts of the High Plains which saw entire years go by with less than 10 inches of precipitation. They essentially became a desert. In fact, in many cases there were several years in a row with less than 10 inches of precipitation. Figure 4 is a map of the number of years with less than 10 inches of precipitation during the 1930 to 1940 time period.

2.4 Length

Even during the longest droughts, there are some periods of increased moisture. So even though the 1930s as a whole were very dry, it would be reasonable to expect at least a few wet years mixed in with all the dry ones.

But in the 30s, almost every year had less precipitation than normal. Figure 5 is a map of the total number of below normal years in this time period. Notice that some areas were below normal all 11 years.

2.5 Breaks

As mentioned previously, it would be reasonable to expect a few wet years mixed in with all the dry years of the 1930s. But even the wet years were not very wet, so the region was not able to take advantage of some very wet periods to recuperate from all of the extremely dry years. The dry years were intensely dry, but the wet years were not nearly as intensely wet. Figure 6 is a map of the departure from normal in terms of a percentage of normal (total precipitation divided by normal precipitation) for the single wettest year in the 1930 to 1940 time period (for example, the wettest year at Boise City was 1930 which was 14% above normal).

2.6 Summary

As can be seen by the data presented above, the dry years of the 1930s were extremely dry, and there were no extremely wet years to help the land recuperate. With this very dry pattern lasting for a decade or more, the 1930s proved to be the most difficult time to be a farmer in the High Plains!

3.0 Precipitation in the 2000s

Since the year 2000, parts of the region have once again turned to a drier than normal pattern. And with 2008 starting out extremely dry in some places, many people have begun to wonder if the area is returning to another "Dust Bowl" climatological period. And although it is very hard to predict the future precipitation over the span of several years, a comparison of the current dry period with the 1930s dry period can be made.

3.1 Extent

The dryness this decade has not been as widespread as in the 30s. Large parts of the region have seen 10% to 20% less precipitation than normal, but some have actually had a surplus. Figure 7 is a map of the precipitation departures from normal in terms of a percentage of normal for 2000 to 2008.

3.2 Intensity

The driest years of this current decade have also not been nearly as intensely dry as the dry years of the 30s. The driest years this decade were generally 30 to 50% below normal. Figure 8 is a map of the departure from normal in terms of a percentage of normal (total precipitation divided by normal precipitation) for the single driest year in the 2000 to 2008 time period.

3.3 Desertification

Another measure which shows how this decade has not been as intensely dry as the 1930s is to look at how many years were not only dry, but extremely dry. Very few locations have seen less than 10 inches of precipitation in a single year this decade, and no location has experienced these desert-like conditions more than once. Figure 9 is a map of the number of years with less than 10 inches of precipitation during the 2000 to 2008 time period.

3.4 Length

Another factor to consider is how many years this decade which have seen below normal precipitation. The comparison here once again shows that not as many years were dry this decade as were in the 1930s, thus the cumulative effect of the dry period has not been as intense. Figure 10 is a map of the total number of below normal years from 2000 to 2008. Notice that no areas have been below normal all 9 years this decade.

3.5 Breaks

The wet years this decade have also been more significantly wet, helping to reduce the cumulative effect of the long term dryness. There have only been a few areas which have not had a very wet year, and most areas have had extremely wet years which more than compensate for the driest year this decade. Figure 11 is a map of the departure from normal in terms of a percentage of normal for the single wettest year in the 2000 to 2008 time period.

3.6 Summary

As can be seen in the various images presented above, the 2000s have been drier than normal in some locations, but not nearly as dry as the 1930s. The extent of the dryness has also not been as widespread as it was in the Dust Bowl.

4.0 Precipitation in 2008

4.1 Early Drought

The year 2008 started off extremely dry in some areas, prompting some to fear a return to the Dust Bowl era. Normally the winter months are very dry, with thunderstorms and significant precipitation returning in March or April and lasting through summer. But in 2008 the spring rains were late to arrive in some areas. Significant rainfall from thunderstorms didn't occur until mid-June or July. From January 1–June 30, 2008 a large part of the region was 40% or more below normal for the period. Figure 12 is a map of the precipitation departures from normal for the first half of 2008 (January 1–June 30).

4.2 Late Rains

Late in the summer the region finally returned to a wet weather pattern. The remnants of three tropical systems (2 Atlantic and 1 Pacific) impacted the area from July through October. Hurricane Dolly brought rains to the area in late July. An unusually large upper level storm system brought cold temperatures and flooding rains to large parts of the area in mid-August, and another system brought heavy rains in mid-October. These systems helped alleviate the effects of the extremely dry start to the year and brought several areas up to above normal for the year as a whole. Figure 13 is a map of the precipitation departures from normal for the entire year (January 1–December 31, 2008):

5.0 Conclusions

To summarize, the 2000s have been dry for parts of the High Plains, most notably the Northwestern Texas Panhandle and the Western Oklahoma Panhandle (along with adjacent areas of Northeast New Mexico). The cumulative effect of the prolonged dryness has impacted not only crops and grasslands, but also area lakes and reservoirs. Because some of the driest areas happen to be in the Canadian River watershed, water levels at Lake Meredith have been dropping for several years. Other factors like water usage from cities also affect lake levels, but the prolonged dryness has been a significant factor for the declining lake levels since 1999. Figure 14 shows the declining water levels at Lake Meredith.

The year 2008 started off extremely dry particularly in the Northwestern Texas Panhandle, the Western Oklahoma Panhandle, and Northeast New Mexico. The first half of the year was as much as 70–80% below normal in some locations. But in large part, the rains returned late in the summer allowing the year as a whole to be not as extremely dry even in the worst affected regions.

One last thing to note is that this region is heavily reliant on warm season thunderstorms for the majority of the annual precipitation (87% of precipitation falls from March thru October in Amarillo). Because of this, the variability in precipitation can be extreme not just from location to location but also from year to year. So one observation point may be very dry, while almost every other location nearby may be wet. But over long periods of time, this variability eventually levels out. What was impressive about the 1930s was that such a widespread area saw very low precipitation amounts for an extended period of time. Although some areas were drier than others, the whole region was uniformly very dry, as if the thunderstorms never came. In the 2000s only a small area has been persistently dry, and the variability noticed in the precipitation departures from normal map (the blotchy nature of the contours) resembles much more what would be expected for an area which receives so much precipitation from localized thunderstorms.

Source: National Weather Service. "Comparison of Dust Bowl Regional Precipitation in 1930s v. 2000s." Available online. URL: http://www.srh.noaa.gov/ama/?n=dust_bowl_verses_today. Accessed November 20, 2009.

1965 Tornadoes: Nathaniel B. Guttman, "General Summary of Tornadoes, 1965" (1966) (excerpt)

The year 1965 was an epic one for tornadoes, including the Palm Sunday tornado that killed 44 in Michigan. The following excerpt is a report to the U.S. Department of Commerce that summarizes the tornado impacts of that year.

1965 will be remembered as the year of the Palm Sunday tornadoes. It was also a year of a record number of reported tornadoes.

During the evening of April 11 and early on the morning of April 12, at least 47 separate twisters ripped through the five-State area of Indiana, Ohio, Michigan, Wisconsin, and Illinois causing 257 fatalities. The first three States were hardest hit with 247 deaths, over 1,500 injuries, and estimated property damage in excess of $200 million. The greatest number of deaths resulting from a single tornado occurred on Palm Sunday; a total of 44 persons were killed in Branch, Hillsdale, Lenawee, and Monroe Counties in Michigan.

Despite the toll of the Palm Sunday outbreak, it was overshadowed historically by the Tri-State tornado that swept through Missouri, Illinois, and Indiana on March 18, 1925. This twister killed 689 persons and injured at least 1,980 others. Another single storm on April 5, 1936 killed 216 persons and injured over 700 at Tupelo, Miss., while one on April 6, 1947 claimed 169 lives in Texas, Oklahoma, and Kansas. Comparable to the Palm Sunday tornadoes were a tornado series in Alabama on March 21, 1932, and another series on March 21–22, 1952 in Arkansas, Missouri, and Tennessee. The former killed 268 persons and injured 1,874 others; the latter killed 208 persons and injured 1,154.

More tornadoes were reported during 1965 than in any previous year. The total of 898 occurred on 180 days spread throughout all 12 months. They killed 299 persons (257 during the Palm Sunday outbreak), injured at least 4,564 others, and caused more than $1 billion worth of property damage. During the year 42 states experienced tornadoes, and from April 1 through September 30 tornadic activity occurred on the average of 3 out of every 4 days. Almost 2/3 of the total number of reported tornadoes, 290 of the yearly total of 299 deaths, and most of the property damage occurred during the 3-month period from April through June.

. . .

The month in which the greatest number of tornadoes was observed in a particular state and the number of tornadoes observed that month are shown by chart. A distinct pattern is evident for most of the country. The region from West Virginia westward through Missouri, and including Michigan, experienced the most tornadoes in April. The next month the maximum occurred in the states from Texas through South Dakota through Wisconsin. The westward progression of the maximum was completed in June when Utah, Colorado, and New Mexico reported the greatest number of tornadoes for any month. Also in June a maximum occurred in northern New England. Montana, North Dakota, and Wyoming were hardest hit by twisters in July, and August was the month of the greatest tornadic activity in southern New England and the Middle Atlantic States. No particular pattern can be discerned in the Southeast.

Normally in April a maximum occurs in the five states from Florida and Georgia through Louisiana. May is usually the month of the greatest number of tornadoes in Oklahoma, Kansas, Missouri, and Arkansas; in June the greatest tornadic activity usually occurs in Minnesota, Iowa, Nebraska, South Dakota, and North Dakota.

In addition to the Palm Sunday outbreak already discussed there were 5 other significant periods of tornadic activity during 1965. The first reported tornado of the year occurred on January 8 in Lafayette Parish, Louisiana, but the first deaths were reported in North Carolina. On March 17 a twister killed 2 persons, injured 100 others, and damaged the New Bern Airport as it crossed Craven, Jones, and Pamlico Counties, N.C.

A family of 6 tornadoes killed 14 persons, injured 683 others, and caused heavy damage mainly in the Minneapolis, Minn., vicinity during the evening of May 6. In November a very unusual number of tornadoes occurred in the Ohio Valley. In Indiana 6 of the 7 twisters during this month were reported on the 26th, while Illinois had 4 tornadoes and Ohio reported

2 on this date. Illinois was the victim of 4 other tornadoes on the 12th that caused 2 deaths, 90 injuries, and extensive property damage. Six tornadoes also occurred in Ohio on the 16th. (Tornado records begun in 1916 show only two tornadoes in this state during November.) The last tornado of 1965 occurred on December 23 in Garza County, Texas.

An eyewitness account of the Austin, Tex., tornado of May 17 was reported by Mr. Nat Henderson, staff writer for the *Austin Statesman*, who stated that the tip of a tiny tornado, estimated at no more than 6 to 8 feet in diameter at its biggest part, passed about 5 feet over his head only seconds before uprooting an elm tree 15 feet away. "Although the distinct funnel roared just above my head," he stated, "it did not even stir a hair. I could not feel the breath of a breeze, but the sound was as deafening as a jet engine warming up near my ear. The only wind was in the funnel itself." The tornado was further described as snow white and not visibly connected to any other cloud.

Source: Nathaniel B. Guttman. "General Summary of Tornadoes, 1965." *Climatological Data: National Summary,* vol. 16, no. 13, 1966. Asheville, N.C.: United States Department of Commerce, pp. 50–51.

RECENT EVENTS

1980 Eruption of Mount St. Helens: R. L. Christianson and D. W. Peterson, "Chronology of the 1980 Eruptive Activity" (1997) (excerpt)

After being dormant for several decades, Mount St. Helens, located in northwestern Washington State, came to life in March 1980 with a series of earthquakes and eruptions of ash. The climax came on May 18, 1980, when the volcano erupted and devastated the surrounding area. The following excerpt gives the precise timing of the events leading up to the eruption.

Initial Period of Seismic and Steam-Blast Activity

On March 27, 1980, Mount St. Helens, in the Cascade Range of southern Washington . . . erupted for the first time in about a century and a quarter. This volcanic eruption, the first in the conterminous United States since the end of a long series of eruptions at Lassen Peak in California that lasted from 1914 to 1917, afforded the first opportunity to study an erupting Cascade volcano since the advent of modem volcano-monitoring techniques. Mount St. Helens itself had not erupted since 1857, at the end of a decades-long period of intermittent activity.

The first major event of the 1980 activity was an earthquake of magnitude 4.0+ at 1547 PSTI on March 20 (day 80:23:47 UT). The swarm of earthquakes increased rapidly to a climax in the late afternoon of March 25, when 24 earthquakes of magnitude 4 and greater occurred during an B-hr period. Overflights on March 24 had indicated no major changes in the appearance of Mount St. Helens other than avalanches of snow and ice. Additional overflights on March 25, during the peak of seismic activity, revealed several new fractures through glaciers high on the mountain and numerous additional large rockfalls and avalanches. None of the new fractures, however, coincided with the larger fractures that later formed across the summit area; the fractures that formed on March 25 probably resulted directly from ground shaking and the accelerated downslope movement of glacial ice on the volcanic edifice. After March 25, seismicity declined somewhat but remained at a high level, with about 30 events per day having a magnitude of 3 or greater (6/day of magnitude 4 or greater).

The first eruption occurred in the early afternoon of March 27. Although extensive cloud cover had hidden the volcano from the air since the morning of March 25, a loud boom was widely heard at 1236 PST on March 27. Aerial observers reported a dark dense column of volcanic ash rising through the clouds, eventually to a height of 2,000 m above the volcano.

With clearing weather later in the afternoon, several changes were conspicuous on the mountain. A new crater about 60–75 m across had formed in the northern part of the old 400-m-wide ice-filled summit crater, and snow on the southeast sector of the volcano was covered by dark ash emitted from the new crater . . . The summit area was bisected by an east-trending fracture nearly 1,500 m long that extended from high on the northwest flank, across the old crater, down the upper northeast flank. Another less continuous fracture system paralleled this master fracture just north of the old crater rim and bounded the south side of a newly uplifted block, or bulge, on the volcano's north flank. These changes clearly had occurred during the period of extremely high seismicity and initial eruption, between observations on the morning of March 25 and the afternoon of March 27. One observer, David Gibney-an aerial spotter of the U.S. Forest Service (oral commun., 1980)-reported seeing the large fractures open and close and the uplifted north flank continue to break and rise during the few hours after the first eruption.

A second explosive eruption, beginning at about 0200 on March 28 and lasting nearly 2 hr, was observed from the air. Ash from that eruption spread for many kilometers to the east of the volcano. By nightfall on March

28, at least a dozen more eruptions had occurred, many of them lasting only a few minutes but some for nearly an hour. Poor weather hampered observations on all of these days. By March 29, however, clear views of the summit revealed the presence of a second larger crater west of the one first seen on March 27. . . . A septum about 10 m wide separated the two craters. Pale-blue flames were first observed on the night of March 29 and were subsequently observed in each of the two craters.

On April 1, a weak burst of harmonic tremor lasted for about 5 min, but stronger tremor bursts the next day were recorded by seismometers as far away as 100 km. Sporadic harmonic tremor continued until April 12, but these tremor bursts could not be correlated directly with the visible character or intensity of eruptive events. The frequency, duration, and intensity of eruptive blasts gradually decreased until April 22, when eruptions temporarily ceased. Eruptions had declined in frequency from an average of about 11hr in March to about II day by the end of this first period of activity.

The eruptions of this first period produced only lithic-crystal ejecta, composed of fragments of preexisting rocks. All of this material was emitted from vents in the new summit craters, which were repeatedly reamed. By April 7, the septum separating the two craters had broken down entirely, and the enlarged single crater grew to about 500 m from west to east and about 300 m north to south. . . . By late in the month, the crater was about 100–250 m deep. (Its rim was highly irregular in elevation.)

As eruptive activity declined slowly through early and mid-April, earthquakes continued at still impressive rates, generally more than 301 day of magnitude 3 or greater. Many of these earthquakes were large enough to be felt strongly on and immediately adjacent to the volcano. All were very shallow, however, and few were felt very far from the volcano. The epicenters were confined to a small area that coincided with the uplift or bulge on the volcano's north flank. These frequent shallow earthquakes triggered numerous avalanches, which were concentrated around the northern sector of the mountain. Most avalanches started on the upper slopes of the volcano where ashfall was heaviest; because they involved dark ash-laden snow, the avalanche deposits stood out prominently and were at first mistaken for mudflows.

. . .

The prominent topographic bulge that had been noted on the north flank of the volcano on the first day of eruption continued for nearly 2 mo to become larger and more conspicuous. Ground ruptures at the 5,400-ft level on the north flank were first noted on April 3 and provided early evidence that the bulge affected a large part of the cone. The first detailed photogrammetric measurements, completed in mid April and compared with contours based on photography of August 1979, revealed the startling dimensions of this bulge. By May 12 (the date of the last contour map predating the climactic eruption of May 18), the high point north of the old crater rim stood 150 m above a corresponding point on the former north slope, and the Goat Rocks area low on the bulge had been displaced northward by 106 m. Geodetic measurements showed that displacement was only slightly upward near the top of the bulge; it was mainly outward-nearly horizontally-at consistent rates of about 2 m/day. No appreciable change occurred in this rate of displacement, even up to 1 1/2 hr before the climactic eruption. Photogrammetric evidence suggests that no appreciable bulging of the north flank occurred before the events of 1980. Comparison of maps made from aerial photographs of 1952 and 1979 indicates that if any bulging did occur before August 1979, the amount was close to the resolution possible with the BO-ft (24.4-m) contour interval of the maps-an order of magnitude less than that which occurred between late March and mid-May 1980.

During late April and early May, the upper part of the bulge changed in general appearance. A high point that had formed in late March (north peak 1) just north of the crater stagnated and subsided during April as a second high point (north peak 2) began to form farther north. . . . An east-west line between these two points appeared to be a fracture zone that delimited areas of differing rates and styles of deformation. In early May, old fractures and disturbed segments of the north and west rims of the summit crater south of the fracture zone were partly filled by snow and drifted volcanic ash and appeared to be parts of a coherent graben block that moved downward with the crater. By contrast, the surface of the actively bulging main part of the north flank continued to break and distort, indicating persistent internal deformation as well as outward bodily displacement.

By mid-May, earthquakes of magnitude 3 or greater continued to occur at a rate of 20–40/ day, including 5–10 of magnitude 4 or greater. Two magnitude-5,O earthquakes occurred, on May 8 and 12. On May 8, the day following a resumption of eruptions, two periods of harmonic tremor were recorded, each lasting only a few minutes. No further harmonic tremor occurred until after the beginning of the May 18 eruption.

In summary, throughout an initial 2-mo period, ejecta consisted entirely of fragmental material derived from the volcanic edifice, most of it generated by shattering and pulverization at shallow levels within the 350-yr-old summit dome. A moderate amount of this ash was distributed 50 km away and some was reported as far as 100 km to the east, but most fell within a few kilometers of the volcano's summit. Evidence for the involvement of water was abundant, and the eruptions apparently all resulted from a steam-blast mechanism that reflected internal heating of the volcano by a shallow intrusion that also produced localized, but very high, seismicity and a rapidly and continuously bulging north flank.

The Climactic Eruption

The situation at Mount St. Helens in the early morning hours of May 18 was much as it had been for the preceding month. There had been no eruptions for 4 days; fumaroles remained active in the summit and north flank areas; seismicity was high but no greater than it had been for many weeks; deformation continued at an awesome rate, but the rate had neither increased nor decreased significantly since detailed measurements began on April 25.

At 0832 PDT (day 139:15:32 UT), with no known immediate precursors, a magnitude-5 + earthquake triggered a rapid series of events. As seen by Keith and Dorothy Stoffel (written commun., 1980) from a small aircraft at low level directly above the summit crater, the earthquake caused avalanching from the walls of the crater and, only a few seconds later, triggered a sudden instability of the north flank. The entire north flank was described as II quivering 11 and appeared to almost liquefy. The slope failed along a surface intersecting the northern of the two high points on the north flank, near the east-west fracture separating the active bulge from the crater block. As the north flank began to slide away from this surface, a small, dark, ash-rich eruption plume rose directly from the base of the scarp and another from the summit crater rose to heights of about 200 m. As virtually the entire upper north flank slid off the cone and became a massive debris avalanche, a blast broke through the remainder of the flank, spewed ash and debris over a sector north of the volcano. . . , overtook the massive avalanche, and devastated an area nearly 30 km from west to east and more than 20 km northward from the former summit of the volcano. In an inner zone extending nearly 10 km from the summit, much of which had been densely forested, virtually no trees remained. Beyond, nearly to the limit of the blast, all standing trees were blown to the ground, and at the blast's outer limit the trees were left standing but thoroughly seared. The devastated area of 600 km 2 was blanketed by a deposit of hot debris carried by the blast.

The sole of the debris avalanche was nearly at the base of the steep volcanic cone on the north side; the avalanche moved down the lower gradients of the volcano's outer flank and was nearly blocked by a ridge 8 km to the north. Part of the avalanche rounded the east end of that ridge and displaced the water from Spirit Lake, raising the bed in its southern part by more than 60 m. The bulk of the avalanche, however, turned westward down the valley of the North Fork Toutle River to form a craggy and hummocky deposit, part of which crossed the ridge to the north, but most of which flowed as far as 23 km down the North Toutle. The total volume of the avalanche in place is about 2.8 km 3, and its length makes it one of the largest on record.

Water incorporated by the avalanche from the North Fork Toutle River and possibly from Spirit Lake combined with melting blocks of ice from the torn-out glaciers of the volcano's north flank and melting snow and ice from the volcano's remaining slopes to produce mudflows that later in the day coursed across the avalanche and down the North Fork Toutle River, sweeping up thousands of logs from timbering operations in the valley and destroying most bridges across the river. The mudflows continued downstream, depositing sediment in the Cowlitz River channel and also obstructing the deep-water navigation channel of the Columbia River. Smaller mudflows were produced from the east flank of the volcano and went down the valleys of Muddy River and Pine Creek into Swift Reservoir. . . . Yet other mudflows and floods went from the volcano's northwest flank down the South Fork Toutle River; smaller floods occurred in the Kalama River on the southwest.

The initial events of the eruption-the rockslide avalanche, the northward blast of ash and debris, and the mudflows-caused most of the casualties and destruction in the immediate region of the volcano. However, within a few minutes a Plinian eruption column . . . began to rise from the position of the fonner summit crater and within less than 10 mm had risen to a height of more than 20 km. Ash from this eruption cloud was rapidly blown east-northeastward, producing lightning and starting hundreds of small forest fires, causing darkness eastward for more than 200 km, and depositing ash for many hundreds of kilometers. Major ash falls occurred as far east as central Montana and ash fell visibly as far eastward as the Great Plains of the Central United States, more than 1,500 km away. As this Plinian eruption column grew, it reamed out the volcanic conduit. The eruptive crater, along with the upper 300 m of the cone that was entirely removed by the initial slide and blast, fonned a great amphitheater 1 1/2 × 3 km across, enclosed by the volcano's former east, south, and west flanks. . . .

The Plinian phase of the eruption continued vigorously for 9 hr and produced numerous ash flows. Some of these were thin flows that spread out over much of the upper surface of the volcano and were generated by fallback from the expanding eruption column in the vicinity of the summit. Most of the flows were directed out through the large northward breach of the crater to form a fan of pumiceous ash flows over the avalanche, extending to Spirit Lake and part way down the North Fork Toutle River valley. Ash flows continued to be emplaced at least until dark on May 18. The hot blast deposits, the avalanche, and these ash flows were frequently disrupted in the vicinity of Spirit Lake and its former drainage into the North Fork Toutle River by large secondary steam-blast eruptions that formed craters as large as 20 m across and drove columns of ash to heights as great as 2,000 m above the surface,

. . .

Period of Subsequent Pyroclastic Activity and Lava Domes

A series of increasingly energetic ash eruptions on the Saturday night following the eruption of May 18 led to the second large eruptive event, on Sunday, May 25, from the vent crater within the amphitheater. That eruption began at about 0230 PDT during a period when winds were blowing in different directions at different altitudes. Although the eruption was an order of magnitude less voluminous than that of May 18, windblown ash was dispersed over wide areas of western Washington and Oregon and affected several metropolitan areas that had not experienced ash falls in the May 18 eruption.

For the next 2 1/2 weeks, the volcano continued to emit large quantities of gas that rose in plumes of steam condensate to altitudes of 3–5 km above sea level. Sulfur emissions, monitored since late March, had remained low until the eruption of May 18. After that eruption, sulfur gases were emitted at a higher rate, but relatively little ash was carried in the gas plumes and appreciably none fell more than a few kilometers beyond the volcano. During this time, no lava appeared at the surface, but there were several night observations of incandescent rock, probably caused by hot gases streaming through the vents from a magma body not far below. Also during this time, small steam blasts continued to erupt through the avalanche and ash-flow fill in the former North Fork Toutle River valley north of the volcano. Some of the craters formed by those eruptions were enlarged considerably by repeated blasts and by the coalescence of smaller craters.

A third magmatic eruption took place on June 12. This eruption was preceded by several hours of harmonic tremor that began around midday and gradually grew in intensity throughout the afternoon. A marked increase in tremor amplitude was noted at 1905 PDT, and an eruption drove an ash column to at least 4 km above sea level by 1910. Tremor amplitude decreased markedly immediately after this brief eruptive pulse and fluctuated at moderate levels for more than 2 hr. The temporary lull was broken by a rapid, large increase in tremor amplitude at 2111, and by 2118 an eruption column had risen to 15 km above sea level. The height of the column fluctuated between 5 and 12 km above sea level until 0043 on June 13, when it decreased abruptly.

Prevailing winds carried the ejected ash south-southwest, allowing centimeter-sized pumice fragments to fall in Cougar, Wash., about 16 km downwind. Portland, Oreg., and Vancouver, Wash., received moderate ash falls beginning at about 2250.

Source: R. L. Christianson and D. W. Peterson. "Chronology of the 1980 Eruptive Activity." In *The 1980 Eruptions of Mount St. Helens, Washington,* edited by Peter W. Lipman and Donal Ray Mullineaux. Seattle: University of Washington Press, 1997. Available online. URL: http://vulcan.wr.usgs.gov/Volcanoes/MSH/Publications/PP1250/ChristiansenPeterson/chronolo gy_1980_activity.html. Accessed November 20, 2009.

1982 Blizzard of St. Louis: National Weather Service Weather Forecast Office, "The St. Louis 'Blizzard' of 30–31 January 1982" (2007)

In 1982, an unexpected snowstorm crippled the city of St. Louis, Missouri. The snow stranded some people in their homes, cars, or places of work for days. Below is the National Weather Service's description of the event.

On January 30th and 31st 1982, a 1-in-70 year snow event occurred from the eastern Ozarks to central Illinois with the heaviest axis of snow blanketing St. Louis, Missouri. The snow began during the evening of January 30th, a Saturday, and ended during the afternoon of Sunday, January 31st. The snow paralyzed the area with government offices, many businesses, and schools cancelling work or class for up to a week after the snow ended. The airport, Amtrak, and bus service were shut down. As many as 4,000 motorists were stranded on highways due to the blizzard-like conditions that were created over the region.

Many people became stranded for days, with hospital and emergency workers working 2 to 3 shifts due to their coworkers inability to make it [to] work. According to the Post Dispatch, one subdivision, Bee Tree Estates in South St. Louis County, was cut off from civilization for five days. Those who owned four-wheel-drive vehicles became the transportation service for the city, escorting nurses, doctors, and law enforcement to work. They also helped deliver necessary supplies to those in need and assisted ambulances, tow trucks, and other vehicles that became stuck in snow drifts.

Residents across the area helped each other dig out from the worst snowstorm since February 20th, 1912 when 15.5 inches of snow was recorded. Mayor Vincent C. Schoemehl Jr., and then County Executive Gene McNary declared snow emergencies. The Missouri National Guard was eventually brought in to help with the disaster and ease the situation in the City of St. Louis.

This snowstorm was remarkable and crippling to the St. Louis metropolitan area, although its claim to fame as being a blizzard is unfortunately untrue. For a blizzard to have occurred, the following conditions must have prevailed for a period of 3 or more consecutive hours:

- Sustained wind or frequent gusts to 35 miles an hour or greater, **and**
- Considerable falling and/or blowing snow that reduces visibility frequently to less than 1/4 mile.

The sustained wind at St. Louis International Airport (Lambert Field) never approached 35 miles per hour, in fact the highest wind speed recorded throughout the event was 26 mph. The 5 hour duration of thunder snow that was reported at Lambert Field was incredible, with snowfall rates of more than 2 inches per hour. The falling snow and wind did create snow drifts up to 6 feet and lower visibilities below one quarter of a mile during the storm.

The storm was not well forecast by the National Weather Service, private meteorologists, local television and radio meteorologists, nor local university meteorologists. The consensus the night before the event was for light snow to occur with a few inches of accumulation. Looking back, who could blame them, the low pressure center never dropped below 998 mb and the concepts and ideas that we as a forecast community have knowledge of now, were not yet known or practiced. So how could such a meager low pressure system produce a band of convective snow with total accumulations exceeding 20 inches?

The late Dr. James Moore and Pamela Blakley from St. Louis University published a paper in Monthly Weather Review (1988), "The Role of Frontogenetical Forcing and Conditional Symmetric Instability in the Midwest Snowstorm of 30–31 January 1982," where it was hypothesized that the instability (conditional symmetric instability) in proximity to the upward vertical motion branch of the ageostrophic direct thermal circulation (located just south of the axis of maximum frontogenesis) acted to focus and intensify the lift, which resulted in heavy convective snowfall.

Source: National Weather Service Weather Forecast Office. "The St. Louis 'Blizzard' of 30–31 January 1982." Available online. URL: http://www.crh.noaa.gov/lsx/?n=01_31_82. Accessed November 20, 2009.

1994 Earthquake near Los Angeles: "USGS Response to an Urban Earthquake—Northridge '94" (2005) (excerpt)

In 1994, an earthquake near Los Angeles caused 57 deaths and billions of dollars in damages. This excerpt from the United States Geological Survey details the events that took place and that organization's response to the disaster.

The Earthquake and Its Impacts

At 4:30 on the morning of January 17, 1994, some 10 million people in the Los Angeles region of southern California were awakened by the shaking of an earthquake. The earthquake, named for its epicenter in the town of Northridge, was a magnitude 6.7 (M = 6.7) shock that proved to be the most costly earthquake in United States history. The shaking heavily damaged communities throughout the San Fernando Valley and Simi Valley, and their surrounding mountains north and west of Los Angeles, causing estimated losses of 20 billion dollars. Fifty-seven people died, more than 9,000 were injured, and more than 20,000 were displaced from their homes by the effects of the quake. Although moderate in size, the earthquake had immense impact on people and structures because it was centered directly beneath a heavily populated and built-up urban region. Thousands of buildings were significantly damaged, and more than 1,600 were later "red-tagged" as unsafe to enter. Another 7,300 buildings were restricted to limited entry ("yellow-tagged"), and many thousands of other structures incurred at least minor damage. The 10–20 seconds of strong shaking collapsed buildings, brought down freeway interchanges, and ruptured gas lines that exploded into fires. Fortuitously, the early morning timing of the earthquake spared many lives that otherwise might have been lost in collapsed parking buildings and on failed freeway structures.

The Early Response—Collecting Information and Organizing Communication

Scientists of the U.S. Geological Survey (USGS) responded quickly to the Northridge earthquake, many arriving on the day of the quake to investigate and report on its geological and engineering effects. Early on January 17, the USGS office at the California Institute of Technology (Caltech) in Pasadena became the center for information processing. The network of seismic instruments for monitoring earthquakes in southern California is operated jointly with the USGS at Caltech and, within minutes of the main shock, scientists were analyzing data and broadcasting information about the quake to the public. Scientists maintained a steady flow of public information over the next few days as details about the earthquake and its effects were gleaned from seismic data and observed by field crews. The USGS maintained communications with emergency-response agencies using a liaison stationed at the nearby Federal Emergency Management Agency (FEMA) Disaster Field Office, while hosting at Caltech a liaison from the California Governor's Office of Emergency Services (OES). The USGS continued data processing, communications, and liaison efforts throughout the following months while disaster cleanup continued and the Los Angeles area was rocked by hundreds of aftershocks.

The Long-Term Response—New Experience with an Urban Earthquake

The long-term response to the earthquake by the four NEHRP agencies (National Science Foundation, National Institute of Standards and Technology, FEMA and the USGS) was based upon four objectives. First, the agencies needed to apply their capabilities immediately by assisting local, State, and Federal jurisdictions to carry out the recovery, reconstruction, and mitigation processes in the aftermath of the quake. Secondly, the agencies needed to commence investigating a sequence of events associated with the earthquake. This sequence of events leads from the earthquake source, through the earth into the built environment, to the resultant economic and social impacts, and finally the response. Studying this continuum helps reveal the lessons the earthquake has to teach. As those lessons were revealed, the third objective was to collaborate in communicating the lessons throughout society. Finally, the lessons needed to be applied both in southern California and, as appropriate, throughout the rest of the United States.

The Tasks of the NEHRP Agencies

By March 14, 1994, the four NEHRP agencies had reached consensus on the highest priority post-earthquake investigations and activities to be funded by emergency disaster appropriations by Congress. In all, 84 specific tasks were assigned to the NEHRP agencies under the March 14 agreement, and an additional 22 tasks were assigned later under appropriations from the President's Discretionary Fund. In general, the tasks were to be completed within a period of about 2 years. The tasks were specifically oriented toward a timely, effective response to the Northridge earthquake, with longer term goals of readying the region and the nation for inevitable, future earthquakes. . . .

Source: United States Geological Survey. "USGS Response to an Urban Earthquake—Northridge '94." Available online. URL: http://pubs.usgs.gov/of/1996/ofr-96-0263. Accessed November 20, 2009.

1997 Red River Flood: North Dakota Department of Emergency Services, "Grand Forks 1997 Flood Recovery: Chronology" (n.d.) (excerpt)

The Red River Flood of 1997 did $3.5 billion in damages along the Red River Valley in North Dakota, Minnesota, and parts of Canada. Below is an account of the major events leading up to the disaster.

April 1997—

April 2: The N.D. Highway Patrol (NDHP) assists NDNG's ice dusting efforts by rerouting traffic to enable helicopters to land on N.D. Highway 18 for sandbag loading. More than 300 tons of sand has been dumped on to river ice.

Warmer temperatures and brisk winds initiate spring runoff. Westbound I-94 between the cities of Casselton and Fargo is closed. Gauging stations along tributaries of the Red River show rapid rises in 24 hours.

USACE emergency dike construction operations at Enderlin go to 24-hours. Other USACE emergency projects underway include: dike construction for the city of Fort Ransom; levee construction for the city of Harwood; levee construction outside floodways for the city of Wahpeton; dike construction and ditch clearing and snow and ice removal work at the English Coulee Diversion in Grand Forks; construction of a dike in the city of Pembina and realignment work on dikes for the city of Drayton.

April 3: Citing imminent flooding in the Red River Basin, NDDEM requests USACE to provide technical, manual and material assistance under the emergency operations portion of U.S. Public Law 84-99. USACE begins work on a $600,000 dike to protect the city of Grafton.

The Red River at Wahpeton reaches its 10-foot flood stage. The Red River at Fargo edges on its 17-foot flood stage. Devils Lake rises to 1,438.03 feet, the highest recorded level in 130 years.

April 4: In 24 hours, the Wild Rice River jumps 7.02 feet to 19.10 feet, far exceeding its flood stage of 10 feet. The Red River at Wahpeton rises to 12.96 feet; flood stage is 10 feet. The Red River at Fargo rises 5.8 feet to 21.6 feet, well above its flood stage of 17 feet.

N.D. Highway 81 is closed after water inundates the bridge at Forest River. The city of Minto is threatened. Valley City officials and volunteers fill 13,800 sandbags in two days.

April 4–5: A blizzard moves into North Dakota late April 4 and early April 5. NWS reports snowfall amounts ranging from 10 to 24 inches throughout the state, adding up to 2.66 inches of moisture to areas already inundated by spring runoff.

The storm creates life-threatening conditions, causes massive power outages and shuts down road systems throughout the state. More than 30,000 North Dakota households are without power. A combination of freezing rain and high winds topples government and commercial radio and television towers, leaving many North Dakotans without access to emergency information. Both flood and winter storm conditions cause the evacuation of hundreds of North Dakotans from their homes.

The Maple River at Mapleton rises to 15.4 becoming the first North Dakota river to exceed its flood-of-record stage. Flood stage is nine feet; flood of record was 15 feet.

The Forest River at Minto isolates about 35 homes and forces the evacuation of six families in Walsh County. ARC opens a shelter at North Dakota State College of Science (NDSCS) in Wahpeton.

USACE levee construction operations in Grafton go to 24-hours. USACE also loans 70 water pumps to eastern North Dakota communities and prepares to send 900,000 sandbags to Grand Forks.

April 6: Governor Schafer requests President Clinton to issue a Presidential Major Disaster Declaration for North Dakota, as a result of severe flooding and the spring ice/snow storm. Major General Keith D. Bjerke, Adjutant General for the NDNG, is designated as the State Coordinating Officer (SCO).

The Red River at Wahpeton crests at an all-time high of 19.2 feet; flood stage is 10 feet.

April 7: Within one day of Governor Schafer's request, President Clinton issues a Major Presidential Disaster Declaration (FEMA-1174-DR-ND) for North Dakota. Lesli Rucker and Pete Bakersky of FEMA's Region VIII office in Denver are named Federal Coordinating Officer (FCO) and Deputy FCO, respectively.

April 8: Thousands of North Dakotans remain without electricity. Ten electric cooperatives report ice and high winds toppled hundreds of transmission towers and 4,300 power poles.

After being closed for up to 48 hours NDDOT and local road crews push through snowdrifts opening all state and interstate systems except N.D. Highway 11 in Wahpeton.

NDNG initiates 27 additional missions in response to the flooding and snow disaster to include: generator deliveries to cities, emergency transportation, dozers to help clear paths for electrical companies, and assistance with floodfight operations.

April 9: Nearly 10,500 rural electric customers still are without power. Utility companies report that power may not be restored to more remote areas of the state for up to seven days because of widespread damages. Utility crews from South Dakota, Minnesota, Montana, Nebraska and Wisconsin assist with efforts to restore power. FEMA makes arrangements to bring 50 generators and operators to North Dakota. NDNG provides 18 generators to communities.

April 10: NWS revises its forecast for the Red River at Fargo. The river is expected to rise from 37.5 to 38 feet. Flood of record was 37.5 feet.

State agencies responding to the disaster meet with 120 members of the N.D. Legislative Assembly to discuss the interagency effort required to address problems created by the flood. FEMA institutes a toll-free registration number for North Dakota storm and flood victims to apply for state and federal disaster assistance.

April 11: Vice President Al Gore visits Fargo flood victims and tours other flood-stricken Red River areas.

The Red River reaches its flood-of-record stage of 37.5 feet at Fargo as residents reinforce sandbags and earthen dikes. Breaches are identified along dikes in the Sheyenne River Diversion in West Fargo.

Reports of injuries continue to mount with several North Dakotans being treated for carbon monoxide poisoning and hypothermia.

As a precautionary measure, the Veterans Administration Hospital in Fargo evacuates one-half of its patients to the Veterans Administration Hospital in St. Paul, Minnesota.

A Disaster Relief Task Force organizes to provide products and equipment for emergency response including generators and sump pumps. Members include Lutheran Social Services, Catholic Family Services, Aid Association of Lutherans, Lutheran Brotherhood, Evangelical Lutheran Church of America and the Missouri Synod of the Lutheran Church.

April 12: Many North Dakotans remain without power preventing operation of sump and water pumps. As a result, basements are flooding. NDNG continues to deliver generators, provide traffic control and evacuate families as part of their floodfight efforts, dubbed "Operation Good Neighbor."

April 14: The SEOC operates around-the-clock. FEMA opens a Disaster Field Office in Bismarck initially staffed by 100 people.

April 15: NWS upgrades its predicted crests for the Red River at Wahpeton, Fargo and Grand Forks. The predicted crest for Wahpeton is upgraded from 18.5 feet to 19.5 feet. The NWS expects the Red River to rise to 50 feet in Grand Forks instead of 49 feet, and to 38 feet in Fargo, approximately one-half foot over original projections.

Sandbag work goes to 24-hour operations in Grand Forks. Approximately 12,000 volunteers filled more than 1.3 million sandbags.

The SEOC and NDAg receive reports of cows suffering from dehydration and weight loss and, as a result, are aborting calves. Power outages prevent operation of electrical pumps used to supply water to livestock. An aerial survey indicates more than 150 head of cattle have drowned in Beaver Creek in Logan County.

April 16: The Governor Schafer announced North Dakotans totally or partially unemployed as a direct result of the disaster may be eligible for disaster unemployment benefits from Job Service North Dakota (JSND).

The Wild Rice River spills out near south Fargo and travels overland toward Rose Creek. NWS forecasts that the Red River at Wahpeton will surpass its 19.2 flood-of-record level set only 10 days prior. The Goose River near Hillsboro rose three feet between April 14 and 15 and is expected to rise another six to seven feet by April 17–19.

Amateur Radio Emergency Services (ARES) provide instantaneous audio and video from all Grand Forks dikes and any trouble spots that arise. The University of North Dakota (UND) closes early so students and staff can join floodfight efforts.

April 17: A sandbag levee in northeast Fargo fails, affecting 23 homes and the Oak Grove School. The Red River at Fargo holds steady at 39.5 feet.

Agriculture officials report approximately 90,000 cattle perished during the blizzard and ice storm compared to 20,000 that died during January and February. With temperatures expected to reach 60 degrees Fahrenheit soon, state and federal officials work to develop strategies for removing the 90,000 carcasses. NDAg, North Dakota State University (NDSU) Extension Service and the FSA open an Agriculture Information Center to assist farmers with problems caused by the winter storm and the spring flood disaster.

Revised flood forecasts cause USACE to raise Grand Forks dikes to 54 feet. Personnel from the U.S. Air Force Base at Grand Forks join Red River floodfight efforts. The Base assigns 500 personnel to that effort.

Approximately 540 NDNG members volunteer to assist with the floodfight "Operation Good Neighbor."

April 18: Approximately 1,000 homes in Grand Forks are evacuated in the early morning hours after Red River floodwaters topple private dikes and flow over a public dike in the Riverside Park area. Emergency management officials report that dikes in the Lincoln Drive and Riverside Park areas are becoming unstable. Evacuation of residents living in those areas begins.

The Red River surpasses its 100-year flood event at Grand Forks with record flows of 145,000 cfs. The velocity of water, doubled from April 17 to April 18 through the river channel at Grand Forks, is compounding threats to dikes already experiencing problems with breaches. NWS reports that the Red River at Grand Forks is at 51.55 feet, gaining more than two feet during the past 24 hours. A revised forecast calls for the river to crest at 53 feet later April 19. NDNG pre-positions equipment, such as pumps and helicopters, to allow for a quick response to communities involved in flood fights.

April 19: In the early morning hours, approximately 10,000 Grand Forks residents leave 3,000 homes as city officials call for a citywide voluntary evacuation and a mandatory evacuation of a 10-block area west of the Red River. Residents of the 118-bed Almonte Living Center are among the evacuees. Earthen and sandbag dikes deteriorate allowing floodwaters through to riverside neighborhoods and downtown areas. The city uses buses to transport flood victims to designated shelters.

The Grand Forks Emergency Operations Center (EOC) relocates to the UND Plant Services Building after floodwaters head toward the EOC.

The only Grand Forks transportation link to Minnesota is lost as officials close the Kennedy Bridge when floodwaters reach 52.9 feet. The U.S. Geological Survey (USGS) reports that flows have decreased from 145,000 cfs to 100,000 cfs as water spreads into Grand Forks. NWS expects the Red River at Grand Forks to reach 54 feet.

Three shelters are open at state universities in Mayville, Valley City and Devils Lake. ARC establishes a shelter at the NDNG Armory in Grand Forks and the Grand Forks Air Force Base. Shelters also are opened at Red River High School and Valley Middle School in Grand Forks. ARC mobilizes its national mass care staff. A FEMA truck carrying 1,500 cots and blankets arrives at the Grand Forks Air Force Base shelter.

Grand Forks city and county officials and floodfighters conduct welfare checks on residents to ensure their safe evacuation. With only two weeks left in the semester, UND officials cancel classes for the remaining spring semester.

With the city of Grand Forks' municipal water system in danger of failing, NDNG sends five water purification units to provide bottled water to evacuees. A Guard helicopter crew transports maintenance personnel to the only radio station on air in Grand Forks, KCNN radio, the only radio station still on-the-air, to ensure its continued operation.

Officials for Pembina and Walsh counties review evacuation procedures and prepare to outline those plans with residents. NWS forecasts call for the Red River to rise at least two feet above flood-of-record levels in Pembina and Drayton. Flood-of-record levels are surpassed in Wahpeton where the river reached 19.5 feet, and in Fargo, where the river reached 39.5 feet. The previous records had been 18.5 feet in Wahpeton, and 37.5 feet in Fargo.

Source: North Dakota Department of Emergency Services, Division of Homeland Security. "Grand Forks 1997 Flood Recovery: Chronology." Available online. URL: http://www.fema.gov/pdf/hazard/archive/grandforks/chronology. pdf. Accessed November 20, 2009.

1999 Oklahoma Tornadoes: National Weather Service. "Storm Data and Unusual Weather Phenomena" (May 5, 1999) (excerpt)

On May 3, 1999, a severe storm resulted in 66 tornadoes in Oklahoma and Kansas. Over the following week, the storm produced a total of 140 tornadoes. The worst of the tornadoes caused widespread damage to parts of Oklahoma City. The following report is from the National Weather Service's bulletin published on the second day of the storm.

A record outbreak of tornadoes struck Oklahoma from late afternoon of May 3, 1999, through early morning of May 4, 1999. To date, 58 tornadoes have been recorded across portions of western and central Oklahoma. Additional tornadoes were reported across eastern Oklahoma from late evening of May 3rd through the early morning of May 4th, and are listed under the eastern Oklahoma portion of Storm Data, provided by the National Weather Service Office in Tulsa, Oklahoma. All direct fatalities (40) and all direct injuries (675) occurred in the Norman National Weather Service warning area. The most notable tornado was rated F5 and formed over Grady County near Amber and tracked northeast for 37 miles eventually into the Oklahoma City metropolitan area. Bridge Creek, Oklahoma City, Moore, Del City, and Midwest City suffered tremendous damage. Thirty-six direct fatalities and 583 direct injuries were recorded. There were many other significant tornadoes as well, including F4 tornadoes in Kingfisher and Logan Counties, and F3 tornadoes in Caddo, Grady, Kingfisher, Logan, and Lincoln Counties. Due to the magnitude of the tornado outbreak, and for easier reference, each tornado has received its own identification. There were 8 tornadic producing thunderstorms, called supercells, and most of them spawned numerous tornadoes, one after another. Occasionally, these thunderstorms spawned tornadoes at the same time. The first tornado producing thunderstorm of the day was labeled storm A, while the last tornado producing thunderstorm of the day was labeled storm I. Tornadoes produced by the same supercell thunderstorm have the same letter and were then numbered chronologically. For example, the 3rd tornado produced by storm B was labeled B3.

Storm A produced 14 tornadoes over a period of about 7 hours and was eventually responsible for the F5 tornado that struck Bridge Creek, Oklahoma City and Moore. The 1st tornado of the outbreak, A1, touched down on US 62, 2 miles north of Interstate 44 in Comanche County at 1641 CST. No damage is believed to have occurred (F0). The 2nd tornado, A2, formed approximately 3 miles west of Elgin in Comanche County. Several witnesses confirmed this tornado, however no damage was observed (F0). The 3rd tornado, A3, touched down in a rural area 3 miles east of Apache in Ceddo County. As the tornado moved northward to near Anadarko, one house was destroyed near the town of Stecker, with its roof ripped off and several walls knocked down (F3). Three person inside the house were injured. Several witnesses reported the 4th tornado, A4, 3 miles northwest of Cyril in Caddo County just west of SH 8. No damage was reported (F0). The 5th tornado, A5, formed 2 miles south of Anadarko in Caddo County. Two witnesses reported the tornado to be brief, and no damage was observed (F0).

The 6th tornado, A6, developed about 3 miles north-northeast of Cement near the Caddo/Grady County border, and quickly intensified to a strong tornado with associated damage rated at the high end of the F3 scale. The tornado tracked northeast for 9 miles before dissipating 2.5 miles west-northwest of downtown Chickasha. Two homes had just a few interior walls standing (F3), one located near US 62 on the northwest side of Chickasha, and several wooden high tension power lines were downed. Several persons were injured south of Verden near the Caddo/Grady County border. The 7th tornado, A7, has been referred to as a satellite tornado, and rotated around A6 for a short period of time, 5 miles west of Chickasha in Grady County. Damage from this satellite tornado was not discovered and was therefore rated F0. The 8th tornado, A8, developed 2.5 miles northwest of downtown Chickasha just north of US 62, and tracked northeast, striking the Chickasha Municipal Airport, resulting in high-end F2 damage to two hangar buildings and destroying several aircraft. An aircraft wing, believed to have originated from this airport was eventually carried airborne approximately 45 miles and dropped in southwest Oklahoma City. Approximately 20 mobile homes near the airport were either damaged or destroyed with several persons injured. The tornado then crossed US 81 about 2 miles north of its intersection with US 62 destroying a large building, then dissipated 4 miles north-northeast of downtown Chickasha

The 9th tornado, A9, was a violent and long-tracked tornado, and eventually produced F5 damage in Bridge Creek, Oklahoma City, and Moore. This tornado developed in Grady County about two miles south-southwest of Amber, and quickly intensified as it crossed State Highway 92. F4 damage was first discovered about 4 miles east-northeast of Amber and extended for 6 1/2 miles, as the tornado continued to move northeast. Two areas of F5 damage were observed. The first was in the Willow Lake Addition, a rural subdivision of mobile homes and some concrete slab homes, in Bridge Creek in far eastern Grady County. Two homes were completely swept from their concrete slabs, and about one dozen automobiles were carried about 1/4 of a mile. All mobile homes in this area in the direct path of the tornado were obliterated, resulting in a high concentration of fatalities. Asphalt pavement about 1-inch thick was also peeled from a section of rural road EW 125. The second area of F5 damage was observed about 1 mile west of the Grady/McClain County line and consisted of a cleanly swept slab home with foundation anchor bolts and another vehicle lofted 1/4 of a mile. The maximum width of damage in Bridge Creek was estimated to be 1 mile. Approximately 200 mobile homes/houses were destroyed, and hundreds of other structures were damaged. The Ridgecrest Baptist Church in Bridge Creek was also destroyed. Twelve persons died in Bridge Creek, nine in mobile homes, and all fatalities and the majority of injuries were concentrated in the Willow Lake Addition, Southern Hills Addition, and Bridge Creek Estates, consisting mostly of mobile homes. Compared to sections of Oklahoma and Cleveland Counties, other counties in the path of this tornado which are more densely populated, eastern Grady County including the Bridge Creek area, is rural and sparsely populated.

The tornado maintained a nearly straight path to the northeast paralleling Interstate 44, as it entered McClain County, except when it made a slight jog to the right and moved directly over the 16th Street overpass in Newcastle where a woman was killed when she was blown out from under the overpass. The tornado continued into northern sections of rural Newcastle and crossed the interstate again just north of the US 62 Newcastle interchange. While this tornado was moving through the northern portion of Newcastle, a satellite tornado (A10) touched down in a field in rural north Newcastle, and caused no damage (F0). Two areas of F4 damage were observed in McClain County, all associated with tornado A9. The first area overlapped the Grady/McClain County line and extended to about 3 miles northwest of Newcastle, ending just west of the 16th St. overpass on Interstate 44, while the other area was observed 2 miles northwest of Newcastle. Thirty-eight tomes and 2 businesses were destroyed in McClain County, and 40 homes were damaged. Damage then diminished to F2 intensity as the tornado crossed the South Canadian River into northern Cleveland County.

The tornado entered Cleveland County between Portland and May and between SW 164th and SW 179th in south Oklahoma City. Damage was rated F2 in this area with a path width averaging 1/2 of a mile. The first major housing development to be struck in Cleveland County was Country Place Estates located just west of Pennsylvania Ave. where about 50 homes were damaged, with 1 dozen of these homes receiving F4 damage. One slab home was cleanly swept from its foundation, and several vehicles were picked up from the subdivision and tossed across Pennsylvania Ave, a distance of approximately 1/4 of a mile. One vehicle was found under a bridge just east of the intersection of Pennsylvania and SW 134th. This particular area of damage has been rated high F4/low F5. Oklahoma City Police indicated that part of an airplane wing, believed to have originated from Chickasha Municipal Airport in Grady County, landed in this area. The tornado then tracked through Eastlake Estates, a densely populated housing development, located north of SW 134th and between Pennsylvania and Western, where 3 fatalities occurred. Entire rows of homes were virtually flattened to piles of rubble. Four adjacent homes on one street were virtually cleaned off their foundations leaving only concrete slabs, which earned an F5 rating. Three other homes in this housing division also received F5 damage, with the remaining destruction rated high F4. Three persons also died in the 600-unit Emerald Springs Apartments on Western Ave. located across the street from Eastlake Estates. One 2-story apartment building on the north end of the apartment complex was virtually flattened, and received an F5 rating. Westmoore High School, located just north of Eastlake Estates, was also heavily damaged. Although a well-attended awards ceremony was being held at the school during the tornado, no one was injured, however dozens of vehicles in the school parking lot were either damaged or destroyed. F4 damage continued northeast into another residential area east of Western Ave. and south of 119th St. The tornado then entered the western city limits of Moore (Cleveland County) along Santa Fe and near NW 12th, and produced damage between 1/2 and 3/4 of a mile wide. Maximum damage, rated high F4/low F5, extended northeast to near Janeway with several large groups of homes flattened. Four persons died in this residential area. F4 damage continued to South Shields just north of the junction with Interstate 35. A woman was also killed when she was blown out from under the Shields overpass of Interstate 35. The tornado appeared to weaken just slightly after crossing interstate 35, however it remained a formidable storm with widespread high F3/low F4 damage observed in Highland Park, a residential area, south of the First Baptist Church on 27th St. in Moore. Escaping with relatively minor damage, and being located near the halfway point of the tornado path, the First Baptist Church in Moore eventually served as the primary coordination center for most tornado relief efforts. The tornado then continued northeast and entered the southern portion of a sparsely populated industrial district. F4 damage continued through this area, to near SE 89th St., the Cleveland/Oklahoma County border.

Moving into Oklahoma County, the tornado curved northward, through the remaining industrial district north of Interstate 240 where 2 businesses were destroyed, with the damage rated F4. Two persons were also killed at a trucking company near the intersection of S. Bryant Ave. and Interstate 240. A freight car, with an approximate weight of 18 tons, was picked up intermittently and blown 3/4 of a mile across an open field, with the body of the freight car being deposited southeast of the intersection of S. Sunnylane Rd. and SE 59th. Gouge marks were observed in the field every 50 to 100 yards, suggesting the freight car had been airborne for at least a short distance. While tornado A9 was moving through southeast Oklahoma City, another tornado (A11) touched down briefly about 1/2 mile south of Interstate 240 (Oklahoma County), near the intersection of SE 80th and Sooner Rd. Damage from tornado A11, rated F0, included fences being blown down and minor roof damage inflicted to a couple of houses. Tornado A9 then entered residential neighborhoods between SE 59th and SE 44th where 1 woman was killed in her house. Crossing SE 44th into Del City (Oklahoma County) the tornado moved through the highly populated Del Aire housing addition killing 6 persons and damaging or destroying hundreds of homes, many with F3/F4 damage. The tornado then crossed Sooner Rd., damaged an entry gate and several costly structures at Tinker Air Force Base, then crossed 29th St. into Midwest City (Oklahoma County), destroying 1 building in the Boeing Complex and damaging 2 others. Widespread F3/F4 damage continued as the tornado moved across Interstate 40 affecting a large business district. Approximately 800 vehicles were damaged at Hudiburg Auto Group, located just south of Interstate 40. Hundreds of the vehicles were moved from their original location, and dozens of vehicles were picked up and tossed northward across Interstate 40 into several motels, a distance of approximately 2 tenths of a mile. Numerous motels and other businesses including Hampton Inn, Comfort Inn, Inn Suites, Clarion Inn, Cracker Barrel, and portions of Rose State College, were destroyed. Some of the damage through this area was rated high F4, however low F5 was considered. The tornado then continued into another residential area located between SE 15th and Reno Ave. where 3 fatalities occurred. High F4 damage was inflicted to 4 homes in this area. Two of these homes were located between SE 12th and SE 11th, near Buena Vista, and the other 2 homes were located on Will Rogers Rd. just south of SE 15th. Damage then diminished rapidly to F0/F1 as the tornado crossed Reno Ave. before dissipating 3 blocks north of Reno Ave. between Sooner Rd. and Air Depot Blvd.

The Oklahoma State Department of Health in Oklahoma City recorded 36 direct fatalities. In addition, 5 persons died of illness or accident during or shortly after the tornado and were not considered in the direct fatality total. Five hundred eighty-three injuries were estimated based on numbers provided from the Department of Health, which were then adjusted to account for persons assumed to be unaccounted for. Injuries which resulted from removing debris, conducting search and rescue efforts, and taking shelter from the tornado, were not considered in the injury total. An estimated 1800 homes were destroyed, and 2500 homes were damaged, resulting in approximately 1 billion dollars in damage.

The 12th tornado, A12, formed about 3 miles southwest of Choctaw in Oklahoma County and produced F2 damage to two homes and lesser damage to many others in the southwest part of Choctaw. The tornado moved into the center of town where a car was thrown over the canopy at a drive-in restaurant. The business strip located on the north side of NE 23rd was especially affected with several businesses destroyed, including Pizza Hut, Sonic, and Tri-City Youth and Family Shelter. Damage was mainly F1 as it moved northeast, except for F2 damage at a nursing home where one woman was injured. The tornado continued to weaken in rural areas and dissipated near the intersection of NE 50th and Triple XXX Road. In total, damage estimates are near 3.2 million dollars, with 8 businesses destroyed, 130 businesses damaged, 14 homes destroyed, and 23 homes damaged. The 13th tornado, A13, formed about 4 miles east-southeast of Jones in Oklahoma County near the intersection of NE 63rd and Triple XXX Rd., and was captured on video. Only minor damage (F0), was observed as the tornado tracked northward along Triple XXX Rd. for 2 miles before dissipating near Britton. The 14th and final tornado (A14) produced by storm A, formed about 3 miles east-northeast of Jones in Oklahoma County just west of Triple XXX Rd and between Hefner and 122nd, then tracked northward for 4 miles before dissipating near Interstate 44. A ground survey concluded F1 damage occurred to several homes south of the intersection of Memorial and Triple XXX Rd.

Storm B was responsible for producing 20 tornadoes in 5 hours. One of these tornadoes, rated F4, caused 1 fatality, and produced a damage path 39 miles long and 1 mile wide. The first tornado, B1, formed about 3 miles south of Roosevelt in Kiowa County near Tom Steed Lake. Touchdown was brief with no damage reported (F0). The 2nd tornado, B2, formed in southwest Caddo County about 12 miles west-northwest of Apache. The tornado was captured on video and remained on the ground for approximately 4 minutes before dissipating. No damage was reported (F0). The 3rd tornado, B3, formed about 8 miles south of Fort Cobb and remained on the ground for 21 minutes covering a distance of 7 miles. Damage, mainly F1, consisted of a destroyed barn 7 miles south of Fort Cobb, a stock trailer which was thrown about 100 yards and a destroyed house garage 6 miles south-southeast of Fort Cobb, and sporadic areas of downed trees and power lines. The 4th tornado, B4, was short-lived and developed about 5 miles west of Anadarko in Caddo County. No significant damage was observed (F0). The 5th tornado, B5, was also short-lived and formed about 4 miles north-northwest of Anadarko in Caddo County. No significant damage was observed (F0). A storm chaser observed the 6th tornado, B6, about 4 miles east-southeast of Gracemont in Caddo County. Touchdown was brief with no significant damage (F0). The 7th tornado, B7, was also reported by a storm chaser, and formed about 9 miles east-southeast of Gracemont in Caddo County. The tornado was brief and produced no significant damage (F0). The 8th and 9th tornadoes, B8 and B9, formed nearly simultaneously. Tornado B8 developed about 8 miles west-southwest of Minco in Grady County and was approximately 300 yards wide. The tornado tracked northeast for 2 miles before dissipating. Tornado B9 formed 5 miles south of Cogar in Caddo County and tracked northward for 5 miles. Tornadoes B8 and B9 moved over rural areas with only extensive tree damage observed, and thus were both rated F1. The 10th tornado, B10, was captured on video and formed about 5 miles west of Minco and tracked northeast for 4 miles before dissipating. Maximum damage, rated F1, consisted of small house moved slightly off its foundation with most of its roof blown off. Trees and road signs were also damaged near the end of the tornado track.

The 11th tornado, B11, formed about 5 miles southwest of Minco in Grady County and was observed by off-duty Storm Prediction Center forecasters. The roof was ripped off a house, which would normally warrant an F2 rating, however the structure was considered somewhat unstable, so an F1 rating was assigned. The 12th tornado, B12, formed about 2.5 miles west-north-west of Union City in Canadian County and was captured on video. No significant damage was observed and thus was rated F0. The 13th tornado, B13, formed about 2 miles north-northeast of Union City in Canadian County and was captured on video. No significant damage was reported (F0). The 14th tornado, B14, formed about 4 miles north-northeast of Union City in Canadian County and was also captured on video. No significant damage was reported (F0). The 15th tornado, B15, was also captured on video as it formed about 3 miles east-southeast of El Reno in Canadian County. No significant damage was reported (F0). The 16th tornado, B16, developed about 6 miles west-northwest of Yukon (Canadian County) and tracked northward for 6 miles before dissipating near Piedmont, also in Canadian County. Two witnesses caught this tornado on video. The majority of damage consisted of mangled and downed trees and downed power poles, however 2 mobile homes sustained heavy damage (F1); a barn was destroyed, and 1 cow was killed. The 17th tornado, B17, developed about 1.5 miles west of Piedmont in Canadian County and tracked northward for 8 miles in a zigzag pattern, ending in far south Kingfisher County. Again the majority of damage consisted of downed trees and power poles, however F2 damage was observed about 4 miles northwest of Piedmont where a garage, attached to the house, was destroyed, and a barn and mobile home were completely demolished with debris from the mobile home, mostly corrugated metal, scattered along a 2-mile stretch of road. F1 damage, consisting of large downed trees and leaning power poles, was observed in southern Kingfisher County. The 18th tornado, B18, developed about 4 miles north-northeast of Piedmont in northeast Canadian County and tracked northward for 10 miles before dissipating about 4 miles northwest of Cashion in Kingfisher County. F1 damage was observed in northeast Canadian County, where large trees were uprooted, and numerous power poles were felled or were leaning. F1 damage was also observed 2 miles west of Cashion in Kingfisher County where telephone/utility poles were downed for approximately 100 feet, and an oil storage tank was knocked off its mount. The 19th tornado, B19, formed about 12 miles south-southwest of Crescent in Logan County and was confirmed by an aerial survey. Damage, rated high F0/low F1, was observed for a distance of 1 mile and consisted mostly of downed power poles and strewn about bales of hay.

Source: National Weather Service. "Storm Data and Unusual Weather Phenomena." May 1999. Available online. URL: http://www.srh.noaa.gov/oun/images/pdf/stormdata/oun199905.pdf. Accessed November 20, 2009.

PLANNING FOR FUTURE DISASTERS

USGS/Cascades Volcano Observatory, "Who's Keeping Watch Over Cascades Volcanoes?" (1997) (excerpt)

More than 20 volcanoes exist in the Cascade Range, which runs from northern California to Canada. The USGS/Cascades Volcano Observatory monitors all of these volcanoes for signs of danger. The following excerpt gives details on this organization's activities.

The USGS Observes, Measures, and Studies Volcanoes in the Cascades
The USGS Cascades Volcano Observatory (CVO) in Vancouver, Washington, was founded in 1980 following the devastating eruption of Mount St. Helens. It is one of three such observatories in the country today (others are in Hawaii and Alaska). Observatory scientists, technicians, and support staff work in partnership with colleagues at other USGS centers, universities, and other agencies to:

- Monitor restless volcanoes and provide timely warning of eruptions
- Assess hazards from volcanoes, including water-related hazards in valleys draining volcanoes
- Share volcano information with emergency-management and planning officials
- Develop new techniques and methods to better monitor and predict behavior of volcanoes
- Study volcanic processes
- Educate public officials, citizens, and the news media about what volcanoes can do

171

Disasters

In the past 200 years, seven volcanoes in the Cascades have erupted, and Cascade eruptions can trigger a variety of hazardous processes. Areas within 10 to 20 miles of erupting volcanoes can be devastated by flows and blasts of hot rock and superheated air, and valleys may be exposed to high concentrations of lethal volcanic gases (also possible during non-eruptive periods). Volcanic ash can rise high into the air to drift with the wind, threatening aircraft and disrupting life on the ground hundreds of miles downwind. Eruptions (and sometimes giant landslides not related to eruptions) can also send floods or torrents of mud and rock hundreds of miles down river valleys. The USGS works to prevent loss of life and property from these catastrophic processes here in the Pacific Northwest and elsewhere.

The USGS Sends Rapid Response Teams to
Areas of Volcanic Crisis

The core of the USGS volcano rapid response team is located at CVO. Team members are capable of responding within 24 hours to threatening volcanic activity anywhere in the US or the rest of the world. USGS staff have the experience that is often needed and requested by the U.S. Agency for International Development (Office of Foreign Disaster Assistance) for crisis situations in many countries. The additional experience gained from work at foreign volcanoes greatly strengthens the USGS program because it provides a training ground where skills are sharpened, experience is broadened, and equipment is tested. USGS scientists then are better prepared to respond to volcano crises at home.

The USGS Provides Information, Products,
and Services about Volcano Hazards

The USGS at CVO offers a range of information about volcano hazards to scientists, planners, emergency-management officials, emergency-response teams, law-enforcement personnel, educators, and citizens.

Products
- "Fact Sheets" about hazards and natural processes related to volcanoes
- Exhibits about individual volcanoes and volcanic hazards
- Resource materials for teachers about volcanoes and geology in general (booklets, videos, rock collections, etc.)
- Reports about scientific studies of hazardous processes and geological history of Cascades volcanoes

- Hazard assessment maps and reports for volcanoes in Washington, Oregon, and California (paper copies and digital files)

Source: USGS/Cascades Volcano Observatory. "Report: Who's Keeping Watch over Cascades Volcanoes?" Available online. URL: http://vulcan.wr.usgs.gov/Volcanoes/Cascades/Publications/OFR97-125/OFR97-125.html. Accessed November 20, 2009.

National Weather Service Weather Forecast Office, "1925 Tri-State Tornado: Now v. Then" (March 2010)

One of the nation's worst disasters, the Tri-State Tornado killed a total of 695 people. The following excerpt gives details about the advances in tornado prediction and the warning systems currently in place to deal with future storms.

Even in today's record books, the resultant toll of 695 fatalities from the Tri-State Tornado remains the largest number of casualties from such a disaster. When searching for an explanation as to why, the answer is clear. From technology to communications and the science of meteorology itself, many things have changed since 1925. Back then, radar and satellite imagery were not even close to invention. In fact, it would take such historical events as World War II and the launch of the U.S. Space Program to bring about the use of these two technological breakthroughs that today's meteorologists could not live without. Communication was also in its primitive stage, as radio was just coming into existence in the larger cities during the 1920's, and television wouldn't make an appearance for another 25 years or so.

When the Tri-State Tornado struck in 1925, there was no such thing as a "Tornado Watch" or "Tornado Warning." People relied on the local newspaper, government mail, or word of mouth to relay a message or communicate current events from one town or family to another. So even if a watch/warning program were in place, the message would have never been disseminated in such a fashion to give people the necessary lead time to seek shelter.

Today, NOAA's National Weather Service (NWS) is a leader in the most effective and sophisticated weather warning system in the world. Thanks to years of research and modern technology, forecasters at NOAA's Storm Prediction Center (SPC) issue forecasts outlining the most likely locations for the development of tornadoes and other severe weather 48 hours in advance, then fine-tune the forecast as the potential for inclement weather draws near. Using GOES satellite imagery, current surface observations, upper-air data, and computer forecast models, the meteorologists at SPC issue Severe Thunderstorm and Tornado Watches when severe weather is expected a few hours out.

From there, the local NWS Weather Forecast Offices (WFO's), such as the office in Paducah, continuously monitor WSR-88D Doppler Radar time-lapse imagery on sophisticated AWIPS Workstations to determine a storm's severe potential. An invaluable resource to the radar operator's final warning decision is the steady stream of reports from a network of trained and dedicated SKYWARN spotters, emergency managers, local law enforcement, and amateur radio "ham" operators.

When severe weather is either spotted or indicated on radar, the WFO radar operator issues a Severe Thunderstorm or Tornado Warning via WarnGen to alert the public to the imminent or existing threat of severe weather. As soon as the warning is disseminated, special tones are broadcast on NOAA Weather Radio in conjunction with the warning message—alerting the public to the impending threat to life and property. Meanwhile, various television and radio stations occasionally interrupt regular programming in order to communicate the NWS warning information to a large segment of the country's population. During the entire process, it takes a tremendous amount of coordination between government and private entities to ensure the best possible warning coverage.

After a severe weather episode, the NWS takes an active role in surveying locales most devastated and compiling information on the storms for research and climatological purposes. Newspapers and broadcasts from radio and television keep local residents updated on storm damage and clean-up efforts. The driving force behind the disaster relief process includes such organizations as the Federal Emergency Management Agency (FEMA) and the American Red Cross. Together, these two agencies bring necessary relief supplies to storm victims, assist in clean-up efforts, and are often instrumental in obtaining state and federal funds to accelerate the clean-up efforts.

Thus, through technological advancements, improved communications, and dedicated scientific research, a death toll of nearly 700 people from such a disaster is highly improbable today—but it is not impossible, especially if the tornado were to strike a highly populated area. Of course, the present warning system is not perfect, as evidenced by sometimes late or missed watches and warnings. However, we have obviously come a long way since the early 1900s! Through a continued cooperation between the NWS, FEMA, the American Red Cross, researchers, emergency managers, spotters, the media, and all concerned entities, the current warning system will undoubtedly experience significant improvements as we journey deeper into the 21st Century.

Following is a description of each facet of the modernized National Weather Service operations and technology mentioned in the preceding text.

AWIPS—Advanced Weather Interactive Processing System. State-of-the-art NWS computer system integrating automated weather observations, satellite imagery, radar data, and numerical model forecasts into forecaster workstations. There are currently over 130 sets of AWIPS Workstations located at numerous Weather Forecast Offices and 13 River Forecast Centers across the United States.

Computer forecast model—A numerical projection of future weather conditions derived by using current weather data in hundreds of mathematical computations. The computations are performed on supercomputers at NOAA's National Centers for Environmental Prediction (NCEP) in Silver Spring, Maryland. Currently, there are several forecast models in existence, including the NGM, NAM/ETA, GFS, and RUC.

GOES—Geostationary Operational Environmental Satellite. A geostationary satellite rotates at the same rate as the earth, remaining over the same spot above the equator. At any given time, there are two GOES satellites in orbit over the Atlantic and Pacific Oceans. These satellites monitor the earth's atmosphere over the entire United States in addition to adjacent land and water masses.

NOAA—National Oceanic and Atmospheric Administration. A branch of the U.S. Department of Commerce, NOAA is the parent organization of the National Weather Service.

NOAA Weather Radio—Continuous 24-hour-a-day VHF broadcasts of weather observations and forecasts directly from National Weather Service offices. A special tone activates an alarm on certain receivers when watches or warnings are issued. With some radios, this alarm can be tailored to sound for specific warnings affecting counties of your choice. Consult your local electronics retailers for more information.

NWS—National Weather Service. Agency of NOAA responsible for providing weather services to the nation. The mission of the NWS, in part, is "to provide weather and flood warnings, public forecasts and advisories for all the United States, its territories, adjacent waters and ocean areas, primarily for the protection of life and property." This mission is carried out by a network of weather offices located throughout the Unites States and its territories along with a highly trained workforce. Through this network, the NWS provides an invaluable service to government agencies, emergency managers, the media, and the general public 24 hours a day.

Severe Thunderstorm Watch—Issued by the SPC when conditions are favorable for severe thunderstorms in and close to the watch area. A watch is generally outlined by a parallelogram and is usually valid for a period of 4 to 7 hours. A severe thunderstorm is defined by wind gusts of 58 mph (50 knots) or greater, 1″ diameter hail or larger, a tornado, or any combination thereof.

Severe Thunderstorm Warning—Issued by the local NWS office when a severe thunderstorm is indicated by radar or reported by trained observers. A warning may cover a part of a county or several counties and is normally valid for 30 minutes to 1 hour in duration. A severe thunderstorm is defined by wind gusts of 58 mph (50 knots) or greater, 1″ diameter hail or larger, a tornado, or any combination thereof.

SKYWARN—A dedicated team of official NWS-trained storm spotters who devote their time and effort to aiding the NWS mission of savings lives via timely warning services. Essential to the warning process, these observers work in conjunction with local emergency officials to relay timely reports of severe weather and tornadoes to local NWS forecast offices. SKYWARN spotters who are licensed in amateur radio operations ("ham" operators) are especially valuable since they bring an alternative means of rapid communication to the warning process.

SPC—Storm Prediction Center. Situated in Norman, Oklahoma, this office is responsible for monitoring and forecasting severe convective weather, as well as winter weather, in the contiguous United States. This includes the issuance of Tornado and Severe Thunderstorm Watches and various outlooks to highlight the degree of severe weather threat.

Surface observations—Information, including such variables as sky condition, present weather, visibility, temperature, humidity, wind, and barometric pressure, analyzed on a map to determine the various weather phenomena occurring at the earth's surface. An integral part of the NWS surface observing program is ASOS, which stands for the Automated Surface Observing System. There are nearly 1000 ASOS units primarily co-located with airports across the United States.

Tornado Watch—Issued by the SPC when conditions are favorable for tornadoes and severe thunderstorms in and close to the watch area. A watch is generally outlined by a parallelogram and is usually valid for a period of 4 to 7 hours. A tornado is defined as a violently rotating column of air usually extending from the base of a cumulonimbus cloud and in contact with the ground. A condensation funnel cloud need not be present, but flying debris near the ground should mark the tornado's lower circulation.

Tornado Warning—Issued by the local NWS office when a tornado is indicated by radar or reported by trained observers. A warning may cover a part of a county or several counties and is normally valid for 15 to 45 minutes in duration. A tornado is defined as a violently rotating column of air usually extending from the base of a cumulonimbus cloud and in contact with the ground. A condensation funnel cloud need not be present, but flying debris near the ground should mark the tornado's lower circulation.

Upper-air data—Information, including such variables as temperature, humidity, and wind, analyzed to determine the weather phenomena occurring in that part of the atmosphere above the earth's surface.

WarnGen—Warning software accompanying AWIPS and used by local NWS offices to issue warnings and statements of inclement weather.

WFO—Weather Forecast Office. Designation of local NWS operational offices, each with its own area of forecast and warning responsibility. For example, WFO Paducah (the NWS office in Paducah, Kentucky) issues forecasts and warnings for a 58 county area, comprising portions of southeast Missouri, southern Illinois, southwest Indiana, and western Kentucky.

WSR-88D—Weather Surveillance Doppler Radar (1988). Whereas conventional radar only detects areas of precipitation, Doppler radar also determines whether atmospheric motion is toward or away from the radar and is useful in detecting rotation within a thunderstorm. To date, over 120 systems have been installed at Weather Forecast Offices with over 30 additional systems at Department of Defense (Air Force) and Department of Transportation (FAA) sites.

Source: National Weather Service Weather Forecast Office. "1925 Tri-State Tornado: Now v. Then." Available online. URL: http://www.crh.noaa.gov/pah/1925/nvt_body.php. Accessed June 30, 2010.

National Oceanic and Atmospheric Administration, "Heat Wave: A Major Summer Killer" (n.d.) (excerpt)

This document details how weather forecasters determine threats from excessive heat and the public announcement system used to warn people when high heat is expected.

Heat is the number one weather-related killer. On average, more than 1,500 people in the U.S. die each year from excessive heat. This number is greater than the 30-year mean annual number of deaths due to tornadoes, hurricanes, floods and lightning combined. In the 40-year period from 1936 through 1975, nearly 20,000 people were killed in the United States by the effects of heat and solar radiation.

In the disastrous heat wave of 1980, more than 1,250 people died. In the heat wave of 1995 more than 700 deaths in the Chicago, Illinois area were attributed to this event. And in August 2003, a record heat wave in Europe claimed an estimated 50,000 lives.

North American summers are hot; most summers see heat waves in one section or another of the United States. East of the Rockies, they tend to combine both high temperature and high humidity although some of the worst have been catastrophically dry. Additional detail on how heat impacts the human body is provided under "The Hazards of Excessive Heat" heading.

NOAA's Watch, Warning, and Advisory Products for Extreme Heat
Each National Weather Service (NWS) Weather Forecast Office (WFO) can issue the following heat-related products as conditions warrant:

> **Excessive Heat Outlook:** when the potential exists for an excessive heat event in the next 3 to 7 days. An outlook is used to indicate that a heat event may develop. It is intended to provide information to those who need considerable lead time to prepare for the event, such as public utilities, emergency management and public health officials.

Excessive Heat Watch: when conditions are favorable for an excessive heat event in the next 12 to 48 hours. A watch is used when the risk of a heat wave has increased, but its occurrence and timing is still uncertain. It is intended to provide enough lead time so those who need to set their plans in motion can do so, such as established individual city excessive heat event mitigation plans.

Excessive Heat Warning/Advisory: when an excessive heat event is expected in the next 36 hours. These products are issued when an excessive heat event is occurring, is imminent, or has a very high probability of occurrence. The warning is used for conditions posing a threat to life or property. An advisory is for less serious conditions that cause significant discomfort or inconvenience and, if caution is not taken, could lead to a threat to life and/or property.

How Forecasters Decide Whether to Issue Excessive Heat Products

National Weather Service Heat Index Based Guidance

The "Heat Index" (HI) is sometimes referred to as the "apparent temperature." The HI, given in degrees F, is a measure of how hot it really feels when relative humidity (RH) is added to the actual air temperature. . . . As an example, if the air temperature is 96°F (found on the top of the table) and the RH is 65% (found on the left of the table), the HI-or how hot it really feels-is 121°F. This is at the intersection of the 96° column and the 65% row.

IMPORTANT: Since HI values were devised for shady, light wind conditions, EXPOSURE TO FULL SUNSHINE CAN INCREASE HI VALUES BY UP TO 15°F. Also, STRONG WINDS, PARTICULARLY WITH VERY HOT, DRY AIR, CAN BE EXTREMELY HAZARDOUS.

Note on the Heat Index Chart shaded zone above 105°F. This corresponds to a level of HI that may cause increasingly severe heat disorders with continued exposure and/or physical activity.

NOAA's Heat Alert Procedures based mainly on Heat Index Values

The National Weather Service will initiate alert procedures when the Heat Index is expected to exceed 105°–110°F (depending on local climate) for at least two consecutive days. The procedures are:

- Include Heat Index values in zone and city forecasts.

- Issue Special Weather Statements and/or Public Information Statements presenting a detailed discussion of: Extent of the hazard including Heat Index values, who is most at risk, and safety rules for reducing the risk.
- Assist state/local health officials in preparing Civil Emergency Messages in severe heat waves. Meteorological information from Special Weather Statements will be included as well as more detailed medical information, advice, and names and telephone numbers of health officials.
- Release all of the above information to the media and over NOAA All-Hazard Weather Radio

Heat Health Watch/Warning System

Recent research has shown that a heat index threshold does not fully account for a variety of factors which impact health including the impact of consecutive stressful days on human health, the time of year, or the location where excessive heat events occur. For example, studies indicate large urban areas are particularly sensitive to heat early in the summer season. Based on this research, NOAA/NWS has supported the implementation of new Heat Health Watch/Warning System (HHWS) that its forecasters use as guidance in producing their daily warning and forecast products. This system was developed in conjunction with researchers at the University of Delaware.

As of summer 2007, about 20 Weather Forecast Offices (WFOs) now utilize the HHWS as additional guidance in their forecast decision-making process. The NWS goal is to expand the HHWS coverage to include approximately 70 vulnerable urban cities across the continental U.S. with mostly populations of 500,000 or more.

The HHWS, tailored for each urban locale, is the first and only meteorological tool based upon the occurrence of certain air masses that have historically been associated with elevated mortality levels. Air masses consider the entire "umbrella" of air over a region, rather than a single meteorological variable such as the heat index. HHWS consider numerous meteorological, seasonal, and social factors, and are based upon actual human health responses. Through the use of, it is possible to predict the likelihood of excess mortality given the synoptic conditions present at specific cities, the number of consecutive days an oppressive air mass is present, and the time of year the event occurs.

Currently, those urban areas with HHWS coverage include Philadelphia, PA; Seattle, WA; Dallas, Fort Worth and Houston. TX; Phoenix and

Yuma, AZ; Baltimore, MD; Washington, D.C.; Chicago, IL; St. Louis, MO; Cincinnati and Dayton, Ohio; New Orleans, Baton Rouge, Lake Charles, Alexandria, Shreveport and Monroe, LA; Memphis, TN; Jackson, Meridian and Tupelo, MS; Little Rock and Pine Bluff, AR; Portland, OR; Minneapolis, MN; San Francisco and San Jose, CA.

The NWS forecaster analyzes the HHWS guidance, as well as heat index values, time of year and expected length of the heat event, collaborate with neighboring WFOs as needed, and then decide which, if any, excessive heat product to issue. If an Outlook, Watch, Warning, or Advisory will be issued, the forecaster will notify the local health department and/or emergency management agency to insure that they are aware of the excessive heat forecast.

The Hazards of Excessive Heat

How Heat Affects the Body Human

Human bodies dissipate heat by varying the rate and depth of blood circulation, by losing water through the skin and sweat glands, and-as the last extremity is reached-by panting, when blood is heated above 98.6 degrees. The heart begins to pump more blood, blood vessels dilate to accommodate the increased flow, and the bundles of tiny capillaries threading through the upper layers of skin are put into operation. The body's blood is circulated closer to the skin's surface, and excess heat drains off into the cooler atmosphere. At the same time, water diffuses through the skin as perspiration. The skin handles about 90 percent of the body's heat dissipating function.

Sweating, by itself, does nothing to cool the body, unless the water is removed by evaporation, and high relative humidity retards evaporation. The evaporation process itself works this way: the heat energy required to evaporate the sweat is extracted from the body, thereby cooling it. Under conditions of high temperature (above 90 degrees) and high relative humidity, the body is doing everything it can to maintain 98.6 degrees inside. The heart is pumping a torrent of blood through dilated circulatory vessels; the sweat glands are pouring liquid—including essential dissolved chemicals, like sodium and chloride onto the surface of the skin.

Too Much Heat

Heat disorders generally have to do with a reduction or collapse of the body's ability to shed heat by circulatory changes and sweating, or a chemical (salt) imbalance caused by too much sweating. When heat gain exceeds the level the body can remove, or when the body cannot compensate for fluids and salt lost through perspiration, the temperature of the body's inner core begins to rise and heat-related illness may develop.

Ranging in severity, heat disorders share one common feature: the individual has overexposed or over exercised for his age and physical condition in the existing thermal environment.

Sunburn, with its ultraviolet radiation burns, can significantly retard the skin's ability to shed excess heat. Studies indicate that, other things being equal, the severity of heat disorders tend to increase with age-heat cramps in a 17-year-old may be heat exhaustion in someone 40, and heat stroke in a person over 60.

Acclimatization has to do with adjusting sweat-salt concentrations, among other things. The idea is to lose enough water to regulate body temperature, with the least possible chemical disturbance.

Cities Pose Special Hazards

The stagnant atmospheric conditions of the heat wave trap pollutants in urban areas and add the stresses of severe pollution to the already dangerous stresses of hot weather, creating a health problem of undiscovered dimensions. A map of heat-related deaths in St. Louis during 1966, for example, shows a heavier concentration in the crowded alleys and towers of the inner city, where air quality would also be poor during a heat wave.

The high inner-city death rates also can be read as poor access to air-conditioned rooms. While air conditioning may be a luxury in normal times, it can be a lifesaver during heat wave conditions.

The cost of cool air moves steadily higher, adding what appears to be a cruel economic side to heat wave fatalities. Indications from the 1978 Texas heat wave suggest that some elderly people on fixed incomes, many of them in buildings that could not be ventilated without air conditioning, found the cost too high, turned off their units, and ultimately succumbed to the stresses of heat

Children, Adults, and Pets Enclosed in Parked Vehicles Are at Great Risk

Each year children die from hyperthermia as a result of being left enclosed in parked vehicles. This can occur even on a mild day. Studies have shown that the temperature inside a parked vehicle can rise rapidly to a dangerous level for children, adults, and pets. Leaving the windows slightly open does not significantly decrease the heating rate. The effects can be more severe on children because their bodies warm at a faster rate than adults.

Excessive Heat Cautions and Safety Tips

Preventing Heat-Related Illness
Elderly persons, small children, chronic invalids, those on certain medications or drugs (especially tranquilizers and anticholinergics), and persons with weight and alcohol problems are particularly susceptible to heat reactions, especially during heat waves in areas where a moderate climate usually prevails.

Heat Wave Safety Tips
Slow down. Strenuous activities should be reduced, eliminated, or rescheduled to the coolest time of the day. Individuals at risk should stay in the coolest available place, not necessarily indoors.

Dress for summer. Lightweight light-colored clothing reflects heat and sunlight, and helps your body maintain normal temperatures.

Put less fuel on your inner fires. Foods (like proteins) that increase metabolic heat production also increase water loss. Drink plenty of water or other non-alcohol fluids. Your body needs water to keep cool. Drink plenty of fluids even if you don't feel thirsty. Persons who (1) have epilepsy or heart, kidney, or liver disease, (2) are on fluid restrictive diets or (3) have a problem with fluid retention should consult a physician before increasing their consumption of fluids. Do not drink alcoholic beverages.

Do not take salt tablets unless specified by a physician.

Spend more time in air-conditioned places. Air conditioning in homes and other buildings markedly reduces danger from the heat. If you cannot afford an air conditioner, spending some time each day (during hot weather) in an air conditioned environment affords some protection.

Don't get too much sun. Sunburn makes the job of heat dissipation that much more difficult.

Never leave persons, especially children, and pets in a closed, parked vehicle.

Know These Heat Disorder Symptoms

SUNBURN: Redness and pain. In severe cases swelling of skin, blisters, fever, headaches. First Aid: Ointments for mild cases if blisters appear and do not break. If breaking occurs, apply dry sterile dressing. Serious, extensive cases should be seen by physician.

HEAT CRAMPS: Painful spasms usually in muscles of legs and abdomen possible. Heavy sweating. First Aid: Firm pressure on cramping muscles, or gentle massage to relieve spasm. Give sips of water. If nausea occurs, discontinue use.

HEAT EXHAUSTION: Heavy sweating, weakness, skin cold, pale and clammy. Pulse thready. Normal temperature possible. Fainting and vomiting. First Aid: Get victim out of sun. Lay down and loosen clothing. Apply cool, wet cloths. Fan or move victim to air conditioned room. Sips of water. If nausea occurs, discontinue use. If vomiting continues, seek immediate medical attention.

HEAT STROKE (or sunstroke): High body temperature (106°F or higher). Hot dry skin. Rapid and strong pulse. Possible unconsciousness. First Aid: HEAT STROKE IS A SEVERE MEDICAL EMERGENCY. SUMMON ENERGENCY MEDICAL ASSISTANCE OR GET THE VICTIM TO A HOSPITAL IMMEDIATELY. DELAY CAN BE FATAL. Move the victim to a cooler environment. Reduce body temperature with cold bath or sponging. Use extreme caution. Remove clothing, use fans and air conditioners. If temperature rises again, repeat process. Do not give fluids. Persons on salt restrictive diets should consult a physician before increasing their salt intake.

*For more information contact your local American Red Cross Chapter. Ask to enroll in a first aid course.

Community Guidance: Preparing for and Responding to Excessive Heat Events

The "Excessive Heat Events Guidebook" was developed by the Environmental Protection Agency (EPA) in 2006, in collaboration with NOAA's National Weather Service (NWS), the Centers for Disease Control and Prevention (CDC), and the U.S. Department of Homeland Security (DHS). This guidebook provides best practices that have been employed to save lives during heat waves in different urban areas, and provides a menu of options that communities can use in developing their own mitigation plans.

Source: National Oceanic and Atmospheric Administration. "Heat Wave: A Major Summer Killer." Available online. URL: http://www.noaawatch.gov/themes/heat.php. Accessed November 20, 2009.

5

International Documents

This chapter contains primary sources of information about the types of natural disasters that are examined in this book. The documents are divided into five sections: Documents of General International Interest, Indian Ocean, China, Africa, and Colombia. Many selections are excerpts of longer documents. The complete document or article can be accessed by using the source information that follows the passage.

DOCUMENTS OF GENERAL INTERNATIONAL INTEREST

The Eruption of Vesuvius: Pliny the Younger, Letter to Tacitus (first century) (excerpt)

A prominent lawyer in ancient Rome, Pliny the Younger (61–ca. 112 C.E.) wrote many letters that provide historical information about the period. In this letter, he provides details concerning the death of his uncle, Pliny the Elder, during the eruption of Mount Vesuvius in 79 C.E. The letter includes eyewitness details of the eruption as well as an account of the last actions of Pliny the Elder, an active author and naturalist who wrote the encyclopedia Naturalis Historia.

I. To Tacitus

You ask me to send you an account of my uncle's death, so that you may be able to give posterity an accurate description of it. I am much obliged to you, for I can see that the immortality of his fame is well assured, if you take in hand to write of it. For although he perished in a disaster which devastated some of the fairest regions of the land, and though he is sure of eternal remembrance like the peoples and cities that fell with him in that memorable calamity, though too he had written a large number of works of lasting value, yet the undying fame of which your writings are assured will secure for his a still further lease of life. For my own part, I think that those people are highly favoured by Providence who are capable either of performing deeds worthy of the historian's pen or of writing histories worthy of being read, but that they are peculiarly favoured who can do both. Among the latter I may class my uncle, thanks to his own writings and to yours. So I am all the more ready to fulfil your injunctions, nay, I am even prepared to beg to be allowed to undertake them.

My uncle was stationed at Misenum, where he was in active command of the fleet, with full powers. On the 23rd of August, about the seventh hour, my mother drew his attention to the fact that a cloud of unusual size and shape had made its appearance. He had taken his sun bath, followed by a cold one, and after a light meal he was lying down and reading. Yet he called for his sandals, and climbed up to a spot from which he could command a good view of the curious phenomenon. Those who were looking at the cloud from some distance could not make out from which mountain it was rising—it was afterwards discovered to have been Mount Vesuvius—but in likeness and form it more closely resembled a pine-tree than anything else, for what corresponded to the trunk was of great length and height, and then spread out into a number of branches, the reason being, I imagine, that while the vapour was fresh, the cloud was borne upwards, but when the vapour became wasted, it lost its motion, or even became dissipated by its own weight, and spread out laterally. At times it looked white, and at other times dirty and spotted, according to the quantity of earth and cinders that were shot up.

To a man of my uncle's learning, the phenomenon appeared one of great importance, which deserved a closer study. He ordered a Liburnian galley to be got ready, and offered to take me with him, if I desired to accompany him, but I replied that I preferred to go on with my studies, and it so happened that he had assigned me some writing to do. He was just leaving the house when he received a written message from Rectina, the wife of Tascus, who was terrified at the peril threatening her—for her villa lay just beneath the mountain, and there were no means of escape save by shipboard—begging him to save her from her perilous position. So he changed his plans, and carried out with the greatest fortitude the ideas which had occurred to him as a student.

He had the galleys launched and went on board himself, in the hope of succouring, not only Rectina, but many others, for there were a number of people living along the shore owing to its delightful situation. He hastened, therefore, towards the place whence others were flying, and steering a direct course, kept the helm straight for the point of danger, so utterly devoid of fear that every movement of the looming portent and every change in its appearance he described and had noted down by his secretary, as soon as his eyes detected it. Already ashes were beginning to fall upon the ships, hotter and in thicker showers as they approached more nearly, with pumice-stones and black flints, charred and cracked by the heat of the flames, while their way was barred by the sudden shoaling of the sea bottom and the litter of the mountain on the shore. He hesitated for a moment whether to turn back, and then, when the helmsman warned him to do so, he exclaimed, "Fortune favours the bold; try to reach Pomponianus." The latter was at Stabiæ, separated by the whole width of the bay, for the sea there pours in upon a gently rounded and curving shore. Although the danger was not yet close upon him, it was none the less clearly seen, and it travelled quickly as it came nearer, so Pomponianus had got his baggage together on shipboard, and had determined upon flight, and was waiting for the wind which was blowing on shore to fall. My uncle sailed in with the wind fair behind him, and embraced Pomponianus, who was in a state of fright, comforting and cheering him at the same time. Then in order to calm his friend's fears by showing how composed he was himself, he ordered the servants to carry him to the bath, and, after his ablutions, he sat down and had dinner in the best of spirits, or with that assumption of good spirits which is quite as remarkable as the reality.

In the meantime broad sheets of flame, which rose high in the air, were breaking out in a number of places on Mount Vesuvius and lighting up the sky, and the glare and brightness seemed all the more striking owing to the darkness of the night. My uncle, in order to allay the fear of his companions, kept declaring that the country people in their terror had left their fires burning, and that the conflagration they saw arose from the blazing and empty villas. Then he betook himself to rest and enjoyed a very deep sleep, for his breathing, which, owing to his bulk, was rather heavy and loud, was heard by those who were waiting at the door of his chamber. But by this time the courtyard leading to the room he occupied was so full of ashes and pumice-stones mingled together, and covered to such a depth, that if he had delayed any longer in the bed chamber there would have been no means of escape. So my uncle was aroused, and came out and joined Pomponianus and the rest who had been keeping watch. They held a consultation whether they should remain indoors or wander forth in the open; for the buildings were beginning to shake with the repeated and intensely severe shocks of earthquake, and seemed to be rocking to and fro as though they had been torn from their foundations. Outside again there was danger to be apprehended from the pumice-stones, though these were light and nearly burnt through, and thus, after weighing the two perils, the latter course was determined upon. With my uncle it was a choice of reasons which prevailed, with the rest a choice of fears.

They placed pillows on their heads and secured them with napkins, as a precaution against the falling bodies. Elsewhere the day had dawned by this time, but there it was still night, and the darkness was blacker and thicker than any ordinary night. This, however, they relieved as best they could by a number of torches and other kinds of lights. They decided to make their way to the shore, and to see from the nearest point whether the sea would enable them to put out, but it was still running high and contrary. A sheet was spread on the ground, and on this my uncle lay, and twice he called for a draught of cold water, which he drank. Then the flames, and the smell of sulphur which gave warning of them, scattered the others in flight and roused him. Leaning on two slaves, he rose to his feet and immediately fell down again, owing, as I think, to his breathing being obstructed by the thickness of the fumes and congestion of the stomach, that organ being naturally weak and narrow, and subject to inflammation. When daylight returned—which was three days after his death—his body was found untouched, uninjured, and covered, dressed just as he had been in life. The corpse suggested a person asleep rather than a dead man.

Meanwhile my mother and I were at Misenum. But that is of no consequence for the purposes of history, nor indeed did you express a wish to be told of anything except of my uncle's death. So I will say no more, except to add that I have given you a full account both of the incidents which I myself witnessed and of those narrated to me immediately afterwards, when, as a rule, one gets the truest account of what has happened. You will pick out what you think will answer your purpose best, for to write a letter is a different thing from writing a history, and to write to a friend is not like writing to all and sundry. Farewell.

II. To Tacitus

You say that the letter which I wrote to you at your request, describing the death of my uncle, has made you anxious to know not only the terrors, but also the distress I suffered while I remained behind at Misenum. I had indeed started to tell you of these, but then broke off. Well, though my mind shudders at the recollection, I will essay the task.

After my uncle had set out I employed the remainder of the time with my studies, for I had stayed behind for that very purpose. Afterwards I had a bath, dined, and then took a brief and restless sleep. For many days previous there had been slight shocks of earthquake, which were not particularly alarming, because they are common enough in Campania. But on that night the shocks were so intense that everything round us seemed not only to be disturbed, but to be tottering to its fall. My mother rushed into my bedchamber, just as I myself was getting up in order to arouse her if she was still sleeping. We sat down in the courtyard of the house, which was of smallish size and lay between the sea and the buildings. I don't know whether my behaviour should be called courageous or rash—for I was only in my eighteenth year—but I called for a volume of Titus Livius, and read it, as though I were perfectly at my ease, and went on making my usual extracts. Then a friend of my uncle's, who had but a little time before come to join him from Spain, on seeing my mother and myself sitting there and me reading, upbraided her for her patience and me for my indifference, but I paid no heed, and pored over my book.

It was now the first hour of the day, but the light was still faint and weak. The buildings all round us were beginning to totter, and, though we were in the open, the courtyard was so narrow that we were greatly afraid, and indeed sure of being overwhelmed by their fall. So that decided us to leave the town. We were followed by a distracted crowd, which, when in a panic, always prefers some one else's judgment to its own as the most prudent course to adopt, and when we set out these people came crowding in masses upon us, and pressed and urged us forward. We came to a halt when we had passed beyond the buildings, and underwent there many wonderful experiences and terrors. For although the ground was perfectly level, the vehicles which we had ordered to be brought with us began to sway to and fro, and though they were wedged with stones, we could not keep them still in their places. Moreover, we saw the sea drawn back upon itself, and, as it were, repelled by the quaking of the earth. The shore certainly was greatly widened, and many marine creatures were stranded on the dry sands. On the other side, the black, fearsome cloud of fiery vapour burst into long, twisting, zigzag flames and gaped asunder, the flames resembling lightning flashes, only they were of greater size. Then indeed my uncle's Spanish friend exclaimed sharply, and with an air of command, to my mother and me, "If your brother and your uncle is still alive, he will be anxious for you to save yourselves; if he is dead, I am sure he wished you to survive him. Come, why do you hesitate to quit this place?" We replied that we could not think of looking after our own safety while we were uncertain of his. He then waited no longer, but tore away as fast as he could and got clear of danger.

Soon afterwards the cloud descended upon the earth, and covered the whole bay; it encircled Capreæ and hid it from sight, and we could no longer see the promontory of Misenum. Then my mother prayed, entreated, and commanded me to fly as best I could, saying that I was young and could escape, while she was old and infirm, and would not fear to die, if only she knew that she had not been the cause of my death. I replied that I would not save myself unless I could save her too, and so, after taking tight hold of her hand, I forced her to quicken her steps. She reluctantly obeyed, accusing herself for retarding my flight. Then the ashes began to fall, but not thickly: I looked back, and a dense blackness was rolling up behind us, which spread itself over the ground and followed like a torrent. "Let us turn aside," I said, "while we can still see, lest we be thrown down in the road and trampled on in the darkness by the thronging crowd." We were considering what to do, when the blackness of night overtook us, not that of a moonless or cloudy night, but the blackness of pent-up places which never see the light. You could hear the wailing of women, the screams of little children, and the shouts of men; some were trying to find their parents, others their children, others their wives, by calling for them and recognising them by their voices alone. Some were commiserating their own lot, others that of their relatives, while some again prayed for death in sheer terror of dying. Many were lifting up their hands to the gods, but more were declaring that now there were no more gods, and that this night would last for ever, and be the end of all the world. Nor were there wanting those who added to the real perils by inventing new and false terrors, for some said that part of Misenum was in ruins and the rest in flames, and though the tale was untrue, it found ready believers.

A gleam of light now appeared, which seemed to us not so much daylight as a token of the approaching fire. The latter remained at a distance, but the darkness came on again, and the ashes once more fell thickly and heavily. We had to keep rising and shaking the latter off us, or we should have been buried by them and crushed by their weight. I might boast that not one groan or cowardly exclamation escaped my lips, despite these perils, had I not believed that I and the world were perishing together—a miserable consolation, indeed, yet one which a mortal creature finds very soothing. At length the blackness became less dense, and dissipated as it were into smoke and cloud; then came the real light of day, and the sun shone out, but as blood-red as it is wont to be at its setting. Our still trembling eyes saw that everything had been transformed, and covered with a deep layer of ashes, like snow. Making our way back to Misenum, we refreshed our bodies as best we could, and passed an anxious, troubled night, hovering between hope and fear. But our fears were uppermost, for the shocks of earthquake still continued, and several persons, driven frantic by dreadful prophecies, made sport of their own calamities and those of others. For our own part, though we had already passed through perils, and expected still more to come, we had no idea even then of leaving the town until we got news of my uncle.

You will not read these details, which are not up to the dignity of history, as though you were about to incorporate them in your writings, and if they seem to you to be hardly worth being made the subject of a letter, you must take the blame yourself, inasmuch as you insisted on having them. Farewell.

Source: Chauncey Wetmore Wells, ed. *A Book of Prose Narratives.* Boston: Ginn and Company, 1914, pp. 169–176.

The Eruption of Krakatoa: Eyewitness Accounts (1833) (excerpt)

In August 1883, the cataclysmic eruption of Krakatoa caused widespread loss of life and destruction on the Indonesian islands of Java and Sumatra. The pyroclastic flows and tsunamis caused by the eruption completely destroyed many villages in the region. One of the most important sources of information concerning the eruption comes from eyewitnesses aboard ships sailing in the area. This excerpt provides the details recorded by captains in ship logs and by passengers who witnessed the events.

On the afternoon of the 26th of August, and through the succeeding night and day till the early morning of the 28th of August, it was evident that the long-continued moderate eruptions (Strombolian stage) which had for some days been growing in intensity, had passed into the paroxysmal (Vesuvian) stage. In order to weigh the evidence which we have concerning the nature of this critical and most interesting period of the eruption of Krakatoa, it may be well to consider what were the facilities for observation possessed by the several individuals from whom the reports concerning the eruptions were obtained.

Situated respectively at a distance of 94 and 100 English miles to the east of Krakatoa are the two important towns of Batavia and Buitenzorg. In both these places, numerous Europeans capable of making accurate observations were resident; there were also self-recording instruments, the tracings of which have proved of the greatest value in these enquiries. At numerous small towns and villages along the Javan and Sumatran coasts of the Strait of Sunda, and in the five lighthouses, two of which were destroyed, European officials were located. Many of these fled during the terrible night of the 26th of August, and others were drowned by the great sea-waves which submerged all the coast-towns on the morning of the 27th. Very admirably has Mr. VERBEEK collected and discussed the reports made by the officials of the coast-towns and villages who survived that night of horrors.

Perhaps, however, the most important evidence of what was actually going on at Krakatoa during the crisis of the eruption is that derived from witnesses on board ships which sailed between Java and Sumatra while the great outburst was in progress, or those that were at the time in the immediate vicinity of either the eastern or western entrance of the Sunda Strait. From many more distant points, however, valuable confirmatory or supplementary evidence has been obtained, for which we are indebted to the captains or passengers of vessels passing through the eastern seas during that period.

Only three European ships appear to have been actually within the Sunda Strait during the height of the eruption on the night of the 26th of August and the early morning of the 27th, and to have escaped destruction, so that those on board could tell the tale of what they witnessed.

The greatest opportunities for observation seem to have been those which were afforded to Captain WATSON of the British ship *Charles Bal,* then on its voyage to Hong Kong. This vessel passed Princes Island at 9 a.m. on Sunday the 26th of August; at noon she was on the south-west side of Krakatoa; and at 4.15 p.m. she reached a point nearly due south of the volcano, and about 10 miles distant from it. The darkness being too great to permit of safe navigation, sail was shortened, and through the whole night the vessel was kept beating about on the east of the volcano, and within a dozen miles from it. At 6 a.m. on the 27th, the Java shore was sighted, and the vessel was enabled to continue her voyage.

The Batavian steamship *Gouverneur-Generaal Loudon,* Commandant T. H. LINDEMAN, left Batavia on the morning of the 26th of August, and reached Anjer at 2 p.m. the same day. Leaving that port at 2.45 p.m., she sailed for Telok Betong, taking a number of coolies and women as passengers, and passing about 30 miles north of Krakatoa, reached her destination at 7.30 p.m. Finding at midnight that it was impossible, on account of the storm which was raging, to communicate with the shore, the vessel steamed out into the bay and anchored. She thus escaped being stranded by the great sea-waves of the early morning, like the unfortunate Government steamer *Berouw,* which was at this time anchored close to the pier-head at Telok Betong. At 7.30 a.m. on the 27th, the steamer *G. G. Loudon* started to return to Anjer, but had to come to anchor at 10 o'clock on account of the rain of pumice, and the storm that was raging. During Tuesday, the 28th, she steamed round the west and south sides of Krakatoa, called at the part of the coast where Anjer formerly stood, and then proceeded to Batavia.

The Dutch barque *Marie,* engaged in the salt-trade, was, during the whole time of the eruption, anchored off Telok Betong. On the morning of the 27th of August, thanks to the precaution of putting out a third anchor, she rode safely, and was able to avoid being stranded by the gigantic sea-waves, which swept on to the land the Government steamer *Berouw,* three schooners, and many smaller craft lying off the same port. The vessel appears to have been at times in imminent danger, but only four of the persons on board of her were drowned.

During the whole of Sunday, the 26th of August, two vessels, the barque *Norham Castle,* Captain O. SAMPSON, and the ship *Sir Robert Sale,* Captain W. T. WOOLDRIDGE, were at the eastern entrance of the Strait, and about 40 miles from Krakatoa. On the morning of Monday, the 27th, both these vessels entered the Strait, but owing to the darkness, neither made much progress till the morning of the 28th, when, falling in with each other, they made their way in company, but with much difficulty, through the Strait.

The Dutch hopper-barge, *Tegal,* which sailed from Batavia for Merak early on Monday, the 27th of August, remained at anchor near the eastern entrance of the Strait during the great darkness, but on Tuesday, the 28th, entered the Strait.

On the morning of Sunday, the 26th of August, the ship *Berbice,* of Greenock, Captain WILLIAM LOGAN, was at the western entrance of Sunda Strait, and about 40 miles from Krakatoa. This vessel remained beating about the entrance till Wednesday, the 29th, when she was able to sail through the Strait.

These are the vessels which, during the crisis of the great eruption, were in the most favourable positions for those on board of them to make observations concerning what was taking place at Krakatoa. The approximate positions of these vessels are shown in the accompanying chart (Fig. 9). Let us now turn our attention to some other vessels which were at greater distances from the scene of eruption, but, from the captains or passengers on board which, valuable information has been received.

The Norwegian barque *Borjild,* Captain AMUNDSEN, was at anchor near Great Kombuis Island, 75 miles east-by-north of Krakatoa, during the 26th and 27th of August.

The British ship *Medea,* Captain THOMSON, was, at 2 p.m. on the 26th, in the vicinity of the last-mentioned vessel, and sailing eastward came to anchor about 89 miles from Krakatoa.

The American barque *William H. Besse,* Captain BAKER, on its way from Manilla to Boston, U.S.A., having called at Batavia, was in the same neighbourhood, and on Wednesday, 29th, and Thursday, 30th of August, was passing through the Strait.

The British steamer *Anerley*, Captain STRACHAN, bound from Singapore to Mauritius, was, on the 26th of August, in Banca Strait, 250 English miles north of Krakatoa. During the 27th the steamer remained at anchor near North Watcher Island, 92 English miles north-east of the volcano.

The Siamese barque *Thoon Kramoom*, Captain ANDERSEN, bound from Bankok to Falmouth, lay, on the 27th and 28th of August, in the Strait of Banca, 230 English miles north of Krakatoa, and, sailing southwards, passed through the Strait of Sunda on the 31st of August.

Several vessels, among which was the barque *Hope*, were lying in Batavia Bay during the great paroxysmal outburst.

The mail steamer *Prinses Wilhelmina*, which passed through the Strait on the 23rd of August, coming from the west, remained at anchor at Batavia during the time of the great eruption.

Among vessels which were at still greater distances from the volcano during the time of the great outburst, the following may be mentioned as those from which information and specimens of the falling pumice and dust have been received:—

The British ship *Bay of Naples*, Captain TIDMARSH, was, during the eruption, about 138 English miles south of Java's First Point, and the barque *Lucia* was about 300 miles to the south-east of Krakatoa.

From the seas to the west of the Strait of Sunda we have information from the steamship *Simla*, Captain M. NICHOLSON, where dust, falling at a distance of about 1,150 English miles from the volcano, was collected, and from the barque *Jonc*, Captain L. REID, at about 600 English miles from the Strait. On board the British ships *Earl of Beaconsfield*, and the *Ardgowan*, Captain ISBISTER, and the German brig *Catherine*, dust fell when they were between 900 and 1,100 English miles from Krakatoa; and on board the British barque *Arabella*, Captain WILLIAMS, when about 1,100 English miles from Krakatoa.

The mail steamer *Prins Frederik*, on its way to Holland, passed near Krakatoa on the 25th of August, and the steamer *Batavia* sailed from Padang to Vlakke Hoek on the evening of the 27th.

The *Prins Hendrik*, a Dutch man-of-war, was ordered to the Strait of Sunda immediately after the eruption, in order to succour the survivors.

H.M.S. *Magpie*, Commander the Hon. F. C. P. VEREKER, was at Sandakang, N. Borneo, at the time of the eruption, and on the 18th of October visited the Strait for the purpose of examining the changes which had taken place. Somewhat later H.M.S. *Merlin*, Commander R. C. BRUNTON, visited the locality, and sent in a report to the Admiralty.

From various ports, accounts have been received, sent by British Consuls and by residents, and many of these have proved to be of great service to the Krakatoa Committee.

The log-books of the different vessels mentioned, and narratives written by the captains and passengers on board of them, taken in conjunction with the reports collected with so much care by Mr. VERBEEK, have afforded the means of compiling the following account of what occurred at Krakatoa during Sunday, the 26th, and Monday, the 27th, August.

The vessels passing through the Strait, as well as the observers on land, all reported a very marked though gradual increase in the violence of the eruption during the three days which preceded Sunday, the 26th of August.

On that day, about 1 p.m., the detonations caused by the explosive action attained such violence as to be heard at Batavia and Buitenzorg, about 100 English miles away.

At 2 p.m. Captain THOMSON, of the *Medea*, then sailing at a point 76 English miles E.N.E. of Krakatoa, saw "a black mass rising up like a smoke, in clouds," to an altitude which has been estimated as being no less than 17 miles. If this estimate be correct, some idea of the violence of the outburst can be formed from the fact that during the eruption of Vesuvius in 1872 the column of steam and dust was propelled to the height of from only 4 to 5 miles.

The great detonations at this time were said to be taking place at intervals of about ten minutes.

By 3 p.m. the sounds produced by the explosions at Krakatoa had so far increased in loudness that they were heard at Bandong and other places 150 miles away; and at 5 p.m. they had become so tremendous that they were heard all over the island of Java, and at many other equally distant localities. At Batavia and Buitenzorg they were, during the whole night, so violent that few people in the district were able to sleep; the noise is described as being like the discharge of artillery close at hand, and as causing rattling of the windows and shaking of pictures, chandeliers, and other hanging bodies. Nearly all observers agree that there was nothing in the nature of earthquake-shocks, but only strong air-vibrations.

Captain WATSON, of the *Charles Bal*, who was only 10 miles south of the volcano during this Sunday afternoon, describes the island as being covered with a dense black cloud; "clouds or something were being propelled from the north-east point with great velocity;" sounds like discharges of artillery at intervals of a second of time, and a crackling noise, probably due to the impact of fragments in the atmosphere, were heard; the whole commotion increasing towards 5 p.m., when it became so intense that the

Captain feared to continue his voyage, and began to shorten sail. From 5 to 6 p.m. a rain of pumice in large pieces, quite warm, fell upon the ship.

Captain WOOLDRIDGE, of the *Sir R. Sale,* viewing the volcano from the north-east at sunset on Sunday evening, the 26th, describes the sky as presenting "a most terrible appearance, the dense mass of clouds being covered with a murky tinge, with fierce flashes of lightning." At 7 p.m., when the dense vapour and dust-clouds rendered it intensely dark, the whole scene was lighted up from time to time by the electrical discharges, and at one time the cloud above the mountain presented "the appearance of an immense pine-tree, with the stem and branches formed with volcanic lightning." The air was loaded with excessively fine ashes, and there was a strong sulphurous smell. Captain O. SAMPSON, of the *Norham Castle,* who was in the same neighbourhood, gives a similar account of what he witnessed. The steamer *G. G. Loudon* passed to the north-west and west of the volcano, within a distance of 20 or 30 miles; it was seen to be "casting forth enormous columns of smoke," and the vessel passed through "a rain of ashes and small bits of stone."

During the night, while the *Charles Bal* remained beating about on the east of Krakatoa, and within about a dozen miles of the island, Captain WATSON records the phenomena of "chains of fire, appearing to ascend" between the volcano and the sky, while on the south-west side there seemed to be "a continual roll of balls of white fire." These appearances were doubtless caused by the discharge of white-hot fragments of lava, and their roll down the sides of the peak of Rakata, which was still standing.

The air at this distance, though the wind was strong at the time, was described by Captain WATSON as being "hot and choking, sulphurous, with a smell as of burning cinders;" masses like "iron-cinders" fell on the ship, and the lead from a bottom of 30 fathoms came up quite warm. From midnight till 4 a.m. explosions continually took place, "the sky, one second intense blackness, the next a blaze of fire."

All these details prove conclusively that Krakatoa had arrived at the paroxysmal phase of eruption. The explosive bursts of vapour beginning on the afternoon of Sunday and continuing at intervals of ten minutes, increased in violence and rapidity, and from sunset till midnight there was an almost continuous roar, which moderated a little towards early morning. Each explosive outburst of steam would have the effect of removing the accumulating pumice from the surface of the melted lava, by blowing it into the atmosphere, and the cauldron of white-hot lava would then have its glowing surface reflected in the clouds of vapour and dust hanging above.

The numerous vents on the low-lying parts of Krakatoa, which were recorded as having been seen by Captain FERZENAAR on the 11th of August, had, doubtless, by this time become more or less united, and the original crater of the old volcano was being rapidly emptied by the great paroxysmal explosions which commenced in the afternoon of the 26th of August.

All the eye-witnesses are in agreement as to the splendour of the electrical phenomena displayed during this paroxysmal outburst. Captain WOOLDRIDGE, viewing the eruption in the afternoon from a distance of 40 miles, speaks of the great vapour-cloud looking like "an immense wall with bursts of forked lightning at times like large serpents rushing through the air." After sunset this dark wall resembled a "blood-red curtain, with the edges of all shades of yellow; the whole of a murky tinge, with fierce flashes of lightning." Captain O. SAMPSON, viewing the volcano from a similar position at the same time, states that Krakatoa "appeared to be alight with flickering flames rising behind a dense black cloud; at the same time balls of fire rested on the mastheads and extremities of the yard-arms."

Captain WATSON states that during the night the mastheads and yard-arms of his ship were "studded with *corposants*," and records the occurrence of "a peculiar pinky flame coming from clouds which seemed to touch the mastheads and yard-arms." From the *G. G. Loudon*, lying in the Bay of Lampong, 40 or 50 English miles north-west of the volcano, it was recorded that "the lightning struck the mainmast-conductor five or six times," and that "the mud-rain which covered the masts, rigging, and decks, was phosphorescent, and on the rigging presented the appearance of St. Elmo's fire. The natives engaged themselves busily in putting this phosphorescent light out with their bands, and were so intent on this occupation that the stokers left the engine-rooms for the purpose, so that the European engineers were left to drive the machinery for themselves. The natives pleaded that if this phosphorescent light, or any portion of it, found its way below, a hole would burst in the ship; not that they feared the ship taking fire, but they thought the light was the work of evil spirits, and that if the ill-omened light found its way below, the evil spirits would triumph in their design to scuttle the ship."

This abundant generation of atmospheric electricity is a familiar phenomenon in all volcanic eruptions on a grand scale. The steam-jets rushing through the orifices of the earth's crust constitute an enormous hydro-electric engine; and the friction of ejected materials striking against one another in their ascent and descent also does much in the way of generating electricity.

Up to late in the afternoon of the 26th of August, the phenomena exhibited by Krakatoa were precisely similar to those witnessed at every

197

great paroxysmal volcanic eruption. But at that time the effects of the somewhat peculiar position of the Krakatoa crater began to be apparent. Lying as it does so close to the sea-level, the work of evisceration by explosive action could not go far without the waters of the ocean finding their way into the heated mass of lava from which the eruption was taking place.

It is often assumed that if a mass of water come into contact with molten lava a terrible outburst of steam, producing a great volcanic eruption, must be the consequence, and some vulcanologists insist that the admission of water by fissures into subterranean reservoirs of lava is the determining cause of all volcanic outbreaks. But careful observation does not give much countenance to this view. Lava-streams have frequently been seen to flow into the sea, and although a considerable generation of steam occurred when the molten mass first came in contact with the water, yet none of the prolonged effects which are popularly supposed to result from the conflict of fire and water were found to occur. The surface of the lava-current becoming rapidly chilled, a layer of slowly conducting rock is formed at its surface, and then the gradual cooling down of the whole mass ensues, without further disturbance.

By the lowering of the mass lying within the old crater-ring of Krakatoa, and the diminution in height of the crater-walls, water would from time to time find a way to the molten lava below; each such influx of water would no doubt lead to the generation of some steam with explosive violence, and the production of small sea-waves which would travel outwards from Krakatoa as a centre. From the reports made by the officials at Anjer and other places on the shores of Java and Sumatra, the production of such waves, which were only a few feet in height, began to be observed about 5.30 p.m. on Sunday, the 26th of August, and continued at irregular intervals all through the night. Towards morning, however, the chilling effects of the water which had from time to time found its way to the molten materials below the volcano began to be felt, and as a result a diminution in the activity of the volcano is recorded.

If, as I shall show when I proceed to discuss the nature of the materials ejected from Krakatoa, the cause of the eruptive action was due to the disengagement of volatile substances *actually contained in those materials,* the checking of the activity, by the influx into the molten mass of vast quantities of cold sea water, would have the same effect as fastening down the safety-valve of a steam-boiler, while the fires below were maintained in full activity.

The constant augmentation of tension beneath Krakatoa, in the end gave rise to a series of tremendous explosions, on a far grander scale than those resulting directly from the influx of the sea-water into the vent; the four principal of these occurred, according to the careful investigations of Mr. VERBEEK, at 5.30, 6.44, 10.2,* and 10.52, Krakatoa time, on the morning of August the 27th. Of these, the third, occurring shortly after 10 o'clock, was by far the most violent, and was productive of the most wide-spread results.

Although no one was near enough to Krakatoa during these paroxysmal outbursts to witness what took place there, a comparison of the condition of the volcano and of the surrounding seas before and after these terrible manifestations of the subterranean forces, leaves little doubt as to the real nature of the action.

Source: George James Symons, ed. *The Eruption of Krakatoa and Subsequent Phenomena*. London: Trubner and Company, 1888, pp. 14–22.

The Eruption of Tambora: Frank G, Carpenter, *Australia, Our Colonies, and Other Islands of the Sea* (1904) (excerpt)

In 1815, one of the largest volcanic eruptions in history occurred at Mount Tambora on Sumbawa Island in Indonesia. Ash and gases ejected into the atmosphere blocked sunlight and caused the Year without a Summer in 1816 in parts of North America and Europe.

Still farther west we coast Sumbawa (sŏom-bä'wå), noted for its volcanoes. The word "Sumbawa" means the land of fire, and this island seems well named, for we can see the steam rising in great clouds from some of its peaks. The crater of Mount Tambora is more than seven miles wide, and so large that a good-sized city might be dropped into it without touching the edges. The crater was caused by an eruption in 1815 when the whole top of the mountain, a mass higher and thicker than Mount Washington, was blown into the air. Before that time Tambora was thirteen thousand feet high. This eruption tore off about eight thousand feet, making so great an explosion that it was heard in Sumatra, a thousand miles away, and also on Ternate, nine hundred miles off in another direction.

Our captain tells us that when the eruption of Tambora occurred, the ocean for miles about was covered with floating timber. Ashes so coated the water that ships could hardly make their way through them, and they so filled the air that it was pitch dark in the daytime for hours after the explosion occurred. At the same time the whirlwinds lashed the sea to a foam; they tore up the largest trees by the roots and carried men, horses, and cattle for miles through the air. A town lying at the foot of Tambora was swallowed up, for the shore sank, and the sea came in and covered the earth to a depth of eight feet, and there it is to this day.

Notwithstanding this, there are still people living on Sumbawa. It has towns and villages, and the natives work away as though they were not in constant danger of another eruption.

As we sail farther westward, we pass Lombok and Bali, other volcanic islands more thickly populated, and thence go on by Madura, an island where great quantities of salt are evaporated from sea water, and then along the north coast of Java with volcanoes in sight all the way, until at last we come to the port for Batavia, the capital of the Dutch East Indies.

Source: Frank G. Carpenter. *Australia, Our Colonies, and Other Islands of the Sea.* New York: American Book Company, 1904, pp. 227–228.

Scientific View of Earthquakes and Volcanoes in 1917

In the following excerpt, the author summarizes the current (as of 1917) understanding of earthquakes and their causes, and he offers a few suggestions for earthquake preparedness. He gives brief overviews of several famous historical earthquakes from around the world. He goes on to explain "the new science" of seismology, or "earthquake survey," and how the seismograph (a relatively new piece of technology at the time) operates.

Earthquakes

There are two minor causes of earthquakes that we may quickly dispose of: (1) The subterranean concussions attending volcanic eruptions produce local quakes, not generally of much violence. (2) The collapse of caverns hollowed out in soluble rocks (limestone, etc.) by underground waters causes quakes, also generally local and unimportant.

The chief cause of earthquakes, and probably the cause of all those of great magnitude, is the jar given to the earth's crust, either by the sudden formation of the fissures that divide the surface shell into the blocks already mentioned, or by the slipping of adjacent blocks along the walls of fissures previously formed. Such slipping is technically described as *faulting*, and the places where it has occurred are called *faults.* They reveal themselves in many places at the earth's surface in the abrupt interruption of exposed strata, or sometimes by crevasses or terraces in the ground. Since ruptures and faults can occur only within the zone of fracture, the source of an earthquake is never more than a few miles beneath the surface.

The breaking of the rock or the slipping of adjacent rock faces upon each other sets up vibrations similar to those produced by the drawing of a violin bow over the strings, and these vibrations are transmitted in all directions through the earth. The part of the earth's surface lying directly over the source, or *focus*, of the disturbance is called the *epicenter* of the earthquake, and here vibrations are most distinctly felt. It is these vibrations, rather than the mere dislocations of the ground, that do most of the damage in destructive earthquakes.

Except where water-waves and fire add to the destruction, the loss of life that occurs in disastrous earthquakes is almost entirely due to the shaking down of buildings. If people lived everywhere in tents, or if houses were constructed like ships—which are so designed that they may be jolted and tossed about by the waves of the ocean without damage—these now dreaded visitations would do comparatively little bodily injury to humanity. A great deal of attention has been devoted to devising methods of construction and selecting building materials adapted to resist the effects of earthquake shocks, but in the countries most subject to these disturbances (except Japan) the knowledge gained in such investigations has not been generally applied. After the frightful Messina disaster of 1908, in which more than 77,000 people perished, the Japanese authority, Omori, declared that at least 99.8 per cent. of this number were the victims of the faulty construction of buildings. Earthquake damage also depends in part upon the kind of ground on which buildings are erected; those standing on rock are generally less affected than those built on soft earth.

Some Famous Earthquakes

An earthquake that destroyed the greater part of the city of Lisbon in 1755 cost the lives of between 30,000 and 40,000 people. A majority of these people were drowned, after they had fled from the city proper, by a huge wave, sweeping in from the ocean. Such waves are frequently generated by earthquake shocks. Sometimes they have their origin in earthquakes far out at sea, which are felt on board ships and which may break submarine cables. They inundate nearby coasts, and, like the huge waves produced by hurricanes, are popularly known as "tidal waves," though they have nothing to do with the tide.

In 1783 numerous towns and villages in Calabria and Sicily were destroyed by an earthquake, with a loss of more than 30,000 lives. This event attracted much attention in scientific circles, and elaborate reports were published upon it.

Probably no other country has suffered so many severe earthquakes as Japan, where the records of modern times show that a destructive shock

201

occurs, on an average, once every two and a half years. In 1891 the densely populated provinces of Mino and Owari, in that empire, were violently shaken, and 7,000 people were killed, while about 17,000 were injured.

The great earthquake that occurred in Assam, India, in 1897, is remarkable for the immense area over which it was destructive, amounting to about 150,000 square miles.

In 1908 the cities of Messina and Reggio, in southern Italy, were completely destroyed by the most disastrous earthquake of which we have any definite record. Official reports place the loss of life at 77,283, but much higher estimates have been published.

The Charleston, S. C., earthquake of 1886, and the San Francisco earthquake of 1906 were the most destructive that have occurred in the United States. In the attendant loss of life these two disasters were insignificant compared with most of the famous earthquakes of history, but in the San Francisco quake the property loss, due chiefly to fire, amounted to more than $200,000,000.

The New Science of Seismology

Seismology, or earthquake science, is mainly a product of the last thirty years, and its remarkable progress is especially due to the invention of delicate instruments *(seismographs)* which make autographic records of the vibrations set up by earthquake shocks. These instruments record not only the shocks that are perceptible to the senses, but also the far more numerous fainter shocks that are not. The seismograph has shattered our faith in *terra firma.* Our globe is trembling somewhere most of the time, and a severe shock at any place upon it is recorded by seismographs all over the world. The seismograph is essentially a heavy mass of metal, suspended in such a manner that its inertia prevents it from partaking readily of the motion of the earth when the latter is shaken. A pen or stylus attached to the suspended weight traces a record of the earth's movements on a sheet of paper carried along by clockwork. Complete apparatus of this kind registers vibrations in three directions—north-south, east-west, and up-down.

An earthquake sends out two principal sets of vibrations. One set travels around the earth's surface (*i.e.,* through the crust), while the other takes a "short cut" through the interior of the globe. The latter set is registered first by a distant seismograph, not only because it has not so far to travel, but also because it is propagated at a much greater speed through the material of the earth's core. Moreover, the vibrations that pass through the earth are of two distinct kinds—longitudinal (moving forward and back) and transverse (moving from side to side)—which travel at different speeds and trace waves of different shapes on the record sheet. The difference in the time of arrival of the various kinds of vibrations gives a clue to the distance of the source from the point of observation. These and other characteristics of seismograph tracings *(seismograms)* make it possible to locate an earthquake with considerable accuracy from the records of stations thousands of miles distant.

Nearly every civilized country now maintains a seismological service, for the sake of coöperating in the worldwide study of earthquakes, and also of determining which regions in each country are most subject to danger from this source. In the United States the "earthquake survey" is conducted by the Weather Bureau.

Volcanoes

A volcano is an opening in the surface of the earth through which heated matter is ejected. The accumulation of ejected material around the opening usually forms a conical hill or mountain. The opening is known as the *vent* or *chimney,* and the cup-shaped enlargement at its upper end is called the *crater.* The act of ejecting material is termed an *eruption,* a volcano being described as *active* while an eruption is in progress.

A few volcanoes are constantly active. Among these is Stromboli, in the Mediterranean, which constitutes a natural lighthouse. In most cases, however, there are long periods of repose between eruptions, during which the volcano is said to be *dormant.* If no eruption has occurred within historic times, or if, for any reason, the volcano is assumed to be incapable of further eruption, it is described as *extinct.*

The rocky material expelled by a volcano is called *lava,* if in a molten form, and *ash, cinders, scoriæ,* etc., if solid. The amount of such material is sometimes enormous. During the eruption of Tambora (or Tomboro), in the Sunda Islands, in 1815, nearly twenty-nine cubic miles are said to have been discharged. A Japanese volcano once ejected so much of the light, spongy material known as *pumice,* that it formed a layer upon the surface of the sea over which it was possible to walk for twenty-three miles.

All violent eruptions include the discharge of an immense volume of steam, forming a huge dust-laden cloud. In many cases this cloud is forked with lightning, and discharges torrents of rain, which, mingling with the dust in the cloud and on the slopes of the volcano, forms a sort of mud-lava. In its hardened form this is called *tuff.*

Contrary to popular belief, there is little real flame connected with a volcanic eruption. The intermittent glow seen over a volcano at night is the reflection of the incandescent lava in the crater upon the overhanging cloud.

The violence of eruptions varies through all grades from quiet outflows of lava to tremendous explosions. The great craters of Kilauea and Mauna Loa, in Hawaii, are examples of the quiet type of volcano. They are characterized by lava lakes, which occasionally well up and overflow. In one discharge Mauna Loa sent forth a stream of lava fifty miles long and in some places three miles wide. The most famous of explosive eruptions was that which blew away a great part of the island of Krakatoa, in the Straits of Sunda, in August, 1883. The noise of this explosion was heard 3,000 miles away, and the concussion was so heavy as to break windows at Batavia, nearly 100 miles distant. The dust from the explosion was hurled upward more than seventeen miles, and was carried around the globe by the winds. Its presence in the atmosphere was responsible for remarkable sunset glows all over the world for months after the eruption. The disturbance in the sea produced enormous waves, which drowned more than 36,000 people on neighboring coasts. Another stupendous explosive eruption was that of Mont Pelé,* in the island of Martinique, in May, 1902. This explosion blew away the side of the mountain and sent down the slope a great cloud of hot gases and glowing dust, which annihilated the city of St. Pierre, killing all but two of its 30,000 inhabitants, and destroying most of the shipping in the harbor. The most frightful of all explosive eruptions is said to have been that of Asama, in Japan, in 1783. Bandaisan, in the same country, after remaining dormant for 1,000 years, was the scene of a gigantic eruption in 1888.

Vesuvius, on the Bay of Naples, probably the best known of all volcanoes, has had many eruptions, generally of an intermediate type of violence, since the memorable one of A.D. 79, which buried the cities of Herculaneum and Pompeii so completely that for 1,600 years the very site of the latter remained unknown. Excavations have since uncovered about half of Pompeii, but comparatively little progress has been made in disinterring Herculaneum. Apart from its occasional violent outbreaks, Vesuvius is in a state of mild activity most of the time.

Between 300 and 400 volcanoes are at present known to be active. More than half of these are on oceanic islands, while most of the others are close to ocean shores. A remarkable belt of volcanoes nearly encircles the Pacific Ocean. The shores of the Atlantic are almost free from volcanoes.

Source: Charles Fitzhugh Talman. "Earthquakes and Volcanoes." *Mentor,* September 15, 1917, pp. 3–6.

1922 Vallenar Earthquake in Chile (November 1922)

The following excerpt was first published less than two weeks after the November 11, 1922, earthquake off the coast of Chile. The earthquake caused a tsunami that traversed the Pacific Ocean and caused damage as far away as Australia. The text provides details of the disaster. It goes on to give a scientific explanation of earthquakes using historical examples from around the world, comparing some of these events to the one in Chile. Finally, the excerpt gives details of Red Cross operations during 1922, explaining that the organization provides relief and help not just to victims of war but to victims of natural disasters such as the recent earthquake and tsunami.

The Earthquake Catastrophe in Chile

It is impossible at this writing to estimate with any degree of accuracy the number of fatalities from the earthquake of November 11 in Chile. Probably a thousand perished, and the fatalities may be largely in excess of that number. The disaster affected a great stretch near the coast; much, if not most, of the damage was inflicted by a tremendous wave which followed a subsidence or break beneath the bottom of the sea; so that, first, enormous quantities of water sank through the crevices, and, secondly, its withdrawal caused an inrush of the ocean. The towns of Coquimbo, Copiapo, and Valenar were seriously damaged and the last was practically destroyed. Numerous small places and country districts were devastated; the length of the territory damaged is put at about 1,200 miles.

Earthquakes are no novelty in Chile. As long ago as 1853 the town of Concepcion was destroyed by an earthquake quite similar to that now recorded, and the whole country along the coast has often suffered from smaller disasters of this kind.

Cause and History of Earthquakes

Of all convulsions of nature an earthquake is undoubtedly the most terrifying, both because of the vastness and mystery of the overwhelming power which produces it and because man can neither escape from it nor protect himself against it. The ancients, as modern barbarians do, ascribed earthquakes to the malevolence of demons or to the anger of outraged gods. All unusual and gigantic phenomena of nature, they thought, were produced by supernatural causes. Thus Herodotus, whose history is one of the great classics of all literature, in two passages mentions eclipses as prodigies or portents of the gods, in both cases unfavorable to the Greeks and foreshadowing their destruction. Science, however, made earlier and more rapid progress in astronomy than in seismology, a term of very old

205

Greek derivation employed by geologists to define the very modern study of earthquakes. For, while the Greeks and Egyptians knew something about the cause of eclipses before the Christian Era, it is only within a few decades that an attempt has been made to formulate the causes of earthquakes. Even now a good deal of explanation of earthquake phenomena is hypothetical. In general, however, it may be said that scientific investigators believe that earthquakes are caused in two ways—either by the explosive pressure of volcanic gases in the molten interior of the earth or by the slipping or displacement of gigantic strata of rock under the earth's surface. In the one case the earthquake is a monstrous explosion, in the other a monstrous landslide.

In most recorded cases the landslide or explosion has taken place near the sea or under its bed, so that it has been accompanied by a violent and death-dealing tidal wave. The earthquake that destroyed Lisbon in 1755 was followed by a tidal wave which swept the shores of Portugal and drowned or dashed to death thousands of human beings. Altogether 40,000 lives were lost in that disaster. Messina was shaken by an earthquake in 1783 and again in 1908, and on the latter occasion a great tidal wave wrought much of the destruction which resulted in the death of 60,000 persons. The recent Chilean earthquake and tidal wave, while terrible and sad enough, are not comparable in magnitude to the Portuguese and Italian disasters, nor probably, in loss of life, to two great earthquakes which have stricken India during the last twenty-five years. It is not surprising that there were times when the Hebrew poet thought man to be a puny thing in the midst of the incalculable forces of nature: "What is man, that thou art mindful of him? and the son of man, that thou visited him? . . . Thou hast made the earth to tremble; thou hast broken it. Heal the breaches thereof; for it shaketh. . . . He looketh on the earth, and it trembleth; he toucheth the hills, and they smoke!"

Answer to the Roll Call

The sixth annual Roll Call of the Red Cross is now under way and will continue until Thanksgiving Day. Once a year the Red Cross appeals to the American public to join in its work. The Roll Call is just what the name indicates; a call to members to renew their membership and to those not members to find out what the Red Cross is and does and then become members. This is not a "drive" in the usual meaning of an attempt to raise contributions and donations; no doubt the National Red Cross welcomes at any time of the year new memberships and special contributions, but its sole direct appeal to the public is to join in membership.

We do not know what the exact figures of membership are at present. A year ago there were about six million members—a much larger membership than had existed before the war.

As probably most of our readers know, the membership fees, amounting to several million dollars, go directly for relief purposes; the beautiful National Red Cross building in Washington was paid for from special contributions made for that purpose; the National officers, or most of them, either serve without pay or are paid from special contributions made outside the membership fees; thus membership fees paid by individuals go intact to carry out the work of the association.

What the Red Cross Does

What is that work, now that the war has long since been ended? This is a question sometimes asked, and easily answered. The watchword of the Red Cross is "Always Ready." One big part of its work is to be ready for emergencies. When such disasters happen as those at San Francisco and Galveston and Tulsa, and now Smyrna, relief and help cannot be improvised in a minute. This is just what the Red Cross is for; to have funds, railway trains, nurses, doctors, medical supplies, food, tents, ready to send with speed to any place where the need is great.

The greatest emergency this century has seen was the emergency of the Great War What the Red Cross did need not now be recapitulated. It can be told only in terms of many millions of money and of arduous and unpaid service of many thousands of men and women.

Since the war the Red Cross has rendered services of vast magnitude in devastated countries and regions; nor is its work in this direction to be confined to the ravages of the past. President Harding, who is the President of the American Red Cross, in announcing the present Roll Call, points out that a fearful emergency exists abroad at this moment. In the Near East, he says, "the lives of millions of unfortunate people even now depend and must continue for a long time to depend on the untiring liberality of more favored communities."

The relief that is to come from this country must be rendered, as President Harding points out, almost entirely through co-operation between the Red Cross, the Near East Relief, and some smaller agencies. It is understood that the Red Cross expects to spend for the Near East at least five million dollars. It could not spend it now if it did not have it now; it would not have it now if it had not been for the membership fees of last year. On the day we go to press the American Red Cross has cabled to Red Cross chapters in Chile offering aid to sufferers from the earthquake.

One other among many activities of the Red Cross may be mentioned, namely, the aid it renders to the returned American soldier. Colonel Forbes, the Director of the Veterans Bureau, speaking for the ex-service man, says: "Whether it be a matter of calling the Bureau's attention to an unrewarded claim, or an ill man needing hospitalization, or of tiding the sick veteran over the time which must elapse before Government aid can be offered, the Red Cross is always on the job with expert service and the necessary goods." These are only the larger divisions of the humane work of the American Red Cross. Its public health activities, its encouragement of sound sanitary systems, its education in first aid, its training of nurses, its work in the schools, are less outstanding, but combined are extremely valuable.

We join with President Harding in urging Americans to renew their allegiance to the Red Cross "in the interests of our common humanity and of the service which we owe to our fellow-men."

Source: [No author.] *The Outlook: An Illustrated Weekly Journal of Current Life*, vol. 132, November 22, 1922. New York: The Outlook Company, pp. 509–510.

Cyclone Nargis: USAID, "Burma—Cyclone" (2008)

On May 5, 2008, Cyclone Nargis made landfall in Myanmar (formerly Burma). The government of Myanmar reported 4,000 deaths and 93,000 displaced persons. USAID published this fact sheet to provide details of the humanitarian assistance needed in the area after the storm.

Burma—Cyclone

Fact Sheet #1, Fiscal Year (FY) 2008   May 5, 2008

BACKGROUND AND KEY DEVELOPMENTS

- On May 2 at 1600 hours local time, Cyclone Nargis made landfall near the mouth of the Irrawaddy River in the Irrawaddy Division of Burma, according to the U.S. Joint Typhoon Warning Center. The cyclone made landfall with a maximum sustained wind speed of approximately 132 miles per hour (mph) and a storm surge of 12 feet. The cyclone then continued east-northeast, affecting Rangoon through May 3 with a maximum sustained wind speed of approximately 90 to 100 mph.

- The Government of Burma (GOB) has reported more than 4,000 deaths and 93,000 displaced, but these figures remain unconfirmed. The U.N. Office for the Coordination of Humanitarian Affairs (OCHA) reports that the most affected areas are Irrawaddy Division, Rangoon Division, Bago Division, Kayin State, Kayah State, and Mon State.

Estimated Numbers at a Glance		Source
Total Dead	4,000	GOB — May 5, 2008
Total Missing	3,000	Reuters — May 5, 2008
Internally Displaced Persons (IDPs)	93,000	GOB — May 5, 2008
Houses Destroyed	20,000	GOB — May 5, 2008

FY 2008 HUMANITARIAN FUNDING PROVIDED TO DATE
USAID/OFDA Assistance to Burma Cyclone................$250,000

CURRENT SITUATION
- OCHA has reported that the GOB is having difficulty contacting islands and low-lying villages due to flooding, blocked roads, and disrupted communications, indicating that the numbers of individuals killed and displaced by the storm will likely increase.
- Initial reports from MRTV, the Burma state television channel, indicate that the cyclone destroyed 20,000 houses on Haing Gyi Island. OCHA has reported that the cyclone destroyed 95 percent of houses in villages along the south coast of the Irrawaddy delta area.
- The GOB has declared a state of emergency in Bago, Irrawaddy, and Rangoon divisions and Kayin and Mon states.

Cyclone-related Damages
- In the city of Rangoon, the cyclone damaged the electrical power grid, which is unlikely to be restored for several days, according to OCHA. The cyclone also disrupted service to telecommunications lines and radio and television stations.
- Throughout affected areas, the storm destroyed buildings, caused widespread flooding, and downed trees and billboards, blocking a number of roads. Local security forces began clearing major roads on May 4. OCHA reports that local residents and monks are clearing smaller roads without assistance.

Humanitarian Needs
- According to OCHA, the most urgent needs of populations in the city of Rangoon include plastic sheeting, water purification tablets, cooking sets, mosquito nets, emergency health kits, food, and possibly fuel supplies. In other areas, determining humanitarian needs is more difficult due to a lack of direct assessment, but immediate needs are likely to include shelter materials and safe drinking water.

GOB and U.N. Response

- The GOB Ministry for Social Welfare, Relief, and Resettlement is coordinating the response to the disaster. The GOB has established an emergency committee headed by the Prime Minister and deployed military units for rescue, rehabilitation, and clean-up efforts in Rangoon. According to media reports, the GOB has also sent rescue teams to the Irrawaddy Division.
- On May 4, the U.N. announced the organization of a U.N. Disaster Assessment and Coordination team, which is standing by to assist the GOB to respond to humanitarian needs, if required.

USAID HUMANITARIAN ASSISTANCE

- On April 27, representatives of USAID/OFDA in Washington, D.C.; Bangkok, Thailand; and Kathmandu, Nepal began tracking the storm. On April 28, USAID/OFDA officials began coordinating storm monitoring efforts with officials from the U.S. Embassy in Rangoon. On May 3, U.S. Embassy staff began conducting informal damage assessments in Rangoon. On May 5, an official from the U.S. Embassy in Rangoon met with other donors to discuss the humanitarian response to the storm.
- On May 5, U.S. Chargé d'Affaires Shari Villarosa declared a disaster in Burma due to the effects of Cyclone Nargis. In response, USAID/OFDA is providing $250,000 to the U.N. Children's Fund (UNICEF), the U.N. World Food Program (WFP), and the Office of the U.N. High Commissioner for Refugees (UNHCR) for emergency food, water and sanitation, and shelter assistance.

USAID HUMANITARIAN ASSISTANCE TO BURMA CYCLONE

Implementing Partner	Activity	Location	Amount
USAID/OFDA Assistance[1]			
UNICEF, WFP, and UNHCR	Water and Sanitation, Emergency Food Assistance, Shelter	Affected Areas	$250,000
TOTAL USAID/OFDA			$250,000

[1] USAID/OFDA funding represents anticipated or actual obligated amounts as of May 5, 2008.

PUBLIC DONATION INFORMATION

- The most effective way people can assist relief efforts is by making cash contributions to humanitarian organizations that are conducting relief operations. Information on organizations responding to the humanitarian situation in Burma may be available at www.relief web.int.

- USAID encourages cash donations because they allow aid professionals to procure the exact items needed (often in the affected region); reduce the burden on scarce resources (such as transportation routes, staff time, warehouse space, etc); can be transferred very quickly and without transportation costs; support the economy of the disaster-stricken region; and ensure culturally, dietary, and environmentally appropriate assistance.
- More information can be found at:
  - USAID: www.usaid.gov—Keyword: Donations
  - The Center for International Disaster Information: www.cidi.org or (703) 276-1914
  - Information on relief activities of the humanitarian community can be found at www.reliefweb.int

Source: USAID. "Burma–Cyclone." Fact Sheet #1, Fiscal Year 2008. Bureau for Democracy, Conflict, and Humanitarian Assistance. Office of U.S. Foreign Disaster Assistance, May 5, 2008.

The 2010 Haiti Earthquake: USAID, "Haiti—Earthquake" (2010)

On January 12, 2010, an earthquake in Haiti caused extensive damage and loss of life in the capital city of Port-au-Prince. Initial reports placed the death toll at over 100,000 people. USAID published this fact sheet shortly after the earthquake to provide details of the disaster.

Haiti—Earthquake

Fact Sheet #1, Fiscal Year (FY) 2010   January 13, 2010

BACKGROUND AND KEY DEVELOPMENTS

- On January 12 at 1653 hours local time, a magnitude 7.0 earthquake struck southern Haiti. According to the U.S. Geological Survey (USGS), the earthquake epicenter was located 10 miles southwest of the capital Port-au-Prince, West Department. Numerous significant aftershocks followed the initial quake. According to the U.N., the earthquake caused extensive infrastructure damage in Port-au-Prince, including to the Haiti presidential palace and the U.N. Stabilization Mission in Haiti (MINUSTAH) headquarters. Residential dwellings and transportation networks also sustained significant damage.

211

- Immediately following the earthquake, U.S. President Barack Obama pledged to provide assistance to Haiti, and U.S. Secretary of State Hillary Clinton reported that the U.S. Government (USG) would provide military and civilian disaster assistance to affected families.
- USAID/OFDA has deployed a Disaster Assistance Response Team (USAID/DART) to Haiti—comprising up to 17 members—and activated a Washington D.C.-based Response Management Team to support the USAID/DART. The USAID/DART will assess humanitarian needs and coordinate assistance with the U.S. Embassy in Port-au-Prince, the international community, and the Government of Haiti (GoH).
- As of 1615 hours local time on January 13, seven members of the USAID/DART, the 72-member Fairfax County Urban Search and Rescue (USAR) team, and four support staff had arrived in Port-au-Prince.
- Two USAID/OFDA-supported heavy USAR teams from Fairfax County, VA, and Los Angeles County, CA, composed of approximately 72 personnel, 6 search and rescue canines, and up to 48 tons of rescue equipment, are also deploying to Haiti. USAID/OFDA expects to support up to two additional heavy USAR teams from Florida. USAID/OFDA has also authorized the deployment of a three-person Americas Support Team (AST) to Haiti. The AST, staffed by additional Fairfax County USAR members and funded by USAID/OFDA, will supplement the U.N. Disaster Assessment Country (UNDAC) team in Haiti. In addition, both the Fairfax County and Los Angeles County Fire Departments are seconding staff members to directly support the UNDAC team.
- On January 13, U.S. Ambassador to Haiti Kenneth H. Merten declared a disaster due to the effects of the earthquake. In response, USAID/OFDA provided an initial $50,000 through the U.S. Embassy in Port-au-Prince for the implementation of an emergency response program. USAID/OFDA plans to provide additional assistance in accordance with the findings of USAID/DART and humanitarian community assessments, as access to affected groups expands over the coming days.

Numbers at a Glance		Source
Estimated Number of Deaths	At least 100,000	GoH — January 13
Estimated Affected Population	Approximately 3 million people	International Media

CURRENT SITUATION

- Preliminary assessments from the International Federation of Red Cross and Red Crescent indicate the earthquake affected approximately 3 million individuals in total.
- Media reports indicate than an unknown number of individuals remain trapped under numerous collapsed buildings, including a hospital in Petionville town, West Department.
- Humanitarian organizations expect a considerable number of affected individuals to require emergency shelter assistance and other relief items. Access to affected populations remains limited at present due to extensive debris and damaged roadways.
- Priority assistance, according to the GoH, includes search and rescue capability, an offshore vessel medical unit, electricity generation capability, and communications equipment to facilitate GoH coordination and response efforts. The GoH has also requested assistance in evacuating patients from a damaged hospital in Petionville to hospitals in Miami, FL.

Emergency Food Assistance

- USAID's Office of Food for Peace (USAID/FFP) has contacted all active P.L. 480 Title II emergency food Assistance implementing partners in Haiti to determine the availability of commodity stocks in-country and the capacity of organizations to reach earthquake victims.
- On January 13, the U.N. World Food Program (WFP) and United Nations Humanitarian Response Depot (UNHRD) planned to airlift 86 metric tons (MT) of high energy biscuits from El Salvador, as well as emergency relief supplies from Panama.
- In addition, WFP approved an Emergency Operation, valued at $500,000, in support of immediate relief efforts.
- On January 13, WFP planned to deploy seven logistics staff and three telecommunications staff to activate the U.N. Logistics and Telecommunications clusters in support of relief efforts. WFP also authorized a $20 million Special Operation to further support logistics augmentation and coordination, as well as emergency telecommunication for the humanitarian community.
- Non-governmental organizations present across affected areas reported the looting of WFP warehouses following the earthquake.

Health

- The American Red Cross reported depleted Red Cross medical supplies in Haiti as of January 13 after distributing the limited supply of medical equipment and supplies in previously positioned stocks.

USG HUMANITARIAN ASSISTANCE

- On January 13. U.S Ambassador to Haiti Kenneth H. Merten declared a disaster due to the effects of the earthquake. In response, USAID/OFDA provided an initial $50,000 through the U.S. Embassy in Port-au-Prince for the implementation of an emergency response program. USAID/OFDA plans to provide additional assistance in accordance with the findings of USAID/DART and humanitarian community assessments, as access to affected groups expands over the coming days.

PUBLIC DONATION INFORMATION

- The most effective way people can assist relief efforts is by making cash contributions to humanitarian organizations that are conducting relief operations. Information on organizations responding to the humanitarian situation in Haiti may be available at www.reliefweb.int.
- USAID encourages cash donations because they allow aid professionals to procure the exact items needed (often in the affected region); reduce the burden on scarce resources (such as transportation routes, staff time, warehouse space, etc); can be transferred very quickly and without transportation costs; support the economy of the disaster-stricken region; and ensure culturally, dietary, and environmentally appropriate assistance.
- More information can be found at:
  - ○ USAID: www.usaid.gov—Keyword. Donations
  - ○ The Center for International Disaster Information: www.cidi.org or (703) 276-1914
  - ○ Information on relief activities of the humanitarian community can be found at www.reliefweb.int

Source: USAID. "Haiti–Earthquake." Fact Sheet #1, Fiscal Year 2010. Bureau for Democracy, Conflict, and Humanitarian Assistance. Office of U.S. Foreign Disaster Assistance, January 13, 2010.

INDIAN OCEAN

2004 Indian Ocean Tsunami Bulletins

The Pacific Tsunami Warning Center distributed the following tsunami bulletins directly after the December 26, 2004, earthquake off the coast of Sumatra. The initial bulletin was published about 15 minutes after the earthquake and communicated that countries in the Pacific Ocean were in no danger. Approximately one hour later, the second bulletin warned of a possible tsunami near the earthquake, but there was no system in place to communicate this information to the endangered populations in the Indian Ocean.

TSUNAMI BULLETIN NUMBER 001
PACIFIC TSUNAMI WARNING CENTER/NOAA/NWS
ISSUED AT 0114Z 26 DEC 2004
THIS BULLETIN IS FOR ALL AREAS OF THE PACIFIC BASIN EXCEPT ALASKA — BRITISH COLUMBIA — WASHINGTON — OREGON — CALIFORNIA.
.... TSUNAMI INFORMATION BULLETIN ....

THIS MESSAGE IS FOR INFORMATION ONLY. THERE IS NO TSU-NAMI WARNING OR WATCH IN EFFECT.

AN EARTHQUAKE HAS OCCURRED WITH THESE PRELIMINARY PARAMETERS
ORIGIN TIME - 0059Z 26 DEC 2004
COORDINATES - 3.4 NORTH 95.7 EAST
LOCATION - OFF W COAST OF NORTHERN SUMATRA
MAGNITUDE - 8.0

EVALUATION
THIS EARTHQUAKE IS LOCATED OUTSIDE THE PACIFIC. NO DESTRUCTIVE TSUNAMI THREAT EXISTS BASED ON HISTORI-CAL EARTHQUAKE AND TSUNAMI DATA.

THIS WILL BE THE ONLY BULLETIN ISSUED FOR THIS EVENT UNLESS ADDITIONAL INFORMATION BECOMES AVAILABLE.

THE WEST COAST/ALASKA TSUNAMI WARNING CENTER WILL ISSUE BULLETINS FOR ALASKA — BRITISH COLUMBIA — WASH-INGTON — OREGON — CALIFORNIA.

TSUNAMI BULLETIN NUMBER 002
PACIFIC TSUNAMI WARNING CENTER/NOAA/NWS
ISSUED AT 0204Z 26 DEC 2004
THIS BULLETIN IS FOR ALL AREAS OF THE PACIFIC BASIN EXCEPT ALASKA — BRITISH COLUMBIA — WASHINGTON — OREGON — CALIFORNIA.

NATURAL DISASTERS

.... TSUNAMI INFORMATION BULLETIN ....

ATTENTION: NOTE REVISED MAGNITUDE.

THIS MESSAGE IS FOR INFORMATION ONLY. THERE IS NO TSU-
NAMI WARNING OR WATCH IN EFFECT.

AN EARTHQUAKE HAS OCCURRED WITH THESE PRELIMINARY
PARAMETERS

ORIGIN TIME	-	0059Z 26 DEC 2004
COORDINATES	-	3.4 NORTH 95.7 EAST
LOCATION	-	OFF W COAST OF NORTHERN SUMATRA
MAGNITUDE	-	8.5

EVALUATION

REVISED MAGNITUDE BASED ON ANALYSIS OF MANTLE WAVES.
THIS EARTHQUAKE IS LOCATED OUTSIDE THE PACIFIC. NO
DESTRUCTIVE TSUNAMI THREAT EXISTS FOR THE PACIFIC BASIN
BASED ON HISTORICAL EARTHQUAKE AND TSUNAMI DATA.

THERE IS THE POSSIBILITY OF A TSUNAMI NEAR THE EPICENTER.

THIS WILL BE THE ONLY BULLETIN ISSUED FOR THIS EVENT
UNLESS ADDITIONAL INFORMATION BECOMES AVAILABLE.

THE WEST COAST/ALASKA TSUNAMI WARNING CENTER WILL
ISSUE BULLETINS FOR ALASKA — BRITISH COLUMBIA — WASH-
INGTON — OREGON — CALIFORNIA.

TSUNAMI BULLETIN NUMBER 003
PACIFIC TSUNAMI WARNING CENTER/NOAA/NWS
ISSUED AT 1535Z 27 DEC 2004
THIS BULLETIN IS FOR ALL AREAS OF THE PACIFIC BASIN EXCEPT
ALASKA — BRITISH COLUMBIA — WASHINGTON — OREGON
— CALIFORNIA.

.... TSUNAMI INFORMATION BULLETIN ....

THIS MESSAGE IS FOR INFORMATION ONLY. THERE IS NO TSU-
NAMI WARNING OR WATCH IN EFFECT.

AN EARTHQUAKE HAS OCCURRED WITH THESE PRELIMINARY
PARAMETERS

ORIGIN TIME   -   0059Z 26 DEC 2004
COORDINATES   -   3.4   NORTH 95.7 EAST
LOCATION   -   OFF W COAST OF NORTHERN SUMATRA
MAGNITUDE   -   9.0

EVALUATION
SOME ENERGY FROM YESTERDAYS TSUNAMI IN THE INDIAN
OCEAN HAS LEAKED INTO THE PACIFIC BASIN . . . PROBABLY
FROM SOUTH OF THE AUSTRALIAN CONTINENT. THIS ENERGY
HAS PRODUCED MINOR SEA LEVEL FLUCTUATIONS AT MANY
PLACES IN THE PACIFIC. FOR EXAMPLE . . .

50 CM CREST-TO-TROUGH AT CALLAO CHILE
19 CM CREST-TO-TROUGH AT IQUIQUE CHILE
13 CM CREST-TO-TROUGH AT PAGO PAGO AMERICAN SAMOA
11 CM CREST-TO-TROUGH AT SUVA FIJI
50 CM CREST-TO-TROUGH AT WAITANGI CHATHAM IS
NEW ZEALAND
65 CM CREST-TO-TROUGH AT JACKSON BAY NEW ZEALAND
18 CM CREST-TO-TROUGH AT PORT VILA VANUATU
06 CM CREST-TO-TROUGH AT HILO HAWAII USA
22 CM CREST-TO-TROUGH AT SAN DIEGO CALIFORNIA USA

HOWEVER . . . AT MANZANILLO MEXICO SEA LEVEL FLUC-
TUATIONS WERE AS MUCH AS 2.6 METERS CREST-TO-TROUGH
PROBABLY DUE TO FOCUSING OF ENERGY BY THE EAST PACIFIC
RISE AS WELL AS LOCAL RESONANCES.

THIS IS TO ADVISE THAT SMALL SEA LEVEL CHANGES COULD
CONTINUE TO BE OBSERVED ACROSS THE PACIFIC OVER THE
NEXT DAY OR TWO UNTIL ALL ENERGY FROM THIS EVENT IS
EVENTUALLY DISSIPATED.

THIS WILL BE THE FINAL BULLETIN ISSUED FOR THIS EVENT
UNLESS ADDITIONAL INFORMATION BECOMES AVAILABLE.

THE WEST COAST/ALASKA TSUNAMI WARNING CENTER WILL
ISSUE BULLETINS FOR ALASKA — BRITISH COLUMBIA — WASH-
INGTON — OREGON — CALIFORNIA.

USAID Projects Associated with the 2004 Sumatra-Andaman Earthquake and Tsunami (February 2007)

This excerpt from a report to U.S. congressional committees provides details concerning USAID projects in Indonesia and Sri Lanka related to the 2004 Indian Ocean tsunami. The report outlines the types of activities underway as well as expenditures and problems related to the progress of the projects.

Results in Brief

Although USAID's signature projects and other activities in Indonesia and Sri Lanka are under way, various difficulties have led the agency to increase initial cost estimates; reduce or cancel some activities in Indonesia; and consider extending project completion dates of signature projects in both countries. In addition, as of February 2007, USAID had not awarded the contract for major signature road construction work in Indonesia. USAID has taken steps to address some of these difficulties, such as revising the length of road to be built in Indonesia and assisting the government in acquiring land and property needed to implement project activities. However, USAID continues to face risks that it may not complete the signature projects within cost and schedule estimates and without further reducing the scope of the work, as currently planned.

- **Indonesia.** As of December 2006, USAID had obligated $186 million (53 percent) and expended $58 million (17 percent) of its $351 million budget for tsunami reconstruction in Indonesia. USAID had obligated $105 million (41 percent) and expended $15 million (6 percent) of its $254 million budget for the signature road construction project. USAID contractors completed the design work for the signature road construction project in Aceh Province and began construction work on some badly damaged road sections. However, in June 2006, because of significant increases in estimated construction costs per mile, USAID reduced the project's scope by over one third—from 150 miles to 91 miles of road construction—and reallocated approximately $9 million from other USAID and State tsunami supplemental appropriations to the road, resulting in a budget that has risen from the original $245 million to $254 million.[7] Based on GAO's analysis, the estimated

---

[7]The government of Japan has agreed to build a 69-mile road that will connect to the U.S. road (91 miles). The total length of the road, 160 miles, differs from the 150-mile road initially planned by USAID because the Japanese road follows a different alignment than the U.S. road would have followed.

per-mile costs as of June 2006 had risen to approximately $2.7 million—a 68 percent increase from USAID's March 2005 projection of $1.6 million per mile—reflecting both higher-than-expected labor and material costs and the availability of more accurate information than in March 2005.[8] For example, the June 2006 estimate indicates that 2.2 million cubic meters of excavation would be required, compared with the March 2005 estimate of 1.5 million cubic meters. Moreover, despite reducing the length of road to be constructed, USAID may extend the project's completion date from September 2009 to February 2010, in part because it did not award a contract for the major road construction work in September 2006, as planned; USAID expects to award the construction contract in May 2007.[9] Further, although USAID is currently assisting the Indonesian government in its land acquisition efforts, the project's completion date may be extended and estimated costs may increase because of ongoing delays by the Indonesian government in acquiring land for the road.[10] For example, as of December 31, 2006, the Indonesian government had acquired only 899 of the needed 3,679 parcels along the road's planned route.[11] In addition, several challenges increase the risk that USAID may not successfully complete the project at planned cost and on schedule. Specifically, delays by the government of Indonesia in acquiring land in a timely manner and community unrest increase USAID's risk of not meeting its schedule, and increasing costs for materials and labor,

---

[8]Based on GAO's analysis, the current $2.7 million cost per-mile estimate does not account for certain administrative and support costs and certain activities performed previously that are not directly related to the costs for constructing 91 miles of road. As such, the $2.7 million per-mile estimate does not directly correlate with the $254 million currently budgeted to the signature road.

[9]Also, in October 2006, the U.S. Ambassador to Indonesia requested that the Department of State ask for additional funds from the Office of Management and Budget to cover anticipated excess costs of constructing two bypasses not included in the revised 86-mile project. The bypasses, totaling approximately 12 miles, are along the section of road being constructed by the government of Japan.

[10]According to the National Research Council, a private, nonprofit institution chartered by Congress that provides science, technology, and health policy advice, extending the completion date for construction projects is likely to increase costs.

[11]The 3,679 parcels account for privately owned land along approximately 65 miles of the 91-mile route between Banda Aceh and Calang. The 899 parcels of privately owned land that have been acquired by the Indonesian government are noncontiguous and account for approximately 16 miles of the road's route. Twenty-six miles along the 91-mile route are owned by the Indonesian government, and 22 miles of this land had been released to USAID as of December 31, 2006.

as well as slippages in the schedule, increase USAID's risk of not meeting its projected costs because longer projects generally cost more.

- **Sri Lanka.** As of December 2006, USAID had obligated $70 million (84 percent) and expended $15 million (18 percent) of its $83 million budget for tsunami reconstruction. USAID had obligated $35 million (73 percent) and expended $5 million (11 percent) of its $48 million budget for the signature project components. The contract completion date for the Sri Lanka signature project, which includes building a bridge and other infrastructure, constructing vocational education facilities, and addressing coastal management issues, is March 2008. However, USAID may extend the signature project completion date. In addition, the projected cost has increased from $35 million to $48 million—about 37 percent—due to increases in the costs of labor and materials. To make up for this shortfall, USAID reallocated approximately $13 million from nonsignature project activities. As in Indonesia, several challenges increase the risk that USAID may not successfully complete all the project components at planned cost and on schedule. Specifically, other factors such as delays in selecting the sites, determining what would be taught, and designing the vocational educational facilities and water treatment facilities, as well as increasing incidents of terrorist-related violence increase USAID's risk of additional costs and slippages in the schedule.

USAID is carrying out measures that it previously established for financial oversight of its reconstruction programs in Indonesia and Sri Lanka, and it has taken steps to enhance its technical oversight capacity. In addition to USAID's standard requiring financial oversight mechanisms, USAID has contracted with the Defense Contract Audit Agency (DCAA) for concurrent audits of its signature road construction project in Aceh Province, Indonesia. Also, USAID's Office of the Inspector General (IG), using funding that was included in the supplemental appropriation, has completed audits showing, among other things, that USAID had not met certain interim milestone dates and was not adequately accounting for funding; other IG audits are ongoing. To ensure technical oversight of its reconstruction projects in both countries, USAID has added experienced staff, including engineers in both Sri Lanka and Indonesia, and enhanced its engineering services through agreements with USACE. However, when USAID began its tsunami reconstruction program in early 2005, the agency did not have disaster reconstruction program guidance, including lessons learned from its prior programs, such as the importance of setting appropriate time frames, conducting thorough cost assessments, and understanding local land tenure systems. In our July 2002 report focusing on USAID's reconstruction efforts in Central America, several USAID missions reported learning numerous lessons, such as the importance of establishing longer implementation periods to complete projects.[12] In response to our May 2006 report, USAID issued guidance, including lessons learned, on implementing disaster recovery activities following hurricanes and tropical storms that struck several Caribbean countries in 2004.[13] Further, as USAID moves forward, the agency faces risks that may affect its ability to complete its signature project activities as planned.

In response to our April 2006 recommendation, the Secretary of State, with information obtained from USAID, has provided some updated cost estimates and schedules in its required reports to Congress. Data in both the June and December 2006 reports included amounts of funds obligated to the affected countries but did not include the amounts of funds signed in agreements with implementing organizations (in USAID's terminology, "subobligated"). Also, the reports do not include USAID's risk information and mitigation strategy for addressing the risks. As a result, Congress lacks funding information that would more clearly reflect the agency's progress; currently, State's required semiannual reports to Congress do not include this information.

Therefore, to ensure that Congress has access to information that clearly reflects both USAID's progress in its tsunami reconstruction programs in Indonesia and Sri Lanka and factors that may slow its progress, we make the following two recommendations regarding the Secretary of State's required semiannual reports to Congress:

- To clearly show USAID's progress in using the appropriated funds for tsunami reconstruction, the reports should include the amounts that USAID obligated to recipient countries for tsunami reconstruction and the amounts that it "subobligated" in transactions with

---

[12]GAO, *Foreign Assistance: Disaster Recovery Program Addressed Intended Purposes, but USAID Needs Greater Flexibility to Improve Its Response Capability,* GAO-02-787 (Washington, D.C.: July 24, 2002).

[13]GAO, *Foreign Assistance: USAID Completed Many Caribbean Disaster Recovery Activities, but Several Challenges Hampered Efforts,* GAO-06-645 (Washington, D.C.: May 26, 2006).

implementing organizations, such as contracts, grants, and cooperative agreements, for specific reconstruction projects.

- To indicate risk of potential changes to the costs, schedules, and
scopes of work of USAID's signature projects in Indonesia and Sri
Lanka, the reports should identify factors that may impact the agency's implementation of the projects and provide strategies for mitigating any impact.

At our request, USAID and the Department of State provided written comments and technical suggestions and clarifications on a draft of this report.
(See app. IV for State's written comments and app. V for USAID's written
comments.) In commenting on a draft of this report, the Department of
State and USAID fully agreed with our recommendation to include cost
data on "subobligated" funds in State's required reports to Congress. USAID
also agreed to separately identify risk and mitigation strategies in State's
reports. We have also incorporated technical suggestions and clarifications
from State and USAID, as appropriate.

Source: United States Government Accountability Office. *USAID Signature Tsunami Reconstruction Efforts in Indonesia and Sri Lanka Exceed Initial Cost and Schedule Estimates, and Face Further Risks,* GAO-07-357, February 2007, pp. 3–7.

CHINA

John R. Freeman, "Flood Problems in China" (1922) (excerpt)

This selection provides details of flooding problems and historical floods in specific regions of China, including details of the flooding in 1851 that changed the course of the Yellow River. The author also offers a history of ancient flood protection practices in China.

Some Localities Having Important Flood Problems

From north to south there are five principal localities within these main
Chinese drainage basins, shown on the map, Fig. 4, in which flood relief
needs to be intensively studied and great works constructed whenever the
Government becomes stabilized and capital gains confidence so that means
can be provided. These localities are, as follows:

(1).—In the flat, level plains of the drainage basin that has its outlet past the
great commercial city of Tientsin. This region suffered terribly in August
and September, 1917. An area of about 12 000 sq. miles was submerged, and

more than 1 000 000 people, mostly farmers, are said to have been driven from their homes. A property loss of more than $25 000 000 is said to have been incurred in this one flood with consequential damage of perhaps $50 000 000.

(2).—In the delta of the Yellow River, or Hoang Ho, along the river's 400-mile course through the delta. During the historic period of 4 200 years, this region has been ravaged time and again by floods in one place and another. These overflows come from breaks in the great dikes and occasionally cause a wide change in the river's source. So terrible have been these visitations that this river is often called "China's Sorrow." The conditions are aggravated by the silt burden, brought from the vast and easily eroded loess deposits of the Provinces just up stream from the delta, which make the Yellow River probably the muddiest great river in the world.

(3).—The Huai River District in Kiang-su Province below Wu-hu. About 5 000 sq. miles of this peculiarly low and level delta region is said to be frequently inundated, with famine often following flood. The land is exceptionally fertile and normally gives two crops per year, so that, for thousands of years, intensive cultivation has been continued in spite of disasters which, sometimes, have brought death to thousands and, at other times, have merely destroyed the second of the two yearly crops.

(4).—Along the Yang-tze River below Ichang. Here, in the Grand Gorges, the river floods are said to rise 80 ft., and below Hankow the flood discharge is said to reach about 3 000 000 sec-ft., which is about 50% larger than the greatest flood discharge of the Mississippi River at New Orleans, La. The Yang-tze River, although turbid, brings down less silt than the Yellow River. Its delta, therefore, grows more slowly. Its whole regimen has become well established, and the flood problems are mainly those of ordinary dikes combined with river training on a mighty and forbidding scale, because of the depth and volume of water. The charts show dozens of spots where the swirl of the flood cuts pools more than 100 ft. deep, and it is no light problem to work out shore protection that will hold so deep a bank of soft fine-grained river silt when attacked by the under-cutting of such a flood. . . .

(5).—In Southwestern China the valleys of the North, East, and West Rivers, in the country around and back from Canton. Near the important City of Wu-chow, the great West River is said to rise 50 and even 80 ft. in extreme floods, and at the rate of 1 ft. per hour.

(6).—The improvement of navigation also presents incidental problems of flood scour, in addition to those of protection from inundation. There are particularly great and difficult problems of river training for improving

223

navigation, and problems of utilization of flood flow for scour of channels for harbor improvements, to be found on the Yang-tze near Shanghai, on the Min River near Foo-chow, and all along each of these five or six great silt-bearing rivers of China proper.

The great depth of the silt deposit in these Chinese deltas gives unlimited scope for the river to scour its bed deeper when confined in a generally straight and narrow way.

(7).—Outside the limits of China proper, to the northeast in Manchuria, there are great flood problems along the Liao River which rises in the Mongolian slopes and brings down silt that forms a harbor bar at New Chwang. On the other side, 17 000 miles away, in a straight line to the southeast, in French Indo-China, the lower course of the Red River presents serious flood problems which the French Government is said to be considering.

Besides the large problems cited, there are many smaller ones in drainage basins from 20 by 100 miles to 40 by 100 miles in extent, that are almost too small to show on the map of all China, to which floods may come once in 10, 20, or 40 years, that are terrible enough to the many people living on little farms or in little villages in their path. Examples of them were found in course at the Grand Canal surveys along the Wen River and in the Red Cross reconnaissance on the Yi River. Doubtless, there are a hundred of these minor areas in fertile and densely populated parts of China, each of which has its own important flood problem which can be solved so as to bring a greater measure of security and happiness, whenever the Government becomes stabilized, a broader community spirit developed, and when the multitude of young Chinese now studying modern engineering methods at home and abroad, are given the means wherewith to work.

Ancient Chinese Flood Protection

The Chinese have had to contend with the problems of flood control by dikes during their whole historic period of about 4 000 years and have developed some wonderfully good technique in many matters of dike building and in the repair of dike breaks. Although in the science of river training, they (and all engineers) still have much to learn, they have, in many great works, shown skill as hydraulic engineers. They have had some of the most difficult problems in the world with which to contend.

One of their most venerated men, of the half-legendary days of 3 000 years ago, was Yu, their great hydraulic engineer, to whose memory many temples were built and who, after a period of trouble, was entrusted with the conservancy of rivers throughout the country. He regulated the waters so wisely, it is said, that with his precepts faithfully followed there was no serious trouble for more than 1 000 years until "the period of the warring States." Tradition says, "Yu labored thirteen years, sparing neither trouble nor fatigue, nor even once entering his own home, though he passed three times before its door." That "he had boats for travel by water, chariots for travel by land, sledges for mud, with relays of men to draw them." "He dug nine great channels to conduct the waters to the sea," and he is supposed to have organized the building of great systems of dikes. In the great fertile Province of Szechwan, many travelers have described great irrigation works and flood channels planned more than 2 000 years ago by an engineer, whose rules are said to be still implicitly followed. Tributary to the Grand Canal, at Tai-tsun-pa, there is a feeder dam, which the speaker has examined, of masonry resting on small piles driven into a soft sand foundation, that was admirably built about 500 years ago. The sea wall, of course cut stone, 25 ft. high, that protects the coast for many miles easterly from Hangchow, is a structure built about 400 years or more ago, of which any modern engineer might be proud. The fact that, during 538 years, the great restless Yellow River was held to its course on a silted bed several feet above the level of the ground on either side, within dikes built of soft friable river silt carefully consolidated by tamping and protected from the erosion of impinging currents in many places only by groynes and revetments of earth bound together with perishable millet stalks, speaks volumes for the skill and resources of these Chinese "Old Masters of River Control."

Preliminary Surveys in Progress

In all the localities previously mentioned, investigation is now being carried on under the Chinese Government; but everywhere . . . this work seems to be mostly in the surveying rather than in the engineering stage. It is far easier to make a tolerable survey than to make a good design and then establish confidence in it; and one gets an impression that the authorities having certain of these matters in charge are staggered by the size, complexity, and uncertainty of their problems, or are floundering in a "Slough of Despond," and that some of them may be forced, by pressure of the public demand for "making the dirt fly," into constructions that will not stand the test of time, and of which they themselves may have doubts. The data collected in these surveys and investigations are few in comparison with the vast territories in China affected by flood.

For investigations within the Hai River Drainage Basin, following the great Tientsin flood of September, 1917, the Chinese Government organized the Chihli River Improvement Commission which, under the guidance of English, American, and Chinese engineers, including three members of the Society, has, for about four years, been making topographic and hydrographic surveys, from which data, works may be planned later.

All along the Grand Canal in Shantung Province, the China Grand Canal Improvement Board, under the supervision of American engineers, members of the Society, has made extensive topographic and hydrographic surveys and a reconnaissance survey of the Yellow River's course and its dikes for 200 miles up stream from the Grand Canal crossing. Plans have been devised for the immediate reconstruction of 253 miles of the Grand Canal, whenever this work can be financed; but from lack of funds, owing to the difficulty of selling Chinese bonds under present disturbed political conditions, the enterprise is now marking time.

Along two or three portions of the 400 miles of the Lower Yellow River within the delta plain, the Provincial Conservancy Boards are making outline surveys of the river's course and of its dikes, but with little attention to accurate hydrography or levels as far as the speaker has been able to learn. Meanwhile, the Chihli River Commission is gauging the Yellow River flow and measuring, at frequent intervals, the percentage of silt that it carries, at the Tientsin-Pukow Railroad Bridge. The daily height of the Yellow River is recorded also at the Peking-Hankow Bridge, 302 miles up stream.

In Kiang-su Province, under the leadership of one of the most remarkable men in China, His Excellency, Chang Chien—classical scholar, Confucian philosopher, captain of industry, and philanthropist, full of the spirit of service to his fellow men—who seeks means for protecting his Province from floods, much of the delta land is being covered by a topographic survey made wholly by Chinese surveyors, that has been in progress for ten years, with ½-m. contours over large critical areas. This is accompanied by occasional, more or lees accurate, river-flow gaugings.

For several years, the Harbor Board of Shanghai, controlled chiefly by the English, American, and French commercial interests, with a view chiefly to the improvement of navigation, has been making an excellent systematic and thorough survey of the neighboring Yang-tze River and its small tributary river, the Whang-poo, on which Shanghai is situated, under A. V. H. von Heidenstam, M. Am. Soc. C. E., assisted by E. C. Stocker, Assoc. M. Am. Soc. C. E.

At Foo-Chow, some good river-training work is said to be in progress, mainly for improving the entrance from the sea to the harbor, in charge of J. R. West, M. Am. Soc. C. E.

In Southwestern China, along the great West River and other rivers near Canton, some excellent preliminary hydrographic studies relative both to navigation and flood problems have been in progress for five years, in charge of Capt. Olivecroner, a Swedish engineer.

As a whole, the outlook for important construction, everywhere, is poor, because of internal political conditions, the impossibility of the present Government securing adequate funds, either by internal taxes or outside loans, and by the lack of community spirit, but the speaker believes that sooner or later China will find a way to work out its own salvation, and that the present outlook on stagnation will suddenly change.

Although China proper presents a range of latitude and longitude about equal to that of the United States east of the Rocky Mountains, this paper will be confined to the first three problems previously mentioned, which are found in the northeastern delta plain of China, north of the Yangtze River, within an area which, on a map of the United States, would about cover a triangle with corners at Buffalo, N. Y., Boston, Mass., and Washington, D. C. First the flood problems of the Yellow River, "China's Sorrow," will be briefly discussed.

Flood Problems of the Yellow River Delta

By far the largest of the Chinese deltas is that of the Yellow River which, with a radius of about 400 miles and an apex angle of about 90°, from its mountain exit, slopes seaward with wonderful uniformity at the rate of 10 in. per mile measured on a straight line, or about 8 in. per mile as measured along the river's winding course. Through millions of years this delta cone has been built up by deposits of fine-grained silt, brought down by the floods from the vast loess deposits of Shansi and other Provinces. The flatness of the surface of this vast delta-cone is the cause of the great width and vast range to which a flood may spread when it escapes from the river's dikes. This flat slope may be due to the extreme fineness of the particles of silt which are mostly derived from the erosion of vast beds of loess, supposed to be accumulations of wind-blown dust, that originally came largely from the vast Gobi Desert. A moderate current can carry this fine-grained material a long way.

The Yellow River is not large in volume, compared with other great rivers of the world, although it is about 2 350 miles long, without counting the minor bends, and drains about 305 000 sq. miles. Within the historic period of about 4 200 years, the Yellow River has meandered and shifted from north to south, and back again, through this delta plain, occupying the nine widely divergent channels. . . .

The courses followed in these migrations appear to have first been brought to the attention of the outside world by the American explorer and geologist, Raphael Pumpelly, about 1865, in a series of small maps. The speaker had the map which is presented on a reduced scale in Fig. 5, prepared in China, in 1919, by tracing on a sheet about 4 ft. square, the several courses laid down on a series of Chinese maps brought to his attention by the engineers of the Chihli River Improvement Commission. These maps were the work of a Chinese historian of about 100 years ago, who based them on the investigation of a Chinese author of about 200 years ago. . . . The Chinese have had some remarkably painstaking scholars and historians.

When a complete topographic map of China on a large scale is prepared, many additional details will probably be secured, because when one travels across the delta, many depressions, sand dunes, abandoned dikes, and other indications of ancient channels are found. Many town and provincial records have been kept for centuries, which note changes and floods with evident care. It is important that all these sources of data be searched and that a complete contour map be made of all the delta, showing all ridges and drainage channels, as a basis for many possible improvements; but the solving of the main problem of flood protection need not wait for this.

Since time immemorial, the river floods have been confined between dikes built of river silt, well tamped into place, which dikes are guarded from erosion, wherever the river threatens, by spur-dikes or groynes commonly built with a facing of loose stone rip-rap over a core of earth and millet stalks, but sometimes built only of bundles of millet stalks, tied together with straw ropes, packed with earth, and pinned down with small stakes. This tuning away of an impinging flood by spur-dikes is made easier by the great width of open ground between the dikes, which area is largely made unavailable for agriculture by the danger of flooding. On some of the higher silt banks of the flood-plain between the dikes, the farmer often sows his seed and takes a chance of loss of harvest if the annual flood arrives early.

. . .

Some Noteworthy Outbreaks

By way of further demonstration of the vast importance of finding a method of solving China's flood problems to the many millions of people inhabiting this vast delta between Kai-feng, Tientsin, and Chin-kiang (possibly more than the total population in the Buffalo-Boston-Washington triangle, for the great city populations in New York, Boston, Buffalo, etc., may be outnumbered in the density of China's farm population), . . . the following cases of serious floods from outbreak of the Yellow River are briefly stated. So recent is the opening up of that vast country to foreign engineering inspection and so imperfect have been the means of transmitting information, that the facts have been little known to the outside world.

The three most noteworthy outbreaks of the Yellow River in recent years of which details are available are as follows:

First.—The great break of 70 years ago, which changed the river's course for 270 miles, as shown in Fig. 5. This break occurred during the summer flood of 1851, but the water seems not to have been completely diverted into the new channel until 1853. The new course of this flood laid waste a stretch of prosperous country about 10 to 20 miles wide, and about 140 miles long, or nearly 2 000 sq. miles, and the escaping waters were not gathered into a narrow channel between dikes for this 140 miles below the break until about a quarter of a century had passed. No record is now known of the thousands of people who perished, or of the great value of the property ruined and the fertile farms destroyed along its path.

This diversion from the course that had been followed for 628 years also caused disaster to many thousands of farmers living along the deserted river bed by depriving them of their previous water supply.

Second.—The break of 35 years ago, which cost more than 1 000 000 lives and a vast property damage, came near causing a permanent diversion and was repaired after 1 ½ years of mighty efforts and great expenditure.

Third.—The break of 1903, about 13 miles below Tsi-nan (or "Chi-nan"), which is described by Capt. W. F. Tyler in his pamphlet . . . on the Yellow River. This outbreak deposited a bed of silt said to be from 3 to 8 ft. in thickness, over much of an inundated area of about 200 sq. miles, and caused such terrible distress as to call for organization of famine relief from foreign sources.

Details of the Great Bank of 1851

It is said that foreigners first heard of this break 6 years after it occurred, and that 11 more years elapsed before anything definite about the cause of the river's change of course was known to the outside world. The first visit of a European to learn what had happened in 1851, was made 17 years after the catastrophe, by an English merchant of scientific education, residing at Shanghai, Mr. Ney Elias, who visited the new course and the scene of the break in October, 1868, and reported his findings to the Royal Geographic Society of England, which published his account. . . .

For many miles up stream from the site of the former course of the Grand Canal, he found the new river flowing irregularly over a strip of country about 10 or 12 miles in width, which had the appearance of a field, inundated and laid waste, rather than that of an ordinary river channel. The strip, thus laid waste, probably was 90 miles long. From 20 to 52 miles up stream, from the break, the river was flowing in a single, well-defined channel, where the shifting currents had deposited low, wide embankments of silt, perhaps 6 to 10 ft. deep, from which protruded half buried houses and temples. The river was flowing between the newly-made banks of silt deposit about 10 ft. high, with its bed at about the previous level of the country. Between the hills of Yu-shan and Chiau-kou, at the northwest corner of the Shantung Mountains, where the escaping waters found and usurped the ancient channel of the Ta-ching Ho, the waters were again gathered together and flowed to the sea, greatly swelling the volume of the Ta-ching River. Some reports state that the silt-laden water raised its deep-cut bed by deposits of silt; other observers, including Mr. Elias, reported that the added flow had cut the Ta-ching wider and deeper. Probably the discrepancy was due to different localities observed. Mr. Elias proceeded down the Ta-ching Ho to where it entered the sea to learn of its possible availability for navigation. The lower 20 miles was found to be flowing through low uninhabited mud flats, deposited from the river's burden of silt. The report by Mr. Elias is an admirably clear description which impresses one with the author's keenness as an observer.

Mr. Elias made a second visit for the purpose of exploring a new connection said to have been established in 1868, between the Yellow River and the Yang-tze *via* the Sha Ho and Huai River, through a break in the south dike 50 miles up stream from the City of Kai-feng. He found, however, that the breach had been repaired and that boat navigation was stopped early in 1870, after more than a year of flow through a breach 1 mile wide, but not deep. Mr. Elias, in his second paper, discusses the merits of turning the Yellow River back into the course it had followed prior to 1851, and calls attention to the great hardship that had been suffered by the large population along the old course of the river, in being deprived of water for irrigation and navigation. A large migration of this river thus brings disaster to millions of people along both the old and the new courses, because this fertile delta is populated in many places so densely that a farm of 3 to 5 acres has to provide the support of a family.

About ten years later, the new course was visited by a prominent English engineer, Mr. G. J. Morrison, who made an outline survey and map. The multitude of shallow shifting irregular streams flooding 10 to 15 miles in width, found by Mr. Elias, had then been gathered into a single stream between dikes, or had gathered itself, into this single, narrow channel by its erosion of its bed slightly below the general level of the plain as now found and shown in the top sections of Fig. 12.

An outline map of the region near the break of 1851, surveyed early in 1919, is given in Fig. 14. The speaker visited this locality in December, 1919, and found the landscape so broad and of such low relief that it was difficult for one in the field to take in the relation of one part to another. Although the chief events are plain, it will require a much closer and more detailed topographic map than is yet available, and a week in the field, map in hand, for tracing out in detail just what happened where the river changed its course. The river, breaking out to the north, found a country about 12 ft. lower in elevation than its former bed, over which it could flow and over this, in ill-defined shifting channels, the flood ultimately reached the Valley of the Ta-ching Ho, within which, the records show, it had flowed about six centuries before, for at least about 80 miles, perhaps then reaching the sea by the Hsiao-ching Ho, instead of by its present course.

As stated previously, the escaping flood of 1851 usurped this Ta-ching channel and, in the course of 50 years, filled some miles in length of its bed with silt almost to the general level of the ancient delta plain, the depth of new silt thus deposited at Lo-kou being about 12 ft., according to borings made for the railroad bridge. Possibly, the "Ta-ching" or "clear-water" river which, although small, sometimes gives brief violent floods, by six centuries of effort had scoured this part of the channel bed below the elevation that it possessed when occupied by the Yellow River from 1194 to 1289.

Source: John R. Freeman. "Flood Problems in China." *Transactions of the American Society of Civil Engineers,* Paper no. 1505, April 1922. New York: The Society, pp. 1,410–1,415, 1,426–1,428.

Great Sichuan Earthquake in China: USAID, "China—Earthquake" (2008)

On May 12, 2008, an earthquake in China caused extensive damage and loss of life in Sichuan Province. Official reports placed the death toll at 69,222 people and estimated 5 million people to be homeless. USAID published this fact sheet to provide details of the disaster.

China—Earthquake

Fact Sheet #6, Fiscal Year (FY) 2008   August 8, 2008

Note: The last fact sheet was dated June 13, 2008

KEY DEVELOPMENTS

- On July 15, a USAID/OFDA acting regional advisor traveled to Beijing to attend a workshop on post-earthquake reconstruction experiences co-hosted by the U.N. and the Government of China (GOC), and meetings related to rural shelter and housing recovery following the earthquake in Sichuan Province. The workshop brought together international experts and GOC officials to discuss the multifaceted issues surrounding recovery, as well as sustainable development practices after a major earthquake, in the context of previous earthquake responses.

- USAID/OFDA is supporting The Asia Foundation (TAF) to promote rural housing reconstruction in earthquake-affected areas of Sichuan Province, and to raise public awareness of disaster risk reduction. USAID/OFDA has supported TAF programming in China since 2006 to enhance private sector participation in and contribution to disaster assistance, as well as to promote multi-sectoral approaches to disaster management.

Numbers at a Glance		Source
Total Dead	69,222	GOC — August 7, 2008
Total Injured	374,638	GOC — August 7, 2008
Total Missing	18,176	GOC — August 7, 2008
Total Homeless (Estimated)	5 million	IFRC[1] — July 29, 2008
Total Displaced (Estimated)	15 million	IFRC — July 29, 2008
Total Affected (Estimated)	46 million	U.N. — July 16, 2008

FY 2008 HUMANITARIAN ASSISTANCE FOR CHINA EARTHQUAKE

USAID/OFDA Assistance to China	$2,672,698
DOD[2] Assistance to China	$2,204,900
Total USG Humanitarian Assistance to China	$4,877,598

CURRENT SITUATION

- On May 12, a magnitude 7.9 earthquake struck Wenchuan County in China's densely populated Sichuan Province, killing nearly 70,000 people and injuring nearly 375,000 others, according to official GOC estimates. As of August 8, the GOC reported spending approximately $9.4 billion on disaster relief and reconstruction.
- As of August 6, the U.S. Geological Survey (USGS) had confirmed more than 950 aftershocks to the May 12 earthquake, the majority of which were between magnitude 4.0 and 4.9. On August 5, a 6.0 magnitude aftershock killed two people and destroyed more than 2,700 houses in Qingchuan county of Sichuan Province.

---

[1] International Federation of the Red Cross and the Red Crescent Societies (IFRC)

[2] U.S. Department of Defense (DOD)

Shelter and Reconstruction

- The earthquake displaced approximately 15 million people according to Sichuan provincial authorities and the IFRC. As of August 7, the GOC had relocated approximately 1.5 million earthquake-affected individuals. To date, the GOC has provided nearly 1.6 million tents, approximately 4.9 million blankets, and more than 14 million articles of clothing to affected populations. As of August 6, the GOC reported that relief workers had constructed 612,400 temporary houses and were in the process of building another 6,700, with 39,500 houses remaining to be built.
- USAID/OFDA is providing $1.2 million to TAF to promote safe rehabilitation of rural shelters and to improve disaster preparedness and mitigation in schools and communities in earthquake-affected areas of Sichuan Province. The program, to be implemented with the GOC's Ministry of Civil Affairs (MOCA), will benefit approximately 1.1 million individuals in affected areas.

Disaster Preparedness

- USAID/OFDA has approved $900,000 to support the second phase of TAF's Public Private Partnerships for Disaster Management in China program. The program, which USAID/OFDA has funded since October 2006, will be extended for an additional two years.
- The program's goal is to promote a partnership between the public and private sectors in disaster management, preparedness, and mitigation. Program participants include the Chinese private sector and member organizations of the American Chamber of Commerce. The program has been developed and implemented in close consultation with MOCA and all other relevant GOC authorities.
- Phase one program activities included training, partnership building, and community awareness campaigns. Phase two will strengthen and expand the public-private partnerships built during the first phase. The second phase of the program will engage more Chinese businesses in disaster preparedness and management initiatives and deepen multi-sectoral collaboration. The program will expand preparedness activities to additional communities, including Chengdu city in earthquake-affected Sichuan Province.

USG HUMANITARIAN ASSISTANCE

- On May 13, U.S. Ambassador Clark T. Randt, Jr., issued a disaster declaration due to the impact of the earthquake and requested disaster assistance from USAID/OFDA. In response, USAID/OFDA provided an initial $500,000 through the U.S. Embassy in Beijing to IFRC. The

IFRC channeled the contribution to the Red Cross Society of China for the local procurement and distribution of emergency relief supplies.

- In response to the GOC's request for international assistance, USAID/OFDA donated specialized search, rescue, and recovery equipment, including 40 crates of saws, hand tools, hydraulic gear, concrete cutters, generators, and personal safety equipment. In addition, USAID/OFDA deployed a nine-person team of experts from Los Angeles (L.A.) County Fire Department and Fairfax County, VA, Fire and Rescue Department, as well as USAID/OFDA support staff.

- On May 23 and 24, the USAID/OFDA-funded team conducted a series of hands-on search and rescue workshops for approximately 40 members of the Public Security and Fire Brigade of Chengdu and the Seismic Disaster Emergency Rescue team of Sichuan Province. The equipment donation and training were aimed at increasing the capacity of GOC earthquake responders as well as facilitating technical cooperation between the two countries during future emergencies.

- In response to the Chinese request for shelter assistance, USAID deployed a shelter expert and a disaster specialist to Beijing and Chengdu in early June in order to provide technical assistance and liaise with relevant GOC departments and international relief organizations. In addition, USGS led a U.S. interagency team in Beijing from June 11 to 13 to assist the China Earthquake Administration with various seismic-related issues, including earthquake monitoring and preparedness activities.

- On May 18, two U.S. Air Force C-17 aircraft arrived in Chengdu with emergency relief supplies, including blankets, plastic sheeting, tents, water containers, and food, valued at nearly $1.3 million, including transportation costs. In response to the GOC's subsequent request for shelter assistance, DOD airlifted 153 medium-sized tents on May 28 to Chengdu. The total value of the shelter commodities, including shipment, was nearly $940,000.

PUBLIC DONATION INFORMATION

- The most effective way people can assist relief efforts is by making cash contributions to humanitarian organizations that are conducting relief operations. Information on identifying such organizations is available in the "How Can I Help" section of www.usaid.gov — Keyword: China Earthquake, or by calling The Center for International Disaster Information (CIDI) at 703-276-1914.

- USAID encourages cash donations because they allow aid professionals to procure the exact items needed (often in the affected region);

USG HUMANITARIAN ASSISTANCE FOR CHINA EARTHQUAKE

Implementing Partner	Activity	Location	Amount
USAID/OFDA Assistance[1]			
IFRC	Emergency Relief Supplies	Affected Areas	$500,000
LA County and Fairfax County Fire Departments	Disaster Support, including Transport	Affected Areas	$796,413
TAF	Shelter, Risk Reduction	Affected Areas	$1,200,000
	Administrative Support		$176,285
TOTAL USAID/OFDA			**$2,672,698**
DOD Assistance[2]			
GOC	Emergency Relief Supplies	Affected Areas	$1,266,300
GOC	Procurement and Transport of Tents	Affected Areas	$938,600
TOTAL DOD			**$2,204,900**
TOTAL USG HUMANITARIAN ASSISTANCE TO CHINA IN FY 2008			**$4,877,598**

[1] USAID/OFDA funding represents anticipated or actual obligated amounts as of August 8, 2008.

[2] The decrease in DOD funding from previous reporting reflects actual or committed amounts rather than preliminary estimates.

reduce the burden on scarce resources (such as transportation routes, staff time, warehouse space, etc.); can be transferred very quickly and without transportation costs; support the economy of the disaster-stricken region; and ensure culturally, dietary, and environmentally appropriate assistance.

- More information can be found at:
  ○ USAID: www.usaid.gov — Keyword: Donations
  ○ The Center for International Disaster Information: www.cidi.org or (703) 276-1914
  ○ Information on relief activities of the humanitarian community can be found at www.reliefweb.int
- In addition to USG contributions, the U.S. private sector has actively responded to the earthquake in China. More information can be found at:
  ○ U.S. Chamber of Commerce Business Civic Leadership Center: www.uschamber.com/bclc/default

° Business Roundtable Partnership for Disaster Response: www. respondtodisaster.com/mambo/

Source: USAID. "China—Earthquake." Fact Sheet #6, Fiscal Year 2008. Bureau for Democracy, Conflict, and Humanitarian Assistance. Office of U.S. Foreign Disaster Assistance, August 8, 2008.

Chinese Government, "China's Actions for Disaster Prevention and Reduction" (2010)

This selection, excerpted from a publication of China's State Council, explains the Chinese government's plan for preparing for natural disasters in that country. This excerpt provides details about the types of natural disasters that occur within China as well as the steps the Chinese government has taken to reduce the risks posed by natural hazards to its citizens.

Foreword

In recent years, natural disasters happened frequently around the world and have caused enormous losses of life and property to human society. They pose a common challenge to all the countries in the world.

China is one of the countries in the world that suffer the most natural disasters. Along with global climate changes and its own economic takeoff and progress in urbanization, China suffers increasing pressure on resources, environment and ecology. The situation in the prevention of and response to natural disasters has become more serious and complicated.

Always placing people first, the Chinese government has all along put the security of people's lives and property on the top of its work, and has listed the disaster prevention and reduction in its economic and social development plan as an important guarantee of sustainable development. In recent years, China has been comprehensively implementing the Scientific Outlook on Development, further strengthened legislation as well as the building of systems and mechanisms on disaster prevention and reduction, committed to building on disaster-prevention capacities, encouraged public contribution, and actively participated in international cooperation in this respect.

The devastating Wenchuan earthquake, which occurred on May 12, 2008, caused massive human casualties and property losses, and caused immeasurable sufferings to the Chinese people. In the wake of the disaster, the Chinese government decided to make May 12 "Disaster Prevention and Reduction

Day," starting in 2009. This document has been written to mark the first anniversary of the Wenchuan earthquake and greet China's first "Disaster Prevention and Reduction Day," with a review of the endeavors the Chinese government and people have made in disaster prevention and reduction.

I. Natural Disasters in China

The natural disasters that China suffers from most have the following characteristics:

1. Diverse types. They include meteorological disasters, earthquakes, geological disasters, marine disasters, biological disasters, and forest and grassland fires. Except for modern volcanic activity, China has suffered from most types of natural disasters.
2. Wide scope of distribution. Natural disasters cause damages in different degrees to all of Chinese provinces (autonomous regions and municipalities directly under the Central Government). More than 70 percent of Chinese cities and more than 50 percent of the Chinese population are living in areas vulnerable to serious earthquakes, or meteorological, geological or marine disasters. Two thirds of China's land are threatened by floods. Tropical cyclones often batter the eastern and southern coasts, and some inland places. Droughts often occur in the northeast, northwest and north, with particularly serious ones common in southwest and south China. Destructive earthquakes with a magnitude of 5 or more on the Richter Scale have struck all the country's provinces (autonomous regions and municipalities). The mountainous and plateau areas, accounting for 69 percent of China's total land territory, suffer frequent landslides, mudrock flows and cliff collapses due to complicated geological conditions.
3. High frequency. Its monsoon climate has a strong impact on China, and causes frequent meteorological disasters. Local or regional droughts occur almost every year, while tropical cyclones, seven times a year on average, batter the east coast. As China lies right in the region where the Eurasian, Pacific and Indian Ocean plates meet, it suffers from frequent earthquakes due to still-active tectonic movements. Most of the quakes shaking China are continental, accounting for one-third of global destructive land quakes. Fires often break out in forests and on grasslands.
4. Huge losses. During the 19 years from 1990 to 2008, on annual average, natural disasters affected about 300 million people, destroyed more than 3 million buildings, and forced the evacuation of more than 9 million people. The direct financial losses caused exceeded 200 billion yuan.

NATURAL DISASTERS

Floods in the Yangtze, Songhua and Nenjiang river valleys in 1998, serious droughts in Sichuan Province and Chongqing Municipality in 2006, devastating floods in the Huaihe River valley in 2007, extreme cold weather and sleet in south China in early 2008, and the earthquake that shook Sichuan, Gansu, Shaanxi and other places on May 12, 2008 all caused tremendous losses.

Now and for a fairly long time to come, the risks of extreme weather phenomena are increasing along with global climate changes. Owing to imbalanced distribution of precipitation, unusual temperature changes and other factors, the occurrences of floods and droughts, hot weather and heat waves, low-temperature rain, snow and sleet, forest and grassland fires, plant diseases, insect and animal pests may grow in number. The probability of strong and extra-strong typhoons, tempests and other disasters is quite high. The tasks of guarding against and preventing such geological disasters as mountain torrents, landslides and mud-rock flows brought about by heavy rains remain weighty. In addition, as a result of the earth's crustal movements, the danger of earthquakes is increasing.

II. Strategic Goals and Tasks for Disaster Reduction

In the National 11th Five-year Plan on Comprehensive Disaster Reduction and other documents issued in recent years, the Chinese government has made clear its medium- and long-term strategic goals during the 11th Five-year Plan period (2006–2010) and the ensuing years; to build a relatively complete working system and operational mechanisms regarding disaster reduction; to greatly enhance the capabilities related to disaster monitoring and early warning, prevention and preparation, emergency handling, disaster relief, and rehabilitation and reconstruction; to notably raise public awareness of disaster reduction and emergency rescue skills; and to significantly reduce human casualties and direct economic losses caused by natural disasters.

The main tasks are as follows:

—To strengthen capability in management over hidden risks of natural disasters and relevant information. Based on a general survey of hidden risks of major natural disasters in key areas, and the overall prevention and reduction capabilities of the nation, China will build a database of the risks of natural disasters, and draw up a national diagram of the situations in high-risk and key areas. It will also build a system for collecting statistics about disasters and the damage inflicted, and a reporting system covering the national, provincial, municipal and county levels. It will improve the mechanisms of prompt news release, check on disaster damage, and the work on information exchange, consultation and announcement. A platform for the sharing and releasing of disaster information will be established, and the analysis, appraisal and application of disaster information be reinforced.

—To strengthen capability in the monitoring, early warning and forecasting of natural disasters. While improving the existing monitoring network, China will increase the monitoring density and launch a satellite remote-sensing monitoring system, thus building a three-dimensional monitoring platform for natural disasters. It will promote the comprehensive utilization and integrated development of monitoring and early warning infrastructures, and improve the supporting systems in the field of disaster warning, forecasting and decision-making.

Particular efforts will be made to strengthen the capability of monitoring, early warning and forecasting extreme weather and serious frequently-occurred disasters. A mechanism to issue disaster risk warnings will be put in place using various channels of communication to accurately and promptly release disaster information.

—To strengthen overall capability to prevent and combat natural disasters. Efforts will be made to carry out various plans concerning disaster prevention and reduction, construct pillar projects, and raise the disaster combat capabilities of large and medium-sized industrial bases, transportation trunk lines, communication hubs and lifeline projects. In line with the national land utilization plan and principle of economical and intensive use of land, the government will make an overall plan in respect of disaster reduction concerning agriculture and rural areas, industry and urban areas, as well as specialized disaster prevention and reduction plans and construction of relevant projects in key areas, so as to enhance the country's overall disaster-prevention capacity in all aspects.

—To strengthen the state capacity for emergency rescue and relief work. A coordinated and efficient disaster emergency management system will be built, characterized by unified command, sound coordination, clear division of work, and level-by-level control with local authorities playing the main role. This will form, by and large, an emergency relief system covering all aspects. The construction of disaster combat and relief materials reserve network, at central and local levels, will be strengthened; the transportation capacity of relief materials will be raised; various backbone or professional rescue contingents will be consolidated; and disaster reduction and relief equipment will be improved. Social mobilization mechanisms are to be improved to give full play to the functions of non-governmental organizations and organizations at the grassroots level as well as volunteers in the sphere of disaster relief.

—To strengthen capability in consolidating flood control in various river valleys. Adhering to the principles of overall planning, sound coordination and comprehensive solution of both root causes and symptoms, a complete flood control and disaster prevention system will be built to guarantee the safety of river valleys. The system will take embankment construction as the basis, backed up by key water control projects on mainstreams and tributaries, flood storage areas and dredged watercourses, as well as levees, lakeside lands returned from farming to water, and water and soil conservation efforts, in addition to non-project measures such as flood and drought control command systems, coordination measures and flood risk management.

—To strengthen capability in comprehensive response to disastrous calamities. By studying the mechanism of occurrence and law of activity of disastrous calamities, and their relations with secondary disasters, China will conduct simulation experiments of massive disaster variations and emergency responses to disastrous calamities. It will build and improve relevant systems and mechanisms, work out policies and emergency response plans against disastrous calamities, and conduct drills to combat them. It will also spread pilot insurance schemes in agriculture and forestry, and introduce, based on national conditions, insurance and reinsurance against devastating disasters. It will strengthen efforts in the construction of projects against huge disasters, and establish an Asian regional disaster research center.

—To strengthen urban and rural community capability in coping with disasters. China will improve the emergency response plans for urban and rural communities, and train the residents against such dangers. Urban and rural emergency facilities will be improved, and model communities will be established throughout the country. Housing safety projects in both rural and urban areas will be built. Shelters will be built where disasters are prone to occur. Community disaster reporters will be nominated, urban and rural residents will be educated to prevent and deal with disasters, and a mechanism to protect disadvantaged community groups will be built.

—To strengthen the scientific and technological support capability in the fields of disaster prevention and reduction. China will fortify the research and development of key technologies, and study and work out national medium- and long-term strategies for scientific and technological development in coping with natural disasters. It will quicken the application of remote-sensing, geographical information system, global positioning system, and network communication technologies. It will invest more funds in scientific and technological development against disasters, support the construction of relevant disciplines and the fostering of skilled people, and build personnel training bases. It will also introduce relevant technological standards, and standardize disaster prevention and reduction operations.

—To strengthen capability in scientific publicity and education concerning disaster reduction. China will heighten the sense of duty of local governments at various levels in disaster reduction. Knowledge related to disaster reduction will be incorporated in school textbooks, and provided to rural residents through activities to bring cultural, scientific, medical services to the rural areas. General or specialized education concerning disaster reduction will be encouraged, and relevant education bases built. A national network platform spreading disaster reduction knowledge will be launched. Popular science books, wall charts and audio-visual products will be published and produced, local experiences in disaster prevention and reduction be publicized, and public awareness and skills raised.

III. Construction of a Legal Framework, Institutional Setup and Working Mechanism Related to Disaster Reduction

China attaches great importance to legislation regarding disaster prevention and reduction and has enacted a number of laws and regulations in this regard, thus gradually institutionalizing disaster reduction efforts. Since the early 1980s, the state has promulgated more than 30 laws and regulations concerning disaster prevention and reduction, including the Emergency Response Law of the People's Republic of China, Law of the People's Republic of China on Water and Soil Conservation, Law of the People's Republic of China on Protection Against and Mitigation of Earthquake Disasters, Water Law of the People's Republic of China, Flood Control Law of the People's Republic of China, Law of the People's Republic of China on Desertification Prevention and Transformation, Meteorology Law of the People's Republic of China, Forestry Law of the People's Republic of China, Grassland Law of the People's Republic of China, Law of the People's Republic of China on the Prevention and Control of

Water Pollution, Law of the People's Republic of China on the Prevention and Control of Pollution from Environmental Noise. Law of the People's Republic of China on the Prevention and Control of Environmental Pollution from Solid Waste, Marine Environment Protection Law of the People's

Republic of China, Fire Control Law of the People's Republic of China, Drought Control Regulations of the People's Republic of China, Hydrology Regulations of the People's Republic of China, Flood Control Regulations of the People's Republic of China, Forest Fire Control Regulations of the People's Republic of China, Grassland Fire Control Regulations of the People's Republic of China, Regulations on Handling Major Animal Epidemic Emergencies, Regulations on the Prevention and Control of Forest Plant Diseases and Insect Pests, Regulations on the Prevention and Control of Geological Disasters, Regulations on the Handling of Destructive Earthquake Emergencies, Regulations on the Administration of Security of Reservoirs and Dams, and Regulations on the Administration of Weather Modification. The state will continue its efforts in the field of legislation as concerns disaster prevention and reduction as the need arises.

The Chinese government has for years persisted in incorporating disaster reduction in the sustainable development strategies at the national and local levels. In "China's Agenda 21," issued in March 1994, the central government clearly defined the relations between disaster reduction and environmental protection at the national level, placing as major concerns the construction of a disaster prevention and reduction system and the reduction of human activities that trigger or worsen natural disasters on its agenda. In April 1998, the state released the Disaster Reduction Plan of the People's Republic of China (1998–2010). It put forward the guidelines, goals, tasks and methods of disaster reduction work, in the form of specialized plans, for the first time in China. In October 2006, the 11th Five-year Plan for the Development of Science and Technology was released, in which the Chinese government included as major tasks the technological development for a public security emergency response system and the enhancement of the nation's capabilities in handling public security disasters and unexpected public incidents. In August 2007, the Chinese government issued the 11th Five-year Plan on Comprehensive Disaster Reduction, requiring local governments to include disaster reduction in their social and economic development plans.

China has adopted a disaster reduction and relief system featuring central leadership, departmental responsibility and disaster administration at different levels with major responsibility on local authorities. Under the unified leadership of the State Council, the central organs coordinating and organizing disaster reduction and relief work are the National Disaster Reduction Committee, State Flood and Drought Control Headquarters,

State Earthquake Control and Rescue Headquarters. State Forest Fire Control Headquarters and National Disaster Control and Relief Coordination Office. Local governments also have set up corresponding coordination offices to handle disaster reduction and relief work. During disaster reduction and relief work, the People's Liberation Army, the Armed Police, militiamen and reservists, as well as policemen play the major role, and often act as task forces. Social groups, non-governmental organizations and volunteers will also join the effort.

With years of experience in disaster reduction and relief work, the Chinese government has established a disaster reduction and relief mechanism geared to the nation's situation. It has established a series of disaster emergency mechanisms, including a disaster emergency response system, disaster information release mechanism, emergency relief materials reserve system, disaster early warning, consultation and information sharing system, major disaster rescue and relief joint coordination mechanism and emergency social mobilization mechanism. Local governments at various levels also have similar working mechanisms.

—Disaster emergency response system. The emergency response system of the central government for unexpected natural disasters is formed of three levels: state overall emergency response plan, state specialized emergency response plan, and departmental emergency response plans. Detailed measures and working regulations are worked out by the relevant government departments in line with the specialized plans and their respective responsibilities. In the wake of a major natural disaster, under the unified leadership of the State Council, the relevant departments with different focuses shall act in coordination and launch emergency response plans to guide disaster control and relief work. The governments of the affected areas shall immediately start emergency response measures and set up a local disaster emergency response command with the heads of the local governments serving as the chief commanders, and leaders of relevant departments as members, to jointly draw up emergency plans and measures, organize field emergency response work, and report disaster details and work progress to governments of higher levels and relevant departments.

—Disaster information release system. Following the principle of "being prompt and precise, open and transparent," the central and local governments are expected to work in earnest on the emergency information release work concerning natural disasters and other emergencies, offering through authorized releases, press releases, interviews and press conferences to the public prompt information on the disasters and their developments, progress of emergency response work, disaster prevention, and knowledge on disaster prevention and other information, thus ensuring the public's rights to know and supervise.

—Relief materials reserve system. China has built a relief materials reserve network based on special storehouses, which has seen year-on-year improvements. The country has now ten such storehouses for daily necessities at the central level, and storage centers for relief supply, flood and forest fire control supplies are continuously being built and improved. Coupled with the reserve relief supply centers established in some provinces, cities and counties, a preliminary disaster control and relief materials reserve system has taken shape. To guarantee the timely purchase of relief supplies, a list of commissioned relief supply manufacturers is established, and emergency purchase agreements signed with them for the supply of relief materials in case of emergency.

—Disaster early warning, consultation and information sharing mechanism. A disaster early warning, consultation and information sharing mechanism has been set up, involving such relevant departments as civil affairs, land and resources, water resources, agriculture, forestry, statistics, seismology, maritime affairs and meteorology. To offer timely and effective support for the decision-making of the central government and local departments in case of emergency, China has initiated the construction of a disaster information database and launched a public platform of national geographical information and a disaster information publishing and sharing system, as well as a platform for national disaster reduction and risk management information.

—Major disaster rescue and relief joint coordination mechanism. In the wake of a major disaster, relevant departments will play their functions and timely dispatch to disaster-hit areas working groups composed of personnel from these departments to gather first-hand information and guide disaster control and relief work on the spot. The groups are also required by the State Council to coordinate with the relevant departments to map out rescue plans, help with disaster relief work and prevent possible secondary disasters.

—Disaster emergency response public mobilization mechanism. A preliminary public mobilization system is now in place, focusing on efforts for rescue, search, first aid, relief, donation and other work. The government also gives full scope to non-government organizations, such as mass organizations, the Red Cross, self-governmental organizations at the grassroots level and individual volunteers in the fields of disaster prevention, emergency rescue, relief and donation work, medical, hygiene and quarantine work, post-disaster reconstruction, psychological support and other aspects.

IV. Enhancement of Disaster-reduction Capability

The Chinese government attaches great importance to the enhancement of disaster-reduction capability. It has made great efforts in undertaking disaster-reduction projects, improving disaster early warning and emergency response, enhancing sci-tech support, strengthening personnel training and disaster reduction work in communities.

1. Carrying out Disaster-reduction Projects and Improving Capability in Comprehensive Prevention of Disasters

In recent years, China has engaged in a series of important disaster-reduction projects, including those concerning flood control, drought combat, earthquake prevention and relief, cyclone control, red tide and other marine disaster prevention, desertification and sandstorm control, and ecological construction.

—Flood control on major rivers. The government has greatly increased its input in harnessing major rivers by way of adopting proactive financial policies and issuance of bonds, which has accelerated the progress of the harnessing of major rivers and lakes. To date, the construction and renovation of the dykes on the middle and lower reaches of the Yangtze River have been completed; construction of standardized dykes on the lower reaches of the Yellow River is in full swing; 19 major flood control projects for the Huaihe River have been, by and large, completed; and pivotal water conservancy projects at the Three Gorges on the Yangtze River, Xiaolangdi on the Yellow River and Linhuaigang on the Huaihe River are playing their full part. China's flood control ability on major rivers has been further improved. Construction work on some major sections of these rivers is now capable of defying the severest flood in 100 years. The flood control capability of small and medium-sized rivers has been continuously improved. The standard for key sea dykes has been raised to withstand the worst flood in 50 years.

—Housing renovation for impoverished rural residents. China attaches great importance to disaster-proof residential construction in rural areas. During the reconstruction of disaster-stricken buildings, technical guidance is given and quality control stressed on site selection, design, construction and acceptance inspection. Housing projects are pushed forward in combination with poverty alleviation efforts. Since 2005, a total of 17.535 billion yuan has been invested nationwide in the renovation and construction of 5.8016 million rural houses for 1.8051 million impoverished households totaling 6.4965 million people.

—Decrepit school building renovation. Since 2001, a school building renovation scheme has been implemented throughout the country. By the end of 2005, a special fund of 9 billion yuan from the state revenues has been allocated for renovating the decrepit buildings of over 40,000 schools. Since 2006, building renovation expenses of all primary and junior high schools in rural areas have been included in the financial support scheme for compulsory education in rural areas.

—Safe school buildings. Starting from 2009, the state will reinforce school buildings nationwide in order to make them meet the earthquake withstanding standard applied for key projects within three years. They should also meet the requirements in preventing and avoiding disasters caused by mountainside landslide, rock collapse, mud-rock flow, tropical heat wave, fire, etc.

—Seepage prevention and reinforcement for unsafe reservoirs. In March 2008, the state issued the Special National Plan on Seepage Prevention and Reinforcement for Unsafe Reservoirs, requiring the completion of renovation of large and medium reservoirs as well as key small-sized ones threatened by floods within three years. In 2008, 4,035 seepage prevention and reservoir reinforcement projects were undertaken, accounting for 65 percent of the total 6,240 targeted reservoirs.

—Drinking water safety in rural areas. During the Tenth Five-year Plan period (2000–2005), a total of 22.3 billion yuan was spent on solving the problem of drinking water for 67 million people in rural areas, thus basically ending the history of a serious shortage of drinking water in rural areas. Since 2006 priorities have been shifted to guaranteeing that all drinking water is safe. From 2006 to 2008, 23.8 billion yuan were invested by the state revenues and 22.6 billion yuan invested by local revenues for providing safe drinking water to an accumulated rural population of 109 million.

—Water and soil erosion control. In the 1980s, key water and soil erosion control projects were launched in areas suffering from serious water and soil erosion, such as those along the Yellow and the Yangtze rivers. During the later period of the Ninth Five-year Plan (1996–2000), efforts were extended to the upper and middle reaches of the seven major rivers (Yangtze, Yellow, Huaihe, Haihe, Songhua, Liaohe, Pearl) and the Taihu Lake. By 2008, key water and soil erosion control projects had covered a total area of 260,000 sq km, with 70 percent of such areas put under control and a silt reduction rate of 40% or more. Soil erosion in the Jialing River area, on the upper reaches of the Yangtze River, has been reduced by one third, while the sediment flow into the Yellow River has been reduced by about 300 million tons per annum.

—Farmland irrigation and drainage. Since the beginning of the Ninth Five-year Plan, China has increased financial input in farmland irrigation and drainage facility construction, focusing on construction of support facilities and water-saving facilities in major irrigation areas.

As a result, farmland irrigation and drainage as well as flood and drought resistance abilities have been improved.

—Ecological construction and environmental improvement. Since the beginning of the 21st century, key ecological construction projects have been carried out, including those concerning natural forest resources protection, reverting farmland to forest, shelter forest construction in the northeast, north and northwest of China, key shelter forest construction on the lower reaches of the Yangtze River, sand-storm sources control in the Beijing and Tianjin areas, desertification control in karst landform areas, wildlife protection and nature reserve construction, coastal shelter forest construction and the restoration of pasturage to natural grassland. All these projects are aimed at checking the rapid expansion of desertification and reduce the damages caused by extreme climatic conditions. Pilot ecological compensation work has been carried out and experimented with six ecological compensation projects, including rational coal resources development in Shanxi Province. Efforts have also been made in the field of ecological construction at the provincial, municipal and county levels, and in building ecological towns and villages with excellent surroundings, especially pushing ahead the construction of 103 key demonstration counties.

—Construction of earthquake-proof buildings and facilities. China has promulgated the Regulations on the Administration of Disaster Prevention of Urban Public Utilities, the Urban Earthquake and Disaster Prevention Planning Standards and the Design Specifications for Earthquake-proof Buildings in Towns (Townships) and Villages. China has published the Zoning Map of China with Seismological Parameters, improved its earthquake-proof evaluation management system for key construction projects, and promoted the implementation of safety guarantees for earthquake-proof houses in rural areas. Construction and reinforcement of 2.45 million such houses have been completed. After the 2008 Wenchuan earthquake in Sichuan Province, the Earth-quake-proof Classification Standards for Construction Projects and the Earthquake-proof Construction Design Specifications have been revised.

—Highway disaster prevention. Since 2006, in view of highways being destroyed by flood or earthquake, the state has implemented highway disaster prevention projects. By 2008, a total of 1.54 billion yuan had been invested to renovate road embankments, roadbeds, bridge structures and flood-proof and drainage facilities with focus on disaster prevention facilities in mountainous and hilly areas. The disaster-prevention capability of China's ordinary highways has also been improved in an all-round way.

2. Building up a Three-dimensional Monitoring System and Enhancing Disaster Monitoring, Early Warning and Forecasting Capability
    China is building a three-dimensional natural disaster monitoring system, including land monitoring, ocean and ocean-bed observation, and space-air-ground observation. A disaster monitoring, early warning and forecasting system has taken initial shape.

—Disaster remote-sensing monitoring system. Small satellites named Constellation A and Constellation B for environmental disaster-reduction monitoring have been launched. A business application system by using the disaster-reduction satellite has taken shape, providing advanced technological support to remote-sensing monitoring, evaluation of and decision-making for disaster reduction.

—Meteorological early warning and forecasting system. Meteorological satellites FY-1, FY-2 and FY-3 have been put into orbit. A new generation of weather radar installations, totaling 146, has been developed. Ninety-one high-altitude meteorological stations equipped with L-band upper-air meteorological sounding system have been established, and 25,420 regional meteorological observation stations are in operation. Special meteorological observation networks have been preliminarily built for studies of atmospheric elements, acid rain, sandstorm, thunder and lightning and agricultural and transportation meteorology. A comparatively complete data forecasting system has been built for early warning of imminent disastrous weather. A meteorological early warning information release platform covering both urban and rural communities has been established, releasing relevant information through radio, television, newspaper, cell-phone and the Internet.

—Hydrological monitoring and flood early warning and forecasting system. A hydrological monitoring network composed of 3,171 hydrological stations, 1,244 gauging stations, 14,602 precipitation stations, 61 hydrological experiment stations and 12,683 groundwater observation wells has been completed. A flood early warning and forecasting system, ground water monitoring system, water resources management system and hydrological data system have been established.

—Earthquake monitoring and forecasting system. China has built 937 fixed seismic stations and over 1,000 mobile seismic stations, enabling China to be capable of quasi-real-time monitoring of earthquakes above 3 on the Richter scale. In addition, 1,300 earthquake precursor observation stations have been established, as well as a mobile observation network composed of over 4,000 mobile observation stations. Seismological forecasting and monitoring systems at both national and provincial levels have taken initial shape. A high-speed seismic data network composed of 700 information nodes has been built. A cell-phone message service to provide timely earthquake reporting has been launched.

—Geological disaster monitoring system. Since 2003, meteorological early warning of geological disasters has been in place. Over 120,000 places with potential geological hazards are now covered by this system. Also practiced in those places is that masses are involved in disaster monitoring and prevention. A special landslide monitoring network at the Three Gorges Reservoir area, and land subsidence monitoring networks in Shanghai, Beijing and Tianjin have been largely completed.

—Environment monitoring and early warning system. Work on environment quality monitoring, pollutant monitoring, environment early warning and forecasting and unexpected environmental accidents monitoring has been carried out for objective observation of the pollution of surface, ground and ocean water, as well as air, noise, solid waste and radiation pollution. The newly launched HJ-1 environmental satellites A and B carry out efficient, macroscopic and real-time ecological monitoring and evaluation. A preliminary air-ground environment monitoring structure has been built. To date, there are altogether 2,399 environment monitoring stations with 49,335 technicians in China.

—Wild animal epidemic sources and disease monitoring and early warning system. A national wild animal epidemic sources and disease monitoring network has been established, with 350 observation stations at the national level, 768 at the provincial level and over 1,400 at the county level in major natural habitats of migratory birds and other wildlife. Thus, a wild animal epidemic sources and disease monitoring and early warning system has been founded, comprising national, provincial and county levels.

—Plant disease and insect pest monitoring and reporting system. A crop pests and disease monitoring and reporting network composed of over 3,000 observation stations and a grassland rodent and insect pest monitoring and reporting network composed of more than 240 observation stations have been established. The categories of crop pests covered by the national monitoring and reporting system have increased from 15 in the early 1990s to 26 at present. The interval of reporting on major plant diseases and insect pests has been reduced from ten days to one week. Also established is a forest pest monitoring and reporting network composed of over 2,500 observation stations at the national, county and township (town) levels. It now covers the most dangerous and frequently occurring forest pests in 35 categories.

—Marine disaster forecasting system. Oceanographic observation instruments, equipment and facilities have been renovated. Offshore observation capacity has been enhanced greatly. Buoy observation and cross-sectional survey abilities have been improved as a whole. A batch of marine observation stations has been constructed or renovated. Upgrading of real-time communication system has been completed at some key observation stations. An observation and evaluation system for sea-air interaction and ocean climate changes has been developed for ocean disaster monitoring closely related to climate changes such as sea level rise, coastal erosion, seawater intrusion and saline tide.

—Forest and grassland fire early warning and monitoring system. The country's three-dimensional monitoring system for forest and grassland fires, including monitoring by satellite remote sensing, air-plane cruise flight, video monitoring, watching on duty and ground detection, has been improved. A graded forest fire early warning and response system and a fire risk evaluation system have been primarily established.

—Sandstorm monitoring and evaluation system. Efforts have been put into building a satellite remote sensing system for sandstorm monitoring and evaluation, as well as a cell-phone message network. Ground observation points have been set up in major sandstorm-stricken areas of north China at the national, provincial, municipal and county levels so as to form a sandstorm monitoring network covering the whole of north China.

3. Establishing an Emergency Rescue and Disaster Relief Response System, and Improving Emergency Handling Capabilities

A disaster relief emergency response system has taken initial shape, with emergency rescue team system, emergency response mechanism and emergency fund appropriation mechanism as its main items. Emergency handling capabilities such as emergency rescue, transportation support, help with daily life, sanitation and epidemic prevention have been greatly enhanced.

—Emergency rescue team system. An emergency rescue team system has taken initial shape, with the public security forces, armed police and armed forces as the main and task force, with special teams such as flood fighting and emergency rescue, earthquake relief, forest fire, maritime search and rescue, mine rescue, and medical care teams as the basic force, with full-time and part-time teams attached to enterprises and public institutions and emergency volunteers as the backup force. The state's land, air search and rescue base construction has been accelerated, and emergency rescue equipment has been further improved.

—Emergency rescue response mechanism. The central government-stipulated responses to unexpected natural disasters are divided into four levels, which are determined by the degree of damage done. The concrete response measures at different levels have been expressly defined, and disaster relief work has been incorporated into a standard management process. The establishment of a disaster rescue emergency response mechanism basically guarantees that people affected by a disaster can receive aid within 24 hours. They are supplied with "food, clothing, clean water, temporary housing, medical treatment and schooling."

—Disaster relief emergency fund appropriation mechanism. A disaster relief fund appropriation mechanism of the central government has been established, including funds for daily life of those affected by natural disasters, funds for severe flood control and drought combat, funds for roads damaged by flood, funds for inland waterway channel rush-repair, funds for medical rescue, funds for culture, education and administration endeavors, and funds for disaster relief in agriculture and forestry. The disaster relief management system characterized by management of disaster relief by levels and funds shared by different levels is being actively promoted. Disaster relief input by local governments must be guaranteed so as to ensure the basic livelihood of people affected by disasters.

4. Establishing a Disaster Reduction Science and Technology Support System, and Enhancing the Scientific and Technological Level of Disaster Reduction

Great importance is attached to the role of science and technology in disaster prevention and reduction and efforts have been made to continuously enhance the scientific and technological level of disaster prevention and reduction by such measures as formulating a special disaster prevention and reduction science and technology development plan, establishing a scientific and technological emergency response mechanism, and undertaking science and technology projects.

—Formulating the National Science and Technology Development Plan for Disaster Prevention and Reduction. In view of the existing problems

251

in natural disaster early warning and forecasting, emergency response, reconstruction, disaster reduction and relief, and information platform, efforts will be made to strengthen top-level design, make overall arrangements, fix weak links, and gradually establish and improve a national science and technology support system for disaster prevention and reduction.
—Strengthening the building of a science and technology emergency response mechanism. The state will set up a national science and technology emergency response mechanism for unexpected public incidents, define the working mechanisms and make arrangements for various links, including the building of a science and technology emergency response system, the enhancement of science and technology support capability, and the application and demonstration of emergency response technology.
—Initiating a batch of disaster prevention and reduction projects. A number of scientific and technological projects, including those concerning meteorology, seismology, geology, oceanography, water conservancy, and agriculture and forestry, has been listed in the National Science and Technology Program, National High-tech R&D Program (863 Program) and key projects supported by the National Natural Science Foundation of China. Financial support will be provided to the basic research programs on disaster prevention and reduction to thoroughly reveal the formation and changing patterns of various types of natural disasters, and the comprehensive risk prevention modes. Research will be carried out in the fields of Asia disastrous calamities overall risk evaluation technology and its applications, disastrous calamity emergency rescue information integration system and demonstrations in China, major natural disaster risk comprehensive rating and evaluation technology in China and the Wenchuan Fault Scientific Drilling Program (WFSD).
—Strengthening the building of scientific and technological research institutes. The National Disaster Reduction Center of the Ministry of Civil Affairs, the International Drought Risk Relief Center and the Satellite Disaster Reduction Application Center of the Ministry of Civil Affairs were established in 2003, 2007 and 2009, respectively. In 2006, the Academy of Disaster Reduction and Emergency Management was jointly established by the Ministry of Civil Affairs and the Ministry of Education.

5. Establishing a Personnel Training System and Improving the Quality of Disaster Relief Personnel

Education of disaster prevention and reduction personnel are incorporated into the national talent development program. A national education system and a training platform for disaster reduction have been gradually established.

—Incorporating disaster reduction into the national education system. Efforts have been made to strengthen personnel training and education, train multi-tiered disaster prevention and reduction personnel by making use of disaster reduction research and the academic advantages of institutions of higher learning; strengthen academic system building for disaster prevention and reduction. Within the framework of the current financial management system, financial support is given to universities and technical colleges that run courses in disaster prevention and reduction, or have established majors in disaster prevention and reduction management and technology, in order to improve the quality of relevant personnel.

—Incorporating disaster prevention and reduction into the training programs for officials. In line of the needs of personnel training, colleges of administration and schools for officials at various levels in China have opened training courses on disaster prevention and reduction, and emergency management. National emergency management personnel training bases are under construction, which will provide disaster prevention and reduction as well as emergency management training to senior civil servants, senior executives of enterprises and public institutions, and senior theoretical workers. A national earthquake emergency rescue training base has been established and put in operation.

—Offering special training courses on disaster emergency management for leading officials. Classes on disaster emergency management for leading officials at the provincial level and classes on unexpected incidents emergency management for officials at the provincial and ministerial level have been held. The participants were leading officials of provinces (autonomous regions, municipalities directly under the central government) and relevant departments of the State Council who were responsible for disaster prevention and reduction, and emergency management work. Since 2005, special training in disaster emergency management for civil servants has been actively carried out, which has effectively helped to improve the overall quality and ability of disaster emergency management personnel at various levels in preventing and dealing with natural disasters and other unexpected incidents. In 2005 and 2006, four special training courses on disaster emergency management for city- and prefecture-level officials were held. Since 2006, four training courses have been held on flood control and drought combat for city- and prefecture-level administrative officials in charge of flood control and drought combat.

—Carrying out emergency rescue training for enterprises and emergency rescue teams. Governments at all levels, in cooperation with related departments, have adopted the method of combining concentrated training with training on their own, to carry out training in disaster prevention and reduction, emergency management for heads of enterprises, management personnel and emergency rescue team members, in order to improve their capability in carrying out rescue efforts, self-protection and coordination in the event of disasters.

6. Carrying out Disaster Reduction Work and Improving Disaster Prevention and Reduction Capability in Communities

Disaster reduction capability building in communities is being carried out in an all-round way. The ability of primary-level communities to fend off disaster risks has been gradually improved.

—Encouraging communities to establish disaster reduction work mechanisms. Promoted by governments at all levels, organizations responsible for disaster reduction work in communities have been gradually established throughout the country, standard disaster reduction rules have been formulated, disaster relief volunteer teams have been organized, and measures for the protection of vulnerable groups including children, senior citizens, the sick and the disabled have been formulated. An effective disaster reduction work mechanism has thereby been established.

—Guiding communities in drawing up plans for emergency response to disasters and carrying out related exercises on a regular basis. According to the General State Emergency Response Plan for Unexpected Public Emergencies, the State Emergency Relief Plan for Natural Disasters and other emergency response plans of local governments, the primary-level governments shall guide communities in formulating communities' emergency relief plans, defining emergency response working procedures, management responsibilities and joint coordination mechanisms in view of the local environment, the pattern of disaster occurrence and characteristics of community residents. With the support and help of related departments of the government, communities frequently organize residents to carry out emergency response exercises in various forms.

—Improving public facilities and equipment for disaster reduction in communities. With government financial support and active public participation, communities may use parks, green land, public squares, sports venues, parking lots, school playgrounds and other open spaces to establish emergency shelters, put up clear emergency safety signs and instruction boards, establish public education places (community disaster reduction classrooms, community libraries, leisure rooms for senior citizens, etc.) and facilities (boards, bulletins, etc.) for disaster reduction, and install necessary fire control and safety facilities for disasters as well as lifesaving appliances, in order to improve public disaster-reduction facilities and equipment.

—Organizing communities to carry out disaster-reduction publicity and education. Proceeding from their respective cultural and regional characteristics, communities may carry out disaster-reduction education in various forms at regular intervals, frequently post disaster-reduction publicity materials in community education venues, and formulate disaster-reduction educational plans based on their specific conditions, so as to enhance the residents' awareness of disaster prevention and reduction and comprehensive disaster-reduction capability.

—Setting up disaster-reduction demonstration communities. In 2007, the state began to establish disaster-reduction demonstration communities. By 2008, 284 communities had been awarded the title of "National Comprehensive Disaster-reduction Demonstration Community" by the state.

Source: Chinese Government's Official Web Portal. "China's Actions for Disaster Prevention and Reduction." Available online. URL: http://www.gov.cn/english/official/2009-05/11/content_1310629.htm. Accessed March 8, 2010.

SAHEL AND HORN OF AFRICA

United States General Accounting Office, *Famine in Africa* (1986) (excerpt)

One of the most important aspects of responding to droughts around the world is the ability to move food quickly to the affected areas. This excerpt examines the problems associated with shipping emergency food during the 1984 crisis in parts of Africa. The excerpt also provides guidelines to improve the efficiency of these processes.

Background

A major problem with the 1984 emergency food program in Burkina Faso, Mali, Mauritania, Senegal, and Somalia was the late arrival of food in-country. Food aid generally arrived later than the times requested by the Agency for International Development's (AID) in-country missions and often after the start of the traditional rainy seasons when its distribution to the most needy normally would be difficult. Because emergency aid arrived after the most opportune time, some food deliveries to needy parts of Burkina

Faso and Mali were delayed or canceled, and significant amounts of food sent to Senegal and Somalia were still in storage at the time of our review in November and December 1984. Major distribution disruptions were avoided in the 1984 program largely because normally heavy rains did not occur during the traditional rainy season due to the continuing drought, and the roads remained accessible in most areas. Nonetheless, potentially significant distribution problems could have resulted because 68 percent of the U.S. emergency aid arrived during traditional rainy seasons.

One reason for the late arrival was the long time lapse between the missions' requests for the food and its arrival in-country. As noted in our report on the 1984 program,[1] it took from 4 to 9 months to deliver emergency food to the five countries after the missions submitted their requests to AID Washington. For the 24 shipments of food to the five countries, approval, procurement, shipment, and inland transport required an average of 6.6 months—program approval, 2 months; obtaining and loading the commodities, 3 months; and ocean and inland transport, 1.6 months.

For the most part, food needs cannot be determined reliably until after crops are harvested, which is usually around October or November for the five countries in our review. The period of time for determining food needs and getting the needed food in-country for those countries where the rainy season is a factor is roughly between December and the first of June, or about 6 months. Therefore, the long program approval and shipping time is a significant constraint to providing food during this period.

The shipping dates indicate that emergency food should have arrived in-country earlier in 1985 than it did in 1984, largely because requests were submitted earlier for the second year of the drought. Still, many of the shipments in our sample were not loaded at U.S. ports until May or June, which would have precluded them from arriving in-country at the most desirable time.

Requests for food aid are received by AID Washington usually from the AID missions and are reviewed by the Bureau for Food for Peace and Voluntary Assistance. Once the Bureau is satisfied with the requests, it submits them for approval to an interagency committee consisting of representatives from AID; the Departments of Agriculture, Commerce, State, and Treasury; and the Office of Management and Budget. After approval by the interagency committee, the requests are submitted to the Department of Agriculture's Kansas City Commodity Office (KCCO) to obtain the commodities and deliver them to a port for shipment to the requesting country. Commodities may be either purchased or obtained from U.S. stocks,

---

[1] *Famine in Attica: Improving Emergency Food Relief Programs* (GAO/NSIAD-86-25) Mar. 4, 1986.

depending on their type and availability. Until June 1985, Agriculture also arranged ocean transportation for commodities provided under government to government programs. Since then, a private contractor under a contract with AID has done so.

Objectives, Scope, and Methodology

We made a survey of emergency food aid approval, procurement, and shipment time frames, focusing primarily on the fiscal year 1985 program. Our objective was to determine how long each function took in comparison with 1984 and to explore actions to reduce the overall time. We performed our work at AID, Agriculture, and KCCO.

We determined the time between receipt of emergency food aid requests by AID Washington and approval by the interagency committee for essentially all requests approved during February and March 1985. We selected these 2 months because improvements were made in record-keeping beginning in February 1985, and thus needed information was more readily available. The approvals included emergency and nonemergency commodities.

We determined the time to obtain and ship commodities for essentially all emergency food aid requested during the first 7 months of fiscal year 1985 for the five countries in our review of the 1984 program. The procurement and waiting time was obtained for a more limited number of shipments under the 1984 program. We also determined the time for a sample of nonemergency or regular program commodities in 1985 for comparison with the emergency program. The number of 1984 emergency shipments and 1985 nonemergency shipments included may not be statistically significant, but we believe they provide indications of the under emergency programs are usually bagged either at the U.S. port of departure or at the port of arrival. Processed commodities include bulgur, flour, cornmeal, and fortified products and are packaged by the vendor.)

Rapid Procurement and Shipment of Large Quantities of Emergency Commodities Is Difficult

AID Handbook 9, containing guidelines and procedures for Public Law 480 programs, indicates that it normally takes 90 days from program approval to arrival of commodity at nearest recipient port. According to the Handbook, if the commodity is needed sooner, arrangements should be made to borrow the same or a similar commodity from a local source to be replaced with a Title II commodity upon delivery. In some circumstances, diversion of another Title II program commodity at a U.S. port or onboard ship may be considered. This involves identifying the correct amount of commodity needed, getting its release from the regular program cooperating sponsor, determining if the vessel involved can enter the desired port and if the commodity can be accommodated at the port, and paying a diversion charge. Since diversion is difficult, expensive, and not often possible, according to the Handbook, it should be considered only after all other sources have proven impracticable.

According to the coordinator of AID's Office of Food for Peace, the 90-day time frame for shipping commodities to destination points referred to in Handbook 9 is unrealistic with the current procedures. He stated that the actual time frame is 2 to 3 months for bulk commodities and 3 to 4 months for processed commodities.

According to KCCO officials, KCCO can acquire and move commodities to U.S. ports for delivery overseas every 60 days. If it takes 60 days to deliver commodities to the ports, then they would have to be shipped and unloaded at an African port in an average of 30 days to be within the 90 days from program approval to arrival at recipient port indicated in Handbook 9.

As shown in table I.1, for 1985 the average procurement time was 58 days, but the combined procurement and waiting at port time was 89 days. Loading and transporting time and unloading time at the recipient port in Africa would be additional. In 1984, this additional time was about 6 to 7 weeks, which if added to the 1985 procurement and waiting time indicates that it requires an average of about 5 months from program approval to delivery of commodities to recipient ports in Africa. (Adding about a month for program approval means that it takes an average of 6 months for program approval and commodity procurement and delivery to recipient ports in Africa. Some individual shipments took much longer than the average time.)

KCCO's acquisition and shipping capabilities are constrained by

- the time required to transport grain owned by the Commodity Credit Corporation from storage sites scattered around the country to U.S. ports,
- the limited number of suppliers of processed commodities and the time required for them to produce processed commodities after contract award, and
- the time required to obtain a ship to transport emergency commodities to Africa.

Orders for emergency and nonemergency food are, for the most part, handled in the same manner. Therefore, the time for procurement and waiting at ports for each was essentially the same for unprocessed commodities—67 days on average for 12 emergency shipments in 1985 versus 70 days for 10 non-emergency shipments. KCCO officials are opposed to establishing special procedures for processing emergency food requests because this (1) would require increased staffing costs, (2) would not necessarily decrease delivery time because vendors have limited production capacity, and (3) other means exist to respond rapidly to emergency requests. The officials stated that long-term emergencies can be handled under the present system if emergency food aid is promptly requested by AID because, after delivery of the initial order to an African port, subsequent deliveries can be made every 30 days from the pipeline established by KCCO's monthly procurement cycle. Also, according to KCCO officials, techniques such as diversions at sea and swapping and pre-positioning of commodities can be used to fill one-time deliveries that are needed in less time than 90 days.

Although an average time of 89 days was required to procure and ship emergency orders during the first 7 months of fiscal year 1985 to the 5 African countries, we did not note an instance where swapping or diversion at sea was used to shorten the delivery time. We saw no instances where this was done except for Ethiopia.

Bulk Commodities Can Be Moved Faster Than Processed Commodities

Bulk commodities can be moved to a U.S. port faster than processed commodities because they are obtained directly from Agriculture surpluses. However, according to KCCO officials, delays are experienced in moving these commodities to U.S. ports because the grain is stored in small elevators at various locations and must be moved by train to an intermediate point for consolidation into the requested amount before it can be transported to a U.S. port for shipment.

The movement of bulk grain from storage elevators to a U.S. port usually takes about 30 days if bagging is not involved. It is difficult to speed up this process because transport from the elevators is limited by the number of railroad cars that can be loaded in a day at the sites. If bagging of bulk commodities is requested, another 15 to 30 days would be required to have the commodities ready for shipment, depending on whether the bagging is done at the U.S. port or the recipient port. KCCO officials estimate that it takes about 2 weeks to award a contract to procure bags marked with the identification of the donor and 20 days for the manufacturer to produce these bags.

It takes longer to acquire and ship processed commodities because they must be procured through the competitive bidding process and processed by the vendor. It takes about 75 days from the time KCCO issues an invitation for bid to the time the commodities are delivered to a U.S. port. Invitations for bid are issued to suppliers on a monthly basis. According to KCCO officials, it takes about 15 days from the issue of an invitation for bid to contract award, about 30 days for the supplier to produce the commodities, and 30 days to deliver them to a U.S. port. Table I.1 shows that the average procurement time for the orders we examined was 69 days, somewhat less than the KCCO estimate.

According to KCCO officials, it would be difficult for suppliers to speed up the production and shipment of emergency commodities. They must have time to obtain the raw materials and to process and package the commodities before shipping them to port. They also require time to gear up production capacity to handle large volumes of processed commodities called for in the contracts, which are usually above normal production levels.

The limit on the number of suppliers of processed commodities and their production capacity can also affect KCCO's ability to fill orders, especially during peak ordering periods. KCCO officials said that there are only five suppliers of bulgur wheat, two suppliers of corn soya milk, and six suppliers of corn meal. The total production capacity of these suppliers for the three commodities is about 92,000 metric tons a month. We noted that for 10 of the 18 months between December 1983 and June 1985, KCCO was not able to fill AID Public Law 480 orders (emergency and nonemergency) for these commodities (especially bulgur and corn meal), because suppliers did not submit bids to cover the total quantities requested in the invitation for bid or the prices offered were unacceptable. For 7 of the 18 months, the shortages were in excess of 1,400 metric tons. KCCO could not procure 17,693 metric tons of bulgur in January 1985 and 16,105 metric tons in March 1985 because of the lack of bids for sufficient quantities.

Actions to Expedite Procurement and Shipment

AID's Office of Food for Peace is giving high priority to speeding up procurement and shipment and has been working with Agriculture and KCCO to accomplish this goal. As a result, KCCO has (1) pre-positioned bulk grain and bags at various locations for rapid movement to U.S. ports when emergency food orders are received and (2) procured processed commodities in advance of anticipated emergency orders on a test basis.

KCCO officials support the pre-positioning of bulk grain and bags to reduce the acquisition time for these items but not of processed commodities because they are too perishable and costly to store. They stated that although pre-positioning can save 30 days for bulk commodities to 45 days for processed commodities in the acquisition and shipping process, the following problems arise.

- Where to store the commodities, since the U.S. port from which they will be shipped is unknown. A central location to all ports or several locations near all ports (the Gulf, Great Lakes, East or West coasts) must be used.

- Whether a ship will be available to move the pre-positioned goods when they are requested; it normally takes about 30 days to acquire a ship.
- What identification markings to use on pre-positioned bags. Although AID and the various private voluntary organizations will accept unmarked bags, they want their own special markings on the bags that they distribute in-country.
- Predicting the type of commodities to pre-position that will be needed in the future in order to avoid useless storage; this will require KCCO to coordinate closely with AID on which commodity items to store.

As of June 1985, KCCO had purchased and pre-positioned a million bags to accompany bulk shipments of grain to be shipped during the African emergency. Also, KCCO pre-positioned about 420,000 metric tons of grain at various locations under a contract to provide for the pre-positioning and rapid movement of grain to U.S. ports during emergencies.

During July and August 1985, Agriculture purchased on AID's behalf 10,000 metric tons of sorghum grits and 3,765 tons of bulgur for use in Ethiopia in advance of an anticipated request from the private voluntary organizations operating in that country. AID estimated that these advance purchases would allow delivery of the food 2 to 3 months quicker than otherwise possible. AID and Agriculture plan to continue to use this process when they see emergencies developing in order to stay ahead of the "emergency curve."

In June 1985, AID sponsored a meeting in Minneapolis among representatives of government, industry, and private voluntary organizations involved in procurement to seek ways to expedite provision of food on an emergency basis. A representative from the shipping industry, however, was not present at the meeting and no followup meetings were planned. Ten proposals and suggestions presented at this meeting were being considered by AID for testing and implementation, including:

- Processing, transporting, loading, and shipping of large orders of food aid to be centralized and coordinated by one entity or organization contracted by the U.S. government.
- Using telex instead of the postal system to invite and transmit bids.
- Using packaging and bags without donor names or identification markings to reduce the time used to manufacture packaging for emergency commodities.

Representatives of private voluntary organizations said they had no objections to not having their organizations' names on bags if the bags have some identification numbers to meet auditing requirements. AID is reconsidering the markings to be used on Food for Peace bags.

The use of telex instead of the postal service for bid invitation and clarification is also being used on an ad hoc basis when deemed necessary for emergency orders. However, Agriculture officials said they do not believe the time necessary for commodity purchases would be significantly reduced by the use of the telex to invite and transmit bids for all emergency procurements.

AID and Agriculture discussed the possibility of testing a competitive vendor to move an order of grain in a continuous operation to a private voluntary organization in a designated country. This testing had been delayed because emergency food requests from the missions had slowed, but AID was planning to try the concept when feasible.

Source: United States General Accounting Office. *Famine in Africa: Improving U.S. Response Time for Emergency Relief,* GAO/NSIAD-86-56, April 1986, pp. 8–16.

USAID, "Responding to Drought through Non-food Interventions" (2008)

This document outlines foreign relief activities by the United States that do not relate to food distribution. These activities include providing drinking water, sanitation, health care, nutritional surveys, agricultural assistance, and other programs designed to promote food security.

Cyclical droughts, such as those currently affecting the Horn of Africa, negatively impact the humanitarian situation among poor populations. Insufficient access to water is the most commonly identified consequence of drought, resulting in shortages of water for human and animal consumption, agriculture, and pasture rejuvenation. As a drought worsens and these shortages of water become prolonged, alleviating the direct negative impacts of the situation on health, food security, and livelihoods of people living in affected areas becomes a primary objective of humanitarian interventions.

As conditions deteriorate, affected populations require interventions that address not only access to water but also related crises including severe and acute malnutrition in children under the age of five, increased mortality from common childhood diseases, poor outcomes for pregnant women and the elderly, threats to longer-term food security, and the depletion of household assets. The following discussion provides illustrative examples of the non-food humanitarian response options in drought situations. These activities should be undertaken in conjunction with appropriate food responses where access or availability of food commodities is problematic.

Water, Sanitation, and Hygiene

Water, sanitation, and hygiene (WASH) programs in drought situations are designed to improve the provision of water for human consumption, help prevent water-borne disease outbreaks such as typhoid and cholera, and boost protection of livelihoods through provision of water for livestock. Examples of these activities are listed below.

Water

Rehabilitation and construction of water sources. During normal times, many communities rely on river systems, water catchments, shallow wells, or scoop holes for their water needs. During droughts, these sources often become unusable or compromised. To address water needs in drought settings, USAID/OFDA supports the rehabilitation of existing water sources or the construction of new water sources in affected communities. These activities may include rehabilitating hand pumps and boreholes, providing spare parts and fuel, digging or drilling new wells and boreholes, digging or rehabilitating traditional water systems, and chlorinating water sources. Any new construction efforts are approached with caution, as consecutive years of drought have often led to decreased water availability that can take several consecutive years of normal rainfall to replenish. Removal of ground water through bore holes and other means can aggravate ongoing environmental degradation and desertification.

Water tankering. Water tankering is extremely expensive, unsustainable, and transfers no capacity to the community. As such, USAID/OFDA supports water tankering only in limited cases, such as to internally displaced person (IDP) camps without adequate access to water through other means.

Community water management activities. USAID/OFDA partners engage communities in the management of new and rehabilitated water systems over the medium and long term. Most often, partners assist communities in establishing volunteer water committees that undertake maintenance of the water system and often address conflict resolution. However, given the limitations of communal management, USAID/OFDA encourages innovative management schemes such as privatization through women's and youth groups.

Sanitation

During droughts, USAID/OFDA promotes the construction and use of family latrines. In limited camp settings, communal latrines have been promoted.

Hygiene Promotion

With any water or sanitation intervention, USAID/OFDA supports the provision of hygiene messages that promote proper transport and storage of water, hand washing, and appropriate sanitation.

Mitigation

During drought conditions, USAID/OFDA supports programming that attempts to mitigate the effects of future droughts while responding to the current crisis.

Nutrition

As drought conditions become prolonged, food security deteriorates among affected populations. Reduced agricultural yield at the household level or diminished household income results in a decreased food basket, compromising the nutritional intake of all family members and particularly affecting children. When the general food basket of a family becomes insufficient in quantity or nutritional diversity, targeted nutrition interventions are required.

Technical assistance in nutrition. Nutritional surveys are the best way to determine the food security of a population. USAID/OFDA provides technical assistance and may financially support nutrition assessment surveys.

Feeding programs. When a population is not getting enough food, either food is not available to purchase or individuals do not have the money to purchase available food. When food is not available to purchase, feeding programs are crucial to address malnutrition resulting from ongoing drought situations. USAID/OFDA supports selective feeding programs such as supplementary and therapeutic feeding. Supplementary feeding programs provide an extra ration of food, while therapeutic feeding provides special high-energy food for severely malnourished children. These programs are targeted interventions that complement USA1D's Office of Food for Peace general ration distributions and supplementary feeding programs. In many drought situations, USAID/OFDA prefers to support community-managed acute malnutrition (CMAM) programs that provide life-saving nutritional support and related medical care to children in their homes, as opposed to hospital or clinic settings. This approach enables USAID/OFDA programs to reach more children and at an earlier stage of malnutrition, allows mothers to remain at home to care for other children, and decreases the risk of transmission of communicable diseases that is likely to occur in clinics or hospitals. However, where appropriate, USAID/OFDA does support traditional therapeutic feeding centers, which provide nutritional and health services in a hospital or clinic environment. In support of all of these activities, USAID/OFDA also supports the procurement of therapeutic and supplementary foods such as ready-to-use therapeutic food (RUTF) and high energy milk.

Training and capacity building. In order to ensure that communities are prepared to deal with future nutrition crises, USAID/OFDA emphasizes the need for communities to be able to recognize and respond to increased levels of malnutrition among their children. USAID/OFDA-funded feeding programs often include training for local health staff on the treatment of severe malnutrition, support for federal health officials and relief staff related to CMAM implementation, development or enhancement of surveillance and sentinel site early warning systems, and nutrition education at the health center and community level.

Most of these interventions focus on the treatment of the malnourished, with some programs aimed at prevention through health education. In addition, to improve the diversity of the diet at the household level and to provide a source of income, USAID/OFDA has funded specific agriculture and livelihood programs, including demonstration gardens and seeds and tools for vegetable gardens, for the beneficiaries of nutrition programs.

Health

As children become increasingly malnourished, they are at increased risk of morbidity and mortality associated with common communicable diseases. Children under the age of five, pregnant women, the elderly, and individuals in ongoing nutrition programs, including those with HIV/AIDS, are at particular risk. The lack of water can further aggravate poor hygiene practices, potentially resulting in diseases of epidemic significance such as cholera and typhoid. As a result, health interventions are critical during droughts. In many cases, drought-affected populations suffer from lack of access to basic health services, resulting in low immunization coverage and lack of basic resources such as trained staff and medical supplies. In such instances, excess morbidity and mortality may result from common diseases such as measles, malaria, pneumonia, and diarrhea, as well as maternal and neonatal complications. In the event that drought conditions lead to population movements, additional needs may arise stemming from overcrowding, lack of appropriate shelter, broken down disease control efforts, and reduced access to health, food, water, sanitation, and hygiene services.

Technical Assistance. USAID/OFDA has funded and works closely with the U.N. World Health Organization (WHO) Disease Control in Emergencies (DCE) branch to provide technical assistance for communicable disease control to partners, including governments in the Horn of Africa. WHO has produced a Communicable Diseases Epidemiologic Profile for the Horn of Africa. The purpose of this profile is to provide public health professionals working in the Horn of Africa with up-to-date information on the major communicable disease threats faced by the static and mobile populations within sub-regions.

Surveillance and early warning for communicable disease control. USAID/OFDA supports national disease early warning and surveillance systems for the prevention and control of communicable diseases. Outbreaks of epidemics can significantly increase mortality in a population suffering from malnutrition and lack of access to basic services. To be effective, these systems should link closely with water, sanitation, and hygiene programs and epidemic response activities to control outbreaks.

Prevention and management of common causes of morbidity and mortality. In drought-affected populations, USAID/OFDA supports the local Ministry of Health and additional partners to increase access of the population to primary health care in order to prevent excess deaths. USAID/OFDA supports the prevention of common communicable diseases such as measles through the provision of vaccines, supplies, and Vitamin A supplements; institution of vector control measures such as the distribution of insecticide-treated nets for malaria prevention; and community education and linkage with water, sanitation, and hygiene programs. USAID/OFDA also supports treatment programs for common communicable diseases and maternal and neonatal complications through training community health workers, traditional birth attendants, midwives, and health educators at the clinic and community level and providing essential medicines to supplement Ministry of Health supplies in affected areas. USAID/OFDA encourages a community-based approach to promote access to health care to the most affected populations.

Agriculture

Subsistence farmers in developing countries are strongly dependent on rainfall to support agricultural activity. Unfortunately, crop yields are not solely determined by the total amount of rainfall received, but also by the timing and distribution of rains. For subsistence farmers, crop production—

rather than an income-generation activity—is the primary means of feeding their families throughout the year, so drought conditions of any magnitude can impact household food security. Several consecutive years of drought can lead families to sell assets such as tools and animals in order to purchase basic necessities. Because crop production and long-term food security are so strongly linked, USAID/OFDA supports agricultural interventions for the most vulnerable households.

Seed distributions, seed fairs, and vouchers. Once USAID/OFDA has determined that vulnerable farmers lack sufficient seeds to plant their fields and are unable to obtain seeds though the market, family connections, or other channels, USAID/OFDA supports seed distributions or seed fair and voucher programs. If seeds are not available on the market, targeted distributions may be used to provide inputs. In many cases, seeds are available on local markets, but the most vulnerable farmers do not have the cash required to access the seeds. For these situations, the seed fair and voucher system is used. This system provides a voucher worth a set amount of money to each beneficiary, who trades it on a specific date for its value in seeds. The farmer is free to choose among different types of crops and seed varieties based on individual household requirements. USAID/OFDA implementing partners assure seed quality.

Agricultural technical assistance. In addition to providing seeds to vulnerable farmers, USAID/OFDA examines the types of crops planted in the affected region, seed quality, and whether any improvements may be made to the varieties offered. For example, if rains are too short for long-cycle sorghum to be planted and harvested prior to the end of the rains, USAID/OFDA might encourage short-cycle varieties to be planted, since such varieties grow to maturity faster, though their yield will be lower. Multiplication of improved drought-resistant varieties is a critical component of mitigation against future droughts.

Livestock

A significant number of vulnerable households in drought-affected areas of Africa are pastoralists or agropastoralists, with a majority of their livelihoods and food security dependent on livestock. As conditions deteriorate in pastoral lands, livestock are unable to find sufficient quantities of fodder and begin to weaken and die from malnutrition or disease. Supplemental grain and fodder decreases in local markets. As a drought intensifies, families begin selling livestock as a coping mechanism. As a result, livestock prices fall and grain prices rise, leaving pastoralists unable to purchase what they need from the sale of their animals. Much like providing seed for subsistence farmers, protecting livestock assets among pastoralist and agro-pastoralist populations is crucial to maintaining medium- and long-term food security. Humanitarian agencies encounter numerous difficulties in responding to drought-affected pastoralists, since populations are nomadic, infrastructure is poor, and security is often problematic.

The carrying capacity of the land determines the number of animals that a region can support. Unpalatable plant species are encroaching in many pastoral areas, competing with the grasses and shrubs that animals eat and reducing the amount of available food. During drought years, available pasture decreases even further. Carrying capacity is important to consider in drought response, since the success of programs like destocking and restocking are based on this principle.

Destocking. Emergency destocking programs provide for the intentional removal of sick or undernourished animals from a region before they die. The programs provide a fair price to farmers for the livestock, based on animal gender and age but not on health. In some cases, the animals are then slaughtered, and the fresh or dried meat is provided to feeding centers in the region. This program can be used to supplement food aid, increase the availability of high protein foods, and provide some supplementary income to vulnerable families. Destocking is most useful early in a crisis since the removal of some animals may reduce the number of animals competing for resources, allowing for the survival of a core group of reproductive animals.

Animal health programs. Improving animal health care may be most important when animals face serious drought conditions. In some situations, simple health interventions such as de-worming at the start of prolonged drought can significantly increase animals' chances of survival. While herd sizes may dramatically decrease during a drought, significant livestock losses can also follow the first rains after drought, when animals already weakened by malnutrition succumb to parasites, dysentery, and disease. Vaccination programs and primary animal health care may prevent some of these losses associated with the onset of rains.

Restocking. Animal restocking is not generally recommended. Providing inputs of animals to a region where the carrying capacity is low will only lead to increased competition for resources and the likely death of the animals provided, which is a waste of money and effort. Except in some isolated cases, it is better to let nature take care of the restocking process.

Provision of water for livestock. The installation of new boreholes can be disastrous for the surrounding environment and should only be pursued in limited cases and with great caution. Livestock owners try to keep animals close to water sources, and the area 60 to 80 kilometers around a new water source often becomes severely degraded. For this reason, water resources should be kept outside of rangelands. In some grazing areas, water pans or small dams might be useful in extending dry-land grazing by four to eight weeks. These projects can be linked to community development and resource management activities. However, smaller water pans are not always naturally replenished in areas with erratic rainfall.

Economic Asset Support

In drought conditions, USAID/OFDA uses a livelihoods-based approach as a means to improve food security. Vulnerable families undertake a variety of subsistence, economic income-generating, and coping mechanism activities in order to ensure household food security. The combination of these activities and the use of existing social, human, and physical assets is known as the collective livelihood of the household.

USAID/OFDA recognizes that a variety of factors contribute to the inability of a family or community to successfully combine these activities to achieve food security. Family members may spend increased time searching for water in lieu of participating in agriculture or herding activities. Households may lose labor assets as family members become too sick or malnourished to work. Individuals may sell productive assets such as tools, livestock, seed, or kitchen utensils as a result of previous or current drought. Diminished environmental assets, such as decreased grazing pasture, also impact household livelihoods. USAID/OFDA addresses the range of factors that affect livelihoods through multi-sectoral interventions, combining asset protection activities with health, nutrition, and water and sanitation programs. Specialized economic asset support interventions include livelihood fairs and alternative income activities.

Livelihood fairs. Much like the seed fair programs, livelihood fairs provide a voucher worth a set amount of money to each beneficiary, who may trade it on a specific date for its value in seeds, tools, livestock, fishing equipment, insecticide-treated nets, or diversified food commodities. This intervention allows the beneficiary to choose his or her own inputs, and fosters diversification of livelihoods. For example, one farmer may choose seeds, a hoe, a chicken, and a bundle of enset, while another may choose a goat and a machete. In some cases, USAID/OFDA partners have reported beneficiaries combining their vouchers to acquire community assets, such as an ox for plowing fields.

Alternative income activities. In drought-prone areas, USAID/OFDA recognizes that subsistence farming is becoming a less viable livelihood strategy over time. This is especially true in the Horn of Africa, where desertification, population increases, and decreasing plot sizes are leaving most subsistence farmers reliant on food assistance for some portion of their annual food consumption. In response, USAID/OFDA supports alternative income-generating activities as one component of an overall livelihood response. Poultry production, bee-keeping, enset production, and raising of small animals are examples of USAID/OFDA pilot alternative income activities implemented in drought situations. Communities identify the most vulnerable households, who in turn receive basic inputs and training in the technical, financial, and management aspects of the specific activity.

Source: USAID. "Responding to Drought through Nonfood Interventions." Bureau for Democracy, Conflict, and Humanitarian Assistance. Office of U.S. Foreign Disaster Assistance, May 2008.

USAID and Famine Early Warning Systems Network, "Food Assistance Outlook Brief" (2010)

USAID and other relief organizations continually track international food security conditions to estimate where relief efforts will most likely be needed. The Food Assistance Outlook Brief *is published by USAID and the Famine Early Warning Systems Network (FEWS-NET). It analyzes the projected food needs of various countries around the world for six months in the future. The following edition of the bulletin outlines the projected emergency food assistance needs for June 2010 in foreign countries where FEWS-NET has a staff presence.*

Projected Food Assistance Needs For July 2010

This section summarizes FEWS NET's most forward-looking analysis of projected external emergency food assistance needs, six months from now, in countries where FEWS NET has a staff presence. Those needs are compared to typical needs at this time of year during the last five years and categorized as Above-average, Average, and Below-average/No need. For more detail on these projections, please visit www.fews.net.

Above-Average Assistance Needs Projected in July 2010

BURKINA FASO: *Early needs following poor season in north and northeast*

Projected start of hunger season (Sahelian zone): *March*

Reduced production, poor pasture growth, and high millet prices will result in reduced household purchasing power, particularly among the poor in areas of the north and northeast. The hunger season will begin in March, as opposed to June in a typical year in these areas.

CHAD: *Elevated malnutrition/ mortality in western Sahelian zone*   **Projected start of pastoral hunger season:** *March*

Given rainfall deficits and dry spells, two million people are expected to require humanitarian assistance between now and November 2010. Areas of the western Sahel (e.g. Kanem, Batha, and Lac) are most affected. Elevated levels of malnutrition and atypical migration have already been reported and the hunger season will begin two to three months early.

DJIBOUTI: *Poor rains and food aid cuts likely to cause extreme food insecurity*   **Projected start of Central/ NW hunger season:** *April*

Poor consecutive rain seasons, followed by poor spatial distribution of the current Heys/Dada rains (Oct–Feb), declining livestock-to-cereal terms of trade, reduced remittances and planned reductions in food aid distribution are expected to drive increasing food insecurity. The country's main hunger season, which usually begins in June, is expected to begin in April.

ETHIOPIA: *High levels of need expected to follow poor* **meher** *harvest*   **Projected start of** *meher* **hunger season:** *March/April*

Significant need is expected to begin in February in SNNPR and in March in eastern *meher* cropping areas, two months earlier than normal. *Deyr* rains have been poor in eastern Somali region and southern Oromia. *Belg* rains have begun early in some areas.

GUATEMALA: *Deteriorating conditions, less response in west*   **Projected start of hunger season (W. Highlands):** *February*

The *postrera* harvest (Oct/Nov) has been below average. Current food insecurity in the dry corridor is expected to persist and conditions in the west are expected to deteriorate, with food stocks depleting, and households turning to purchases, as early as February.

HAITI: *Massive earthquake has far-reaching food security implications*   **Projected start of hunger season:** *Not relevant*

The January 12 earthquake resulted in heavy mortality and a massive disruption of livelihoods in Port au Prince and nearby cities. In addition to large acute needs, medium and longer term food security will be affected by potential increases in food prices and reduced labor opportunities.

KENYA: *Good short-rains harvest expected*

Projected start of main pastoral hunger season: *August*

Short-rains harvests have been normal in many areas. Assistance needs will likely decline by June but remain above the recent five year average.

MALI: *Above-average need expected in northern pastoral areas*

Projected start of hunger season (Gao): *March/April*

Cereal production has been average-to-good and pastoral conditions in Timbuktu are satisfactory. However, in some areas of eastern Gao, agricultural production and pasture regeneration have been poor. The hunger season will start two to three months early in these areas.

MOZAMBIQUE: *Extended dry spell likely to impact main harvest*

Projected start of hunger season: *August/September*

An extended dry spell in central/southern Mozambique during December/January has seriously affected the crops of 785,000 households. If rainfall does not improve, assistance needs could begin as early as July, with the hunger season beginning in August/September, 1–2 months early.

NIGER: *Large-scale humanitarian assistance needs expected*

Projected start of hunger season (cropping areas): *March*

Due to poorly distributed rains and an early end of season, food insecurity during 2010 will be quite severe, particularly in the agropastoral belt. Atypical food-insecurity-related migration has begun among eastern farming households whose production has been particularly limited. Rains in pastoral areas have also been poor for the second consecutive year. Hunger seasons will begin two to three months earlier than normal.

SOMALIA: *Needs will not decline as much as anticipated*

Projected start of agricultural hunger season: *April/May*

Though the recent post-*deyr* assessment suggested a drop in assistance needs, a deepening drought in the central and northern regions, the suspension of the food assistance, and unseasonable increases in staple food prices, mean that this decline will be less substantial than expected.

SOUTHERN SUDAN: *Poor rainfall, expanding conflict*

Projected start of hunger season: *March*

Needs are likely to rapidly increase in January–February, due to poor rainfall, widespread yield reductions and ongoing conflict. The main hunger season is expected to begin in March–April rather than in May–June.

Below-Average Assistance Needs or No Assistance Needs Projected in
July 2010
AFGHANISTAN, MALAWI (none), MAURITANIA, NIGERIA (none),
TANZANIA (none), UGANDA, ZAMBIA (none), ZIMBABWE

Source: USAID Famine Early Warning Systems Network. "Food Assistance Outlook Brief." January 11, 2010. Available online. URL: http://www.fews.net/docs/Publications/FAOB_020810_ext.pdf. Accessed March 8, 2010.

Famine in Ethiopia: FEWS NET Ethiopia, "Ethiopia Food Security Outlook Update, May 2010" (2010) (excerpt)

Although a quarter-century has passed since the 1984–85 famine in Ethiopia, it is still mired in a deadly cycle of drought and famine. Mapping famine vulnerability in Ethiopia is a practice that can save lives. The Famine Early Warning Systems Network (FEWS), funded by USAID, collaborates with other U.S. organizations and international partners to provide famine vulnerability data for Ethiopia—data that relief organizations can use to plan food aid or other relief programs. Each month, FEWS publishes a food security update detailing the conditions in areas susceptible to famine. The May 2010 update, excerpted below, details the number of people in Ethiopia needing emergency food services and outlines the projected status of food security in specific regions of Ethiopia for the months to follow. The report links the status of food security to natural and man-made causes including rain and armed conflict.

Improvements expected in SNNP/Somali regions following good *belg* rains.

Key Messages

- An estimated 5.23 million people will continue to require emergency food assistance up to December 2010 according to the Joint Government and Humanitarian Partners' Humanitarian Requirement Document released on 2 February 2010.

- Food security is expected to improve in large parts of Southern Nations Nationalities and Peoples' Region with the green harvests in May. In the southern pastoral and agropastoral parts of the country, improvements are also expected following the *belg* harvest in June/July and improved pasture and water availability.

- Despite planned assistance and the improved availability of pasture and water in the eastern marginal *meher* producing parts of the country, food security is expected to deteriorate further between June and September as stocks from the previous *meher* harvest decline and prices of staples rise.

273

Updated food security outlook through September 2010
An estimated 5.23 million people will continue to require emergency food assistance up to December 2010 according to the Joint Government and Humanitarian Partners' Humanitarian Requirement Document released on 2 February 2010. Assistance for this population is expected, and will have a significant impact on food insecurity, though high and extreme food insecurity still likely during the Outlook period. . . . A significant proportion of the people who require emergency food assistance are those in SNNPR and in the southern pastoral and agropastoral parts of the country. Food security is expected to improve in these areas with the start of the *belg* harvest in June/July in most of SNNPR and in the agropastoral zones of southern Somali Region, and the Bale and Borena lowlands of Oromia Region. In eastern marginal *meher* cropping areas (Much of Tigray, eastern Amhara, eastern Oromia) however, where the 2010 harvest is not expected until October, food security is likely to deteriorate between June and September. In addition people being supported through the emergency relief program, 7.8 million chronically food insecure people are currently receiving cash and food under the productive safety net program. Additionally, a total of 171,772 refugees in five regions of the country (Gambella 21,180; Benishangul 3,738; Somali 99,124; Tigray 44,885; and Oromia 2,845) remain in need of food and non-food assistance between May and September.

. . . .

Normal to above normal *Belg* (February to May) rains have been received in most **eastern *meher* producing parts of the country,** including Wag Hamra zone, East Belesa and Beyeda in North Gondar zone, Abergelle and Tselemt woredas (those along the Tekeze river) in Western and Central Tigray zones, and some woredas in the lowlands of East and West Hararghe zones in Oromia region. In these areas, the performance of the 2009 *kiremt* (June to September) rains was very poor and has resulted in extreme levels of food insecurity. The recent rains have brought much needed relief in most areas by improving water and pasture availability and have also been beneficial for land preparation and planting of long cycle *meher* crops. Despite these improvements, the period June to September is typically the lean season in these areas as the stocks from the *meher* season decline and staple prices rise. This year, the lean season has started sooner than average and will be particularly difficult due to poor performance of the *meher* crops in 2009. Affected people in these areas remain highly and extremely food insecure and will continue to heavily rely on assistance from both the Productive Safety Net and the emergency relief programs during the June to September period. In addition, it is possible that the number of safety net beneficiaries will be expanded through the risk finance mechanism later in the year. There are also areas in the eastern *meher* producing parts of the country that have not received sufficient rain and continue to report water shortages such as Samre Seharti, Tahitay Adiabo, Lailay Adiabo, Tanqua Abergele and Erob in Tigray Region and East Belesa in Amhara region.

There are some woredas in Southern Tigray, North Wello, South Wello, North Shewa of Amhara region and East and West Hararghe zones of Oromiya region that produce *belg* crops. The dependence on *belg* production in these areas is high, especially in some woredas of North and South Wello. In the northern *belg* crop producing parts of the country, farmers planted using unseasonable rains in January. Though these initial plantings failed due to the late start of the *belg* rains, February replanting went well and a good harvest is possible if there is no crop damage from excessive rainfall. Thus, improvement in food security is anticipated in these *belg* producing parts of the country following the harvest in June/July.

. . . .

The southern parts of Afar region and the northern two zones (Jijiga and Shinile) of Somali Region are expected to continue to be moderately food insecure following the normal to above-normal *sugum/belg* (April to June) rains and subsequent improvements in livestock body conditions, milk production and livestock prices. Ongoing food distribution both from the PSNP and the emergency program has also contributed to the improvement of food security and the stabilization of staple food prices. The **northern parts of Afar region** however, will continue to be highly food insecure because of the repeated poor performance of seasonal rains and the below average performance of the current *sugum* (March to May) rains. There were some rains during the second half of April that improved availability of water, especially in woredas such as Afdera, Kori, Bidu, Teru and Erebti that experience chronic water shortages. But the rains have been below average and are insufficient for regeneration of browse and pasture. . . . Livestock have not returned to their areas of origin and milk availability continues to be low. Improvement in food security in these parts of the region is expected only after the *karma* rains (July–September), assuming that the rains will perform normally.

In Southern Somali Region, lowlands of Bale, Guji and Borena of Oromia Region and South Omo Zone of SNNPR, *Gu/Ganna* (April to

June) rains have been normal to above normal thus far. Pasture and water availability have significantly improved, leading to good livestock body conditions, and normal conception and terms of trade. Milk availability has improved but remains below normal due to low births as a result poor performance of consecutive rainy seasons (Gu/Ganna and Dayr/Hagaya) in 2009, particularly for larger ruminants (camels and cattle). Prices of cereals continue to be stable with ongoing food aid distribution, though this is less true in areas that are affected by security related market access problems (i.e., Korahe, Dagehabur, Warder, Fik and Gode zones). Following the repeated poor performance of seasonal rains, most southern pastoral and agropastoral areas continue to be highly dependent on relief food assistance. Finally, although recent flooding along the Shabelle river has caused temporary displacement and destroyed farmlands, people along the river practice flood recession crop production which is expected to be harvested in August/September.

In the Southern Nations Nationalities and Peoples' Region, the performance of the current *belg* rains has been favorable; especially in the *belg* dependent parts of Gamo Gofa, Wolayita, Kembata, Sidama and Hadiya zones and southern Special Woredas of Konso, Derashe, Amaro and Burji. . . . Furthermore, improved availability sweet potatoes and other root crops, regeneration of pasture, replenishment of water sources, and ongoing food and cash distributions under the PSNP and relief programs have significantly improved food insecurity. Green maize and haricot bean harvests from May onwards are expected to further improve the food security situation in these areas. Household income from agricultural labor and sale of livestock and livestock products will remain normal following the expected normal *belg* harvest. Despite the changes in the food security situation, excessive rains across most of the zones in the region have caused localized flooding in many areas. Hailstorms and landslides have also been reported in some areas and there has been a widespread outbreak of armyworm. The extent of crop and pasture damage from these shocks has yet to be assessed. However, these hazards are likely to adversely affect food security by damaging crops (including coffee which is at its flowering stage), livestock, and rural infrastructure. Despite these hazards, most of *belg* producing parts of SNNPR is expected to continue to be moderately food insecure. However, the chronically food insecure *meher* cropping highlands in SNNPR, including Guraghe, Siltie, Sidama, and Gamo Gofa zones and Alaba Special Woreda, will continue to face high levels of food insecurity through September due to early depletion of stocks from the reduced 2009 *meher* harvest, limited availability of *enset*, as a result of repeated shocks in the last three years, and increased cereal prices during the lean season.

Most of **Gambella region** continues to be highly food insecure following the poor performance of the 2009 main season *(meher)* crops, lack of flood recession crop production, reduced availability of fish due to the low level of the rivers, and clan conflicts within the region, as well as across the border with Sudan, that displaced 11,500 households in Akobo, Lare Itang and Jor woredas. Most of Gambella region relies on *meher* season long cycle crops that are typically planted in April and harvested in August, with green harvest beginning in July. This year, the rains in April were late by about three weeks. Although the rains started late, performance since then has been good and land preparation and planting are ongoing, though there is an expanding army worm outbreak in Gog, Akobo, and Gambella Zuria woredas that is damaging maize seedlings. Assuming that the rains continue to perform normally, and the recent army worm outbreak is controlled, food security is expected to improve beginning July with the start of the green maize harvest. Improvements in food security are not expected for those who are displaced due to the conflicts since they don't have access to their land.

Source: Famine Early Warning Systems Network. "Ethiopia Food Security Outlook Update, May 2010." Available online. URL: http://www.fews.net/docs/Publications/ethiopia_FSOU_05_2010_final.pdf. Accessed July 7, 2010.

COLOMBIA

Eruption of Nevado del Ruiz, 1985: USGS, "Mobile Response Team Saves Lives in Volcano Crises" (2005) (excerpt)

In 1985, the eruption and lahars of Nevado del Ruiz in Colombia killed more than 23,000 people in Armero and other villages. In the months that followed this disaster, the U.S. Agency for International Development's Office of Foreign Disaster Assistance (OFDA) asked the U.S. Geological Survey (USGS) to collaborate on the creation of a program to mitigate fatalities and economic losses resulting from volcano crises. Together, these two agencies established the Volcano Disaster Assistance Program (VDAP) in 1986. The history, purpose, and activities of VDAP are described in the following bulletin published by the USGS in 2005, including a short case study of the program's actions

277

during its first major mission in response to the anticipated eruption of Mount Pinatubo in the Philippines in 1991.

The world's only volcano crisis response team, organized and operated by the USGS, can be quickly mobilized to assess and monitor hazards at volcanoes threatening to erupt. Since 1986, the team has responded to more than a dozen volcano crises as part of the Volcano Disaster Assistance Program (VDAP), a cooperative effort with the Office of Foreign Disaster Assistance of the U.S. Agency for International Development. The work of USGS scientists with VDAP has helped save countless lives, and the valuable lessons learned are being used to reduce risks from volcano hazards in the United States.

On April 2, 1991, after being dormant for 500 years, Mount Pinatubo volcano in the Philippines awoke with a series of steam explosions and earthquakes. Ten weeks later, on the morning of June 15th, Pinatubo exploded in a climactic volcanic eruption. Fiery avalanches of hot ash (pyroclastic flows) roared down the flanks of the volcano, and giant mudflows of ash (lahars) swept more than 30 miles down valleys. Cities and towns near Pinatubo were devastated by falling ash. Ash fall also inundated the two largest U.S. military bases in the Philippines. On Clark Air Force Base, which was home to more than 15,000 American servicemen and dependents, many buildings collapsed under the weight of rain-saturated ash. Facilities at the U.S. Naval Station at Subic Bay, 25 miles from Pinatubo, were also severely damaged.

. . . .

Despite the enormity of the devastation wrought by this explosive eruption, quick work by earth scientists helped keep the death toll low. Shortly after Mount Pinatubo's reawakening, U.S. Geological Survey (USGS) scientists with the Volcano Disaster Assistance Program's crisis response team arrived at Clark Air Force Base. Once on the scene, they joined scientists from the Philippine Institute of Volcanology and Seismology, who had begun monitoring the volcano. This joint team worked quickly to evaluate the threat from Pinatubo, installing instruments to detect earthquakes and swelling on the mountain and mapping volcanic deposits in order to understand the volcano's eruptive history. Their evaluation enabled them to alert people in areas at risk and also provide critical advice to the Philippine Government and to U.S. military commanders at Clark Air Force Base and Subic Bay.

When data from monitoring instruments indicated that a large eruption was imminent, the scientists issued timely warnings that resulted in safe evacuation of more than 75,000 people before the volcano's climactic June

15 eruption. In addition to the thousands of lives saved, hundreds of millions of dollars in military aircraft and hardware were moved out of harm's way.

Volcanoes produce a wide variety of hazards that can kill people and destroy property. Large explosive eruptions can endanger people and property hundreds of miles away, affect global climate, and cause widespread economic losses. For example, the drifting ash cloud from the June 15, 1991, eruption of Mount Pinatubo damaged more than 20 passenger jetliners (including those of American air carriers), most of which were flying more than 600 miles from the volcano.

Since 1980, volcanic activity worldwide has killed more than 29,000 people, forced more than 1,000,000 to flee from their homes, and caused billions of dollars in economic losses. On average, about 10 eruptions a year cause significant damage and casualties, and eruptions powerful enough to cause major disasters happen several times a decade.

There are more than 1,500 potentially active volcanoes in the world, about 550 of which have erupted in historical times. Moreover, most of the truly devastating and strongest explosive eruptions since 1800 have occurred at volcanoes with no historical record of previous eruptions.

Despite the threat posed by volcanoes, only about 20 of the 550 historically active volcanoes in the world are monitored adequately, and fewer than one-third are monitored at all. As growing human populations encroach further into areas of greater volcano hazard, the potential for deadly disasters increases.

In the early 1980's, scientists at the USGS Cascades Volcano Observatory (CVO) in Vancouver, Washington, recognized that it was not economically feasible to fully monitor all potentially active volcanoes in the Pacific Northwest. To meet this problem, the USGS developed a suite of portable monitoring instruments that could be quickly deployed to a reawakening volcano. These instruments are used to detect and analyze earthquakes, ground deformation, mudflows, and volcanic gas emissions. The data from these instruments are supplemented by additional information from global positioning (GPS) satellites, weather radar, and other equipment.

In 1985, the eruption of Nevado del Ruiz volcano, Colombia, triggered giant, fast-moving lahars that killed more than 23,000 people. Following this tragedy, the U.S. Agency for International Development's Office of Foreign Disaster Assistance (OFDA) asked the USGS to help create a program to reduce fatalities and economic losses in countries experiencing a volcano crisis. Toward this goal the two agencies jointly established the Volcano Disaster Assistance Program (VDAP).

VDAP consists of a small core group of scientists at CVO, a larger group of other contributing USGS scientists, and portable volcano-monitoring equipment ready for rapid deployment. The VDAP crisis-response team is mobilized and sent overseas only when the U.S. State Department receives an official request from a country with a restless volcano. Once on site, the VDAP team works with local scientists and technicians to help them provide timely information and analysis to emergency managers and public officials. VDAP also conducts training exercises and workshops in volcano-hazards response with foreign scientists and emergency-management officials.

. . . .

Since the 1991 eruption of Mount Pinatubo, VDAP has responded to volcano crises in Central and South America, the Caribbean, Africa, Asia, and the South Pacific. Most recently, VDAP teams have been providing assistance to Mexico and to the Caribbean island of Montserrat.

. . . .

When majestic, 17,887-foot-high Popocatépetl Volcano, near Mexico City, began to erupt intermittently in December 1994, the Mexican Government requested the aid of a scientific team from the Volcano Disaster Assistance Program (VDAP). "El Popo" has erupted violently in the past, and a major explosive eruption today would put millions of people at risk. The VDAP team worked with local scientists . . . from the Centro Nacional de Prevencion de Desastres and the Universidad Nacional Autonoma de Mexico to set up a monitoring network on the volcano. Popocatépetl is now being intensively monitored. Should the volcano's current weak activity escalate, the effective forecasts and warnings made possible in part with VDAP's assistance may save countless lives and help lessen a massive, wide-ranging refugee crisis.

Although highly visible, the activities of VDAP are only one aspect of the USGS Volcano Hazards Program. The new techniques developed and the experience gained by scientists with VDAP in volcano crises overseas prepare the USGS to better protect people's lives and property from future eruptions of volcanoes in the United States.

Source: USGS. "Mobile Response Team Saves Lives in Volcano Crises." Fact Sheet 064-97, 2005. Available online. URL: http://pubs.usgs.gov/fs/1997/fs064-97. Accessed July 7, 2010.

Hawaiian Volcano Observatory, "Lessons Learned from the Armero, Colombia Tragedy" (2009) (excerpt)

On November 13, 1985, the volcanic eruption of Nevado del Ruiz in Colombia produced lahars that killed most residents in the city of Armero, even though the mudflows took two hours to reach the city. This excerpt provides insights concerning the breakdown in communication that led to the tragedy.

Late in the evening on November 13, 1985, most of those living in the Colombian town of Armero, on the shores of the river Lagunillas, were in bed. The nearby volcano, Nevado del Ruiz, had been quiet for the past couple of months, and the mayor and the town priest had assured the people of Armero that they were safe for the night.

Unfortunately, a storm had been brewing over the area, and the explosive eruption that occurred on the mountain, obscured by rain that night, went unnoticed by Armero residents. Those who experienced the eruption had no way of relaying information quickly and efficiently to Armero, the place most in danger.

Nevado del Ruiz is a stratovolcano, akin to Washington's Mount St. Helens, topped by glaciers, rising 5,389 m (17,784 ft) above sea level. Such volcanoes are especially dangerous, because heat from the eruptions melts the ice to create lahars, or mudflows of volcanic debris. These are not slow, cumbersome mudflows, but fast, deep, and destructive walls of debris and water.

The volcano first began to stir about a year earlier, its reawakening marked by a swarm of earthquakes. Fumarolic activity in the summit crater began around the same time. A visiting UN geologist advised the installation of monitoring equipment and the creation of a hazard map and evacuation plans. But Colombian scientists lacked the expertise, government support, and equipment necessary to effectively monitor the volcano and relay information to public authorities. They requested aid-both equipment and scientists-from foreign countries and were sent a few seismographs without proper instructions on how to operate the instruments and analyze the data.

In addition, the Colombian government was preoccupied with civil matters. In Bogotá, Colombia's capital, guerrilla warfare had broken out, and the Colombian President sent troops to quell the rebellion. To the government, the unstable political situation was more pressing than the volcanic activity. It did not help that the information about threats from Nevada del Ruiz provided to officials by various visiting and Colombian scientists was often contradictory and vague.

A minor explosive eruption on September 11 and a calmer political situation refocused officials on volcanic matters. City, county, and federal officials started meeting with scientists to discuss hazard map creation, evacuation plans, and possible eruption times and outcomes. Progress finally started at the federal and county levels, but the little town of Armero, built on old lahar deposits 45 km (28 mi) from the volcano, still went mostly unnoticed by government officials.

Before an effective line of communication and evacuation plan could be created, Nevado del Ruiz erupted again and sent lahars racing north and east through the deep river valleys on Nevado del Ruiz's flanks. With little warning, the river Lagunillas heaved its muddy contents onto the flatlands, directly into the city of Armero.

Within minutes, 23,000 people-most of the town's inhabitants-were killed, entombed within a concrete-like mixture of mud, vegetation, buildings, and everything else swept away by the lahars. Sadly, the lahars reached Armero approximately two hours after the eruption-plenty of time for the people to have evacuated to higher ground, had they been notified more quickly.

The 1985 eruption of Colombia's Nevado del Ruiz is the second most deadly volcanic eruption of the 20th century, resulting in the deaths of more than 25,000 people. Communities worldwide learned valuable lessons from this calamity. These include proper equipment and training for monitoring scientists to understand and help others understand what is happening.

Effective communication between all parties is one of the most essential components of living around an active volcano. When Mount Pinatubo in the Philippines awoke in 1991, the entire volcanological community responded, thus averting a disaster like the Armero tragedy. Every day, we continue to learn more about the volcanoes of our Earth. Looking back into the past is another way to catch a glimpse of the future.

Source: Hawaiian Volcano Observatory. "Lessons Learned from the Armero, Colombia, Tragedy." Available online. URL: http://hvo.wr.usgs.gov/volcanowatch/2009/09_10_29.html. Accessed March 8, 2010.

PART III

Research Tools

6

How to Research
Natural Disasters

Scientists are continually discovering new information and drawing new conclusions about natural disasters. Each week, new hazards, large or small, occur around the world while relief and rebuilding efforts continue in response to previous disasters. Occasionally, scientists or officials implement new procedures or technologies to help predict or mitigate hazards. For instance, when the Sumatra earthquake and tsunami occurred in 2004, no tsunami warning system was in place for the Indian Ocean region. Following that disaster, scientists put the Indian Ocean Tsunami Warning System into place.

Because environmental changes and scientific advancements are ongoing, keeping abreast of current events and issues is vital to research in this field. The guidelines and tips in the following sections provide useful starting points.

GETTING STARTED

Whether you want to learn about a specific type of natural hazard or a particular event that occurred, information is plentiful. A wide array of print and electronic sources exist to provide the facts, data, and resources that you seek. This plentiful supply of information is both a blessing and a curse. On one hand, nearly any information you need is out there, waiting for you to locate and use it. On the other hand, the sea of information can easily swamp you, drowning you in facts that are irrelevant, unreliable, or outdated. For this reason, making a solid research plan is essential.

Select a Topic

Begin by choosing your research topic. Perhaps you live on the Gulf Coast and you want to learn about hurricane hazards for your region. Perhaps a science teacher has asked you to write a report explaining what a stratovolcano is and describing a specific historical eruption. Or perhaps you watch a news report on flooding along the Mississippi River, and you want to research the use of levees on rivers. Each of these topics relates to natural disasters. Moreover, each topic represents an opportunity to gain a wealth of knowledge regarding the connections between natural hazards, humans, and the environment.

Identify Your Research Purpose

Once you have selected your research topic, you should formulate a thesis about this topic. A *thesis* makes a claim about the topic that your research project will prove or explain. For instance, suppose your research topic is the Indian Ocean Tsunami Warning System. What is your purpose for researching this topic? A thesis statement expresses this purpose. For example, your thesis may be, "The 2004 Sumatra earthquake and tsunami highlighted the need for a tsunami warning system in the Indian Ocean." Your purpose for researching the topic, then, would be to find information that explains how the earthquake brought about the need for the early warning system.

Another thesis about the same topic might be, "The Indian Ocean Tsunami Warning System consists of a network of sensors to detect tsunamis and a communications network to notify people of the need for evacuation." In this case, your research purpose is to explain and describe the workings of the warning system.

Whatever your thesis is, identifying your research purpose helps you focus your energies on useful, relevant information. Reviewing your thesis statement during your research process helps you stay on track.

FINDING INFORMATION

A good way to start your research is to review the information on your topic that you have at hand, whether this is something as brief as a writing assignment or thesis statement or something more substantial such as a textbook chapter or introductory book on the topic. Then, determine the types of information sources that would best fit your research purpose. Do you need historical facts, such as what the city of Galveston, Texas, was like just before and just after the hurricane in 1900? If so, then you may want to begin by searching for historical newspaper accounts or essays and by finding contemporary accounts written by scholars who have researched the historical event. Or maybe you need information on a cutting-edge technology such as geothermal drilling. In this case, you may want to begin by searching for recently published articles in scientific, technological, and news publications and Web sites. The annotated bibliography in this book can give you a head start on your initial research, as it lists both print and electronic sources of information on the history of natural disasters, specific natural disasters, and general information and overviews. When you find sources that seem useful, scan the introduction and the first few paragraphs of each chapter to determine whether the source relates to your research purpose. Don't forget to check the back of books for glossaries, time lines, maps, bibliographies, and other useful resources.

United States Topics

If you are researching a natural hazard, a specific disaster event, or other topic within the United States, you can often find valuable information from U.S. state and national government sources. Most of these are available online, including:

- State Offices and Agencies of Emergency Management, http://www. fema.gov/about/contact/statedr.shtm. This site provides links to the emergency management office in each U.S. state and territory. Clicking through to a state's Web site will take you to information on the state's current disasters (if any), state hazards, news links, and much more.
- FEMA Federal Disaster Declarations, http://www.fema.gov/news/disasters.fema. This site is a database of information on U.S. disasters, both natural and human-caused. You can search by year, by state, by region, and by type of disaster (an array of choices ranging from chemical/biological to winter storm).
- U.S. Department of Homeland Security, http://www.dhs.gov. This site provides links to numerous government sites relating to disaster preparedness, response, and recovery, including the Office of Emergency Communications, the National Response Framework, laws and regulations, and fact sheets on natural and human-caused disasters.
- U.S. Geological Survey, http://www.usgs.gov. This is the flagship government Web site on natural hazards and disasters relating to geology (as opposed to the oceans and atmosphere). The site contains extensive information on natural hazards and specific natural disasters, both current and historical, including maps and graphics.
- National Oceanic and Atmospheric Administration, http://www.noaa. gov. This is the flagship government Web site on the exploration and research of the oceans and weather of our planet. Use the search box to

find fact sheets on the tsunami warning system, hurricanes, and other water- and weather-related hazards and disasters. Includes links to government organizations such as the National Hurricane Center.

- Science.gov: Earth and Ocean Sciences, http://www.science.gov/browse/w_119.htm. This is an alphabetical, annotated, clickable list of Web sites, databases, and organizations that provide information on disasters and related topics. Examples of links include CDC Tsunamis and Geodata.gov.

International Topics

If you are researching hazards or events outside the United States, you can find valuable information on the sites of international organizations, most of which are available online and in English, including:

- EM-DAT: The International Disaster Database, http://www.emdat.be. This site contains statistics and other data on the occurrence and effects of more than 18,000 mass disasters worldwide from 1900 to the present. Includes country profiles, disaster lists, disaster profiles, reference maps, and more.
- Center for International Disaster Information, http://www.cidi.org. This site provides situation reports on current and recent disasters worldwide and has a searchable database of current and historical disasters worldwide.
- Global Disaster Alert and Coordination System, http://www.gdacs.org. This site posts up-to-the-minute alerts for current disaster events around the world and includes damage maps and links to news articles.

Conducting Research Online

The Internet is your gateway to an abundance of electronic resources, including Web sites, databases, journals, encyclopedias, graphical data, bibliographies, and more. Computer software is another type of electronic resource.

Search engines. A search engine is one of the most useful online tools you can use. It is a service that allows you to search for information on the Internet. It uses special software that allows it to browse the Internet, reading Web sites and indexing the information on them. The search engine allows Internet users to type one or more keywords into a search box, and then it finds and lists the Web pages that contain those key words. You can choose among many different search engines. Some commonly used search engines include the ones listed below. These are the most respected search engines because of the volume of sites they have indexed and because they are user friendly. Some of them have additional capabilities such as finding and displaying maps.

- AltaVista (http://www.altavista.com)
- AOL Search (http://search.aol.com)
- Ask (http://www.ask.com)
- Bing (http://www.bing.com)
- Google (http://google.com)
- HotBot (http://hotbot.com)
- Yahoo (http://yahoo.com)

One advantage of using a search engine is that, with the click of a button, you can locate dozens, hundreds, or even thousands of Web pages relating to your search terms. The sheer volume of listings, however, can easily become a disadvantage. It is up to you to sort through the search results to find trustworthy, updated, and relevant Web pages. The tips in the following list will help you get the most useful results from searching and using Web sites.

TIPS FOR USING SEARCH ENGINES

- **Choosing keywords.** Remember that the more specific your search term is, the more useful your results are likely to be. For instance, using the search word *flood* will return a list of results for products, services, maps, towns, definitions, articles, and more. Not all of these types of information will be relevant to your research purpose. To narrow the search results to a more useful selection, narrow your search term by adding words to it. For example, instead of searching *flood,* search *Mississippi floods* or *China floods* or *flood safety.*

- **Using quotation marks.** Another search technique is to enclose the words of your search term in quotation marks. Doing this tells the search engine that you want to find Web pages that use the exact phrasing within the quotation marks, not a random combination of the words. For example, searching *Yellow River floods* will return a list of pages about the Yellow River, flood management, flood facts, and Yellow River flooding. In contrast, searching *"Yellow River floods"* returns a list of pages that contain that exact phrase.

- **Using Boolean operators.** Some search engines support the use of Boolean operators, which are the words *and, or,* and *not.* These words help you fine-tune your search in three ways. By typing *and* between the words of your search term, you tell the search engine to list only pages that have *all* your search terms. By typing *or* between search words, you tell the search engine to include results that include one search term or another. By typing *not* before a word in the search term (e.g., *Yellow River **not** floods*), you tell the search engine to exclude results that use the word *floods.*

Conducting Research in a Library

A library is a rich source of both electronic and print resources. Electronic resources include those available on the Internet as well as the library's own databases and software programs. Print resources include books, magazines, journals, newspapers, government documents, and other sources that are in printed versus electronic format. Conducting your research in a library allows you to find specific sources that you know about. In addition, you can browse subject areas in a database or browse related books on a shelf to discover resources that will help you accomplish your research purpose. For a successful research trip to a library, be prepared to use the library's research tools to find items you need.

RESEARCH TOOLS AVAILABLE IN A LIBRARY

To find what you need in a library, head for the reference area. Here you will find print or electronic catalogs of the library's holdings. You will also find dictionaries and encyclopedias to consult for a better understanding of words and terms that you encounter in your research. The list below explains some of the research tools available in a library.

- **Catalog of holdings.** A library's catalog is usually available in an online database, or set of records. You can search the database by keyword, author, title, or subject. Most libraries have computers set aside for the sole purpose of searching the catalog. Some libraries allow patrons to log into the database from their computers at home via the Internet. The record of each book or other holding tells you where the item is located in the library and whether the item is currently available for checkout. Often the record contains the publication data for the item and perhaps a summary of its contents.
- **Printed periodical indexes.** In printed periodical indexes, you can search for articles by title, author, or subject by turning to specific parts of a book or by using specific volumes of a set.

- **Online periodical databases.** Besides the database of its holdings, a library may have databases such as EBSCOhost, ProQuest, and InfoTrac. These are indexes of articles in hundreds of magazines, journals, and newspapers. You can search periodical databases by keyword, author, or title. The citation of each article contains the publication information for the article so that you can locate the article. Besides providing a citation, the database record may also include a summary or abstract of the article. In some cases, you can download the entire article.

- **Periodicals, brochures, and documents.** A library subscribes to numerous periodicals, journals, newsletters, and other items published on a monthly, bimonthly, or other schedule. In addition, the library holds documents published by city, state, and federal governments. Recent issues and copies of these kinds of documents are usually available in printed form. Older copies may be stored electronically.

TIPS FOR RESEARCHING IN A LIBRARY

Searching databases. Whether you are searching the library's catalog or a periodical database, first decide whether you need to search by title, author, keyword, or subject. Search by title if you know the exact title of a work that you need. If the title begins with a, an, or the, omit this word when you type the title into the search field. Search by author's name to find a list of all the library's holdings of an author's works. Use the format of last name, first name, like this: Davis, Lee. Searching by author's name is useful when you know that the writer has written at least one work relevant to your research, and you want to find out if he or she has written other works that also may be useful.

When doing a keyword search, decide whether a general or a specific term would yield the most helpful results. General search terms are useful for finding out the range of information available to you. A broad search term such as *natural disasters* will yield results that discuss natural disasters in general as well as results that discuss one or more specific natural disasters. You'll need to scroll through the results, checking the data on each item to see if it is useful to your research purpose. Use narrower search terms to find sources on specific topics. For example, hurricanes is more specific than natural disasters, and Hurricane Katrina is yet more specific. The more specific your search term, the more specific your results will be.

Search by subject to get a list of items that the library has assigned to a particular subject heading. Libraries use the subject headings established by the Library of Congress. A subject search of natural disasters may return a list of 50 or 60 items, depending on the size of the library. Beneath this subject heading, the holdings are further categorized in narrower subjects, such as Natural Disasters United States and Natural Disasters Safety Measures and Natural Disasters Maps. Browsing these subject listings can help you find a list of sources that directly relate to your research purpose.

Consulting librarians. Libraries have a reference desk or a help desk staffed by one or more librarians who can help you. If you are not sure how to choose a search term, if you cannot locate a specific item, or if you run into another problem, ask for help. Librarians are trained in the use of research methods and are a rich source of information, whatever your research topic is.

EVALUATING INFORMATION SOURCES

Whether you locate a source online or in print, evaluate it to determine its reliability and usefulness. Here are questions to consider pertaining to the source's author, publisher, publication date, intended audience, and content.

Author. What are the author's qualifications for writing this work? Skim the biographical paragraph on the author, usually found on the back cover or inside flap of the dust jacket. What is the writer's educational background? Has she or he published other books? Does she or he have other related experience in the topic? Is she or he associated with a professional organization? If this is a Web page, are the author's name and credentials provided? If the author of a work has no professional education or affiliations related to the topic, this is an indicator that the author may not be a reliable source of information.

Publisher. Established publishers have rigorous standards for the research, writing, and editing of a work to ensure accuracy and balance of coverage. As a general rule, works published by university presses may be considered reliable sources. Other publishers have earned a reputation for publishing quality works. Generally, librarians at a school or public library consider the reputation of a publisher before purchasing books. Remember, too, that works published by certain types of publishers can be biased depending on the publisher's agenda and mission. A book published by a human rights watch organization, for instance, may place the lion's share of responsibility for a famine on government policies within the affected country, while a book on the same famine written by a university professor may show less bias against the government by attributing the famine to causes including governmental policies, policies of the international aid community, and agricultural practices within the country. When in doubt about the reliability or possible bias of a publisher, discuss your concern with a librarian.

If the source is a Web page or Web site, check the domain of the site. Web sites published by universities (those ending in .edu) and those published by a state or federal government agency (those ending in .gov) are generally trustworthy and well-researched sources. Web sites created by reputable authors, teachers, and other professionals are usually reliable sources as well. These sites are often affiliated with a school or professional organization. Beware of Web sites created by individuals with no professional or educational affiliation, especially if they do not identify the sources of their information. Consider blogs to be the writer's personal ideas and opinions, not a trustworthy research source, unless the blog is officially sponsored by a trustworthy organization. It is becoming increasingly common for reputable organizations to provide tutorials, analyses, historical recaps, and other useful information in the form of blog entries to attract readers to their Web sites.

Date of Publication or Revision. When was the work published, revised, or updated? More recent publication dates are generally preferable to older ones. However, consider the type of information. General information in older sources can be just as accurate and reliable as general information in recent sources. The advantage of recently published sources is that they contain the most up-to-date information on current events. For instance, a book published in 2005 may explain what an earthquake is and give facts about the December 2004 earthquake in Indonesia. But it will not contain information on the January 2010 earthquake in Haiti.

Intended Audience. Who is the intended audience of the work? Is it a general audience or a specialized audience? Is the book too simple or too advanced for your research purpose? Is it too technical? Or is it a solid match for your research needs? Note that books written for younger audiences tend to simplify information for first-time learners and may omit many or most controversial details associated with the event. In contrast, books intended for high school and college readers are more likely to address the full spectrum of information related to a topic, including controversies and hot-button issues. Articles written by journalists for the general public typically combine explanations of basic facts along with references to relevant controversial issues.

Content. Are the sources of facts, statistics, and other data clearly documented in footnotes or a bibliography? Trustworthy works clearly document their research sources. For instance, graphical information such as charts and graphs should include a source notation and an indication of when the data was gathered or the time period the data represents. Scholarly works, particularly those that present controversial issues, should include a bibliography of research sources that the author used to prepare arguments or information in the book. In addition to checking the author's documentation of data and information sources, ask, does the writer use a balanced, objective tone? Or are the paragraphs peppered with emotional, judgmental, or unfair word choices? If the writer shows bias for or against the topic, it is usually wise to skip the source in favor of a more objective source. However, if you are researching the opposing sides of a controversial topic (such as geothermal drilling or the Three Gorges Dam), then you may find it useful to seek out biased sources in order to understand the strengths and weaknesses of opposing viewpoints. The important point is that you, the researcher, are aware of whether a source is biased or not and use the source's information responsibly.

INTERPRETING GRAPHICAL DATA

During your research process, you will find information presented in graphical formats such as graphs, charts, and diagrams. You will find graphical data in chapter 7 of this book, "Facts and Figures." Knowing how to read these graphical sources of data will help you conduct thorough, accurate research. Watch for the following commonly used graphical forms.

- A **bar graph** (or bar chart) uses bars to represent different numerical values. Compare the lengths (or heights) of the bars to draw conclusions about the values represented.

- A **pie chart** shows percentages as "slices of pie" or wedges in a circle. Compare the sizes of the slices and read the data supplied to draw conclusions.

- A **line chart** plots data along a time line, connecting the dots to form a line. The trajectory of the line up, down, and straight across shows the change of the data over time.

- A **table** organizes information into rows and columns for the purpose of comparing or contrasting the facts or data.

- A **diagram** is a drawing of a subject, such as a volcano, with the parts labeled.

- A **map** shows the layout of a place as seen from overhead. The content of the map depends on the map's purpose. For example, climate maps give information, usually color-coded, about the climate or precipitation of a region. Physical maps show mountains, rivers, lakes, and relief, or differences in land elevation (as in mountains and valleys).

DOCUMENTING YOUR SOURCES

With the wide array of sources available to you, it is important to document each source that you use. For Web sites, note the site's name, the title of the page you used, the URL, and the date you accessed the page. For books, note the author, title, and publication information. In addition, record the page numbers that correspond to facts, charts, quotations, or other information you noted. For articles, note the author, article title, journal or magazine name, volume number, and page numbers of the article. In addition, write down the page numbers of quotations or data that you may use in your research project.

Documenting your sources during the research process is vital for two reasons. First, it allows you to find the source easily if you need to consult it again. Second, it provides you with the data you need to create the bibliography or works-cited page of your research paper.

PLAGIARISM

Documenting your sources is an essential step in preparing and presenting your research and ideas, but don't stop there. As you takes notes during your research process and as you write the drafts of your paper, keep one vital question alive in your mind: Are these my own exact words, or did they come from someone else? Doing so can help you avoid plagiarism, or the presentation of someone else's words or ideas as your own. Simple ways to avoid plagiarism include these tips:

- Use quotation marks around any series of words, any sentence, or any series of sentences that you borrow from a source. The time to insert the quotation marks is immediately as you take notes from a source. Don't expect to remember later which words you borrowed and which are your own. When you incorporate phrases or sentences from your notes into your paper, include the quotation marks. By continually paying attention to the use of quotation marks, you are able to keep track of which words are borrowed and which are your own.

- Know how to quote sources responsibly. Often, you will take down a long quote in your notes and then use only part of that quote in your paper. Remember that any sequence of words from the quote—even if it is just a few words—should remain in quotation marks in your paper. Simply adding an original sentence of your own between two sentences from the source does not mean you can omit the quotation marks. Likewise, using half of a sentence from your source and writing the remainder of

295

the sentence with your own words does not mean that you can omit the quotation marks. In this case, place quotation marks around the borrowed words even when they do not make up a complete sentence on their own.

• Acknowledge the source of original ideas as well as direct quotations. Plagiarism is not limited to presenting someone else's words as your own; it also includes presenting someone else's ideas as your own. If, in your written work, you include a notable idea, perspective, opinion, or other product of thought from one of your sources, acknowledge whose product of thought it is. One way of doing so is by starting the sentence "According to (insert name), . . ." and go on to describe the idea or opinion in your own words. You can also use phrases such as "(Insert name)'s idea that . . ." or "(Insert name)'s perspective on the controversy is that. . . ." In this way, you can present crucial ideas while acknowledging where they came from, and then expand upon them with your original ideas and interpretations.

Using these simple tips can help you incorporate words and ideas from others into your original written work in a way that showcases your ability to research and present information professionally.

7

Facts and Figures

INTERNATIONAL

1.1 Natural Disasters and Deaths by Continent,1900–2010

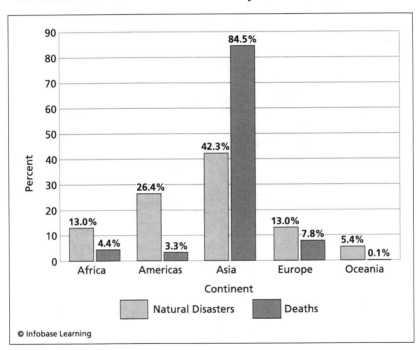

The majority of natural disasters occurs in Asia and the Americas. Though only 42 percent of natural disasters occur in Asia, that continent experiences more than 84 percent of the deaths attributed to natural disasters.

Source: EM-DAT: The Public International Disaster Database. Available online. URL: http://www.emdat.be/advanced-search. Accessed May 24, 2010.

1.2 Natural Disasters and Deaths by Disaster Type, 1900–2010

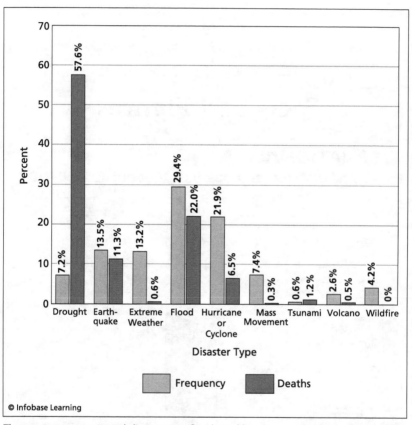

The most common natural disasters are floods and hurricanes, which account for more than 51 percent of all natural disasters. Droughts occur much less frequently but cause more than 57 percent of deaths from natural disasters.

Source: EM-DAT: The Public International Disaster Database. Available online. URL: http://www.emdat.be/advanced-search. Accessed May 24, 2010.

1.3 Natural Disaster Data, 1900–2010

Continent	Disaster Type	Number of Events	Killed	Total Affected	Damages (thousands U.S. $)
Africa	Drought	261	844,145	295,346,829	$5,41 9,593
	Earthquake	74	21,072	1,694,094	$12,129,699
	Extreme Weather	52	1,142	1,386,906	$716,372
	Flood	494	15,944	36,983,036	$4,407,266
	Hurricane or Cyclone	100	3,401	14,995,592	$3,079,430
	Mass Movement*	29	1,299	48,630	$0
	Tsunami	4	312	109,913	$230,000
	Volcano	17	2,218	511,353	$9,000
	Wildfire	24	217	25,822	$440,000
Africa Total		**1,055**	**889,750**	**351,102,175**	**$26,431,360**
Americas	Drought	121	77	65,214,341	$20,711,139
	Earthquake	253	438,550	32,060,254	$84,109,906
	Extreme Weather	364	14,662	5,991,091	$81,407,010
	Flood	515	43,673	46,269,274	$59,269,430
	Hurricane or Cyclone	530	85,993	46,488,783	$402,032,632
	Mass Movement	168	21,779	5,472,334	$2,221,727
	Tsunami	7	380	2,472	$900
	Volcano	76	67,851	1,551,770	$2,168,697
	Wildfire	115	1,516	1,170,748	$19,368,900
Americas Total		**2,149**	**674,481**	**204,221,067**	**$671,290,341**
Asia	Drought	144	9,663,389	1,648,500,427	$28, 140,159
	Earthquake	583	1,559,335	126,549,463	$309,753,474
	Extreme Weather	285	26,406	243,996,964	$31,255,182
	Flood	988	4,413,840	2,130,426,906	$207,441,430
	Hurricane or Cyclone	934	1,237,288	560,084,367	$151,642,503
	Mass Movement	313	20,654	5,701,091	$2,040,838
	Tsunami	**25**	**235,285**	**2,382,682**	**$10,266,000**

(table continues)

(continued)

Continent	Disaster Type	Number of Events	Killed	Total Affected	Damages (thousands U.S. $)
Asia	Volcano	86	21,462	2,802,262	$708,351
	Wildfire	80	690	3,255,820	$11,633,500
Asia Total		3,438	17,178,349	4,723,699,982	$752,881,437
Europe	Drought	37	1,200,002	15,482,969	$20,061,309
	Earthquake	151	275,924	5,443,816	$61,734,076
	Extreme Weather	336	84,712	5,369,129	$65,073,055
	Flood	327	5,091	9,214,805	$74,957,041
	Hurricane or Cyclone	22	201	94,682	$1,817,360
	Mass Movement	76	16,798	48,805	$3,111,489
	Tsunami	4	2,376	2	$0
	Volcano	11	783	26,224	$44,300
	Wildfire	92	426	1,287,800	$11,018,811
Europe Total		1,056	1,586,313	36,968,232	$237,817,441
Oceania	Drought	19	660	8,027,635	$10,703,000
	Earthquake	38	428	89,167	$1,379,419
	Extreme Weather	36	655	4,788,678	$3,535,128
	Flood	70	230	276,845	$5,784,354
	Hurricane or Cyclone	194	1,721	2,294,553	$7,461,364
	Mass Movement	18	562	21,315	$2,466
	Tsunami	10	2,798	20,843	$159,500
	Volcano	23	3,665	259,900	$110,000
	Wildfire	31	496	97,140	$2,622,844
Oceania Total		439	11,215	15,876,076	$31,758,075
Grand Total		8,137	20,340,108	5,331,867,532	$1,720,178,654

* Mass movement refers to a landslide or avalanche

This table shows the number of events, deaths, and affected population in thousands and shows damages in U.S. dollars for all natural disasters that occurred between 1900 and 2010. The data is grouped by continent and disaster type.

Source: EM-DAT: The Public International Disaster Database. Available online. URL: http://www.emdat.be/advanced-search. Accessed May 24, 2010.

1.4 Major Tsunamis in the Past 250 Years

Location	Date	Magnitude	Maximum Run-up (feet)
Lisbon	November 1, 1755	8.5	59
Chile	November 7, 1837	8.5	26
Chile	August 13, 1868	9.1	59
Krakatau	August 27, 1883	8.5	118
Kamchatka	February 3, 1923	8.3	26
Aleutians	April 1, 1946	7.4	138
Kamchatka	November 4, 1952	9	59
Aleutians	March 9, 1957	9.1	49
Chile	May 22, 1960	9.5	59
Alaska	March 28, 1964	9.2	223
Sumatra	December 26, 2004	9.3	112

A tsunami occurs when a large earthquake in the ocean causes an abnormally high run-up of sea levels along coastal areas in the form of large waves. In the past 250 years, there have been 11 major tsunamis caused by earthquakes ranging in magnitude from 7.4 to 9.5. The largest run-up of 223 feet occurred in Alaska in 1964, and the deadliest tsunami occurred near Sumatra in 2004.

Source: Tsunami Laboratory, Institute of Computational Mathematics and Mathematical Geophysics, Novosibirsk, Russia, 2005.

1.5 Major Historical Floods, 1900–2010

Country	Date	Fatalities
China	July 1931	3,700,000
China	July 1959	2,000,000
China	July 1939	500,000
China	1935	142,000
China	1911	100,000
China	July 1949	57,000
Guatemala	October 1949	40,000
China	August 1954	30,000
Venezuela	December 15, 1999	30,000
Bangladesh	July 1974	28,700

Floods account for some of the worst natural disasters in terms of deaths. Of the worst floods in the past century, seven of 10 have occurred in China, and those floods caused more than 4.7 million deaths.

Source: EM-DAT: The Public International Disaster Database. Available online. URL: http://www. emdat.be/advanced-search. Accessed May 24, 2010.

1.6 Major Historical Droughts, 1900–2010

Country	Date	Fatalities
China	1928	3,000,000
Bangladesh	1943	1,900,000
India	1942	1,500,000
India	1965	1,500,000
India	1900	1,250,000
Soviet Union	1921	1,200,000
China	1920	500,000
Ethiopia	May 1983	300,000
Sudan	April 1983	150,000
Ethiopia	December 1973	100,000

China experienced the worst drought in history in 1928, when 3 million people died. The decade of the 1920s was especially deadly, with droughts in China, Bangladesh, India, and the Soviet Union that killed more than 1 million people each and an additional drought in China that killed half a million people.

Source: EM-DAT: The Public International Disaster Database. Available online. URL: http://www.emdat.be/advanced-search. Accessed May 24, 2010.

1.7 Major Historical Earthquakes, 1900–2010

Country	Date	Fatalities
China	July 27, 1976	242,000
Haiti	January 12, 2010	222,570
China	May 22, 1927	200,000
China	December 16, 1920	180,000
Indonesia (tsunami)	December 26, 2004	165,708
Japan	September 1, 1923	143,000
Soviet Union	October 5, 1948	110,000
China	May 12, 2008	87,476
Italy	December 28, 1908	75,000
Pakistan	October 8, 2005	73,338

Four of the 10 most deadly earthquakes occurred between 2000 and 2010. The most devastating recent disaster occurred in Haiti in 2010 and officially killed 222,570 people. The 2004 tsunami in Indonesia, which was caused by an earthquake, also caused significant fatalities, with 165,708 deaths officially recorded.

Source: EM-DAT: The Public International Disaster Database. Available online. URL: http://www. emdat.be/advanced-search. Accessed June 11, 2010.

UNITED STATES
2.1 Presidential Disaster Declarations
by Type, 1953–2009

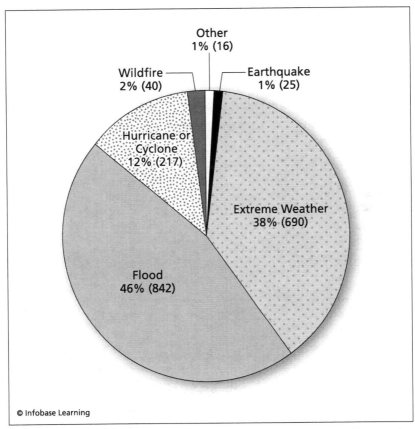

Other
1% (16)

Wildfire
2% (40)

Earthquake
1% (25)

Hurricane or
Cyclone
12% (217)

Extreme Weather
38% (690)

Flood
46% (842)

© Infobase Learning

A total of 1,830 major presidential disasters were declared between 1953 (the year that presidential disasters began to be serially numbered) and 2009. The most common types of disasters were floods and extreme weather, which includes freezing weather, heat waves, ice and snow storms, and tornadoes.

Source: Public Entity Risk Institute, University of Delaware. Available online. URL: http://www.peripresdecusa.org/mainframe.htm. Accessed June 11, 2010.

2.2 U.S. Economic Losses from Major Natural Disasters, 1900–2009

Date	Disaster Type	Name	Damages (millions U.S. $)	Fatalities	Total Affected
8/29/2005	Hurricane	Katrina	$125,000	1,833	500,000
9/12/2008	Hurricane	Hurricane Ike	$30,000	82	200,000
1/17/1994	Earthquake	Northridge Earthquake	$30,000	60	27,000
8/24/1992	Hurricane	Andrew	$26,500	44	250,055
9/15/2004	Hurricane	Ivan	$18,000	52	not available
8/13/2004	Hurricane	Charley	$16,000	10	30,000
9/23/2005	Hurricane	Rita	$16,000	10	300,000
10/24/2005	Hurricane	Hurricane 'Wilma'	$14,300	4	30,000
6/24/1993	Flood	Great Flood of 1993	$12,000	48	31,000
9/5/2004	Hurricane	Frances	$11,000	47	5,000,000
6/9/2008	Flood	2008 Midwest Floods	$10,000	24	11,000,148
9/25/2004	Hurricane	Jeanne	$8,000	6	40,000
9/25/1989	Hurricane	Hugo	$7,000	51	25,000
9/13/1999	Hurricane	Floyd	$7,000	70	3,000,010
9/1/2008	Hurricane	Hurricane 'Gustav'	$7,000	43	2,100,000

The most costly natural disaster in U.S. history occurred in 2005 when Hurricane Katrina caused an estimated $125 billion in damages. Compared to the $30 billion in damages each caused by Hurricane Ike and the Northridge earthquake, Hurricane Katrina caused more than four times as much damage.

Source: EM-DAT: The Public International Disaster Database. Available online. URL: http://www.emdat.be/advanced-search. Accessed June 11, 2010.

2.3 Major U.S. Hurricane Damages, 1900–2010

Date	Name	Damages (millions U.S. $)
8/29/2005	Katrina	$125,000
9/12/2008	Ike	$30,000
8/24/1992	Andrew	$26,500
9/15/2004	Ivan	$18,000
8/13/2004	Charley	$16,000
9/23/2005	Rita	$16,000
10/24/2005	Wilma	$14,300
9/5/2004	Frances	$11,000
9/25/2004	Jeanne	$8,000
9/25/1989	Hugo	$7,000
9/13/1999	Floyd	$7,000
9/1/2008	Gustav	$7,000
6/5/2001	Allison	$6,000
9/11/1992	Iniki	$5,000
9/5/1996	Fran	$3,400
9/18/2003	Isabel	$3,370
8/17/1983	Alicia	$3,000
10/4/1995	Opal	$3,000
9/12/1979	Frederic	$2,300
7/10/2005	Dennis	$2,230
6/18/1972	Agnes	$2,100
9/14/1995	Marilyn	$2,100
10/3/2002	Lili	$2,000
10/27/1985	Juan	$1,500
8/18/1991	Bob	$1,500
8/19/1998	Bonnie	$1,500
9/7/1965	Betsy	$1,420
8/17/1969	Camille	$1,420
9/20/1998	Georges	$1,205
7/23/2008	Dolly	$1,200
8/30/1985	Elena	$1,100
Total		$176,145

Between 1900 and 2010, there were 29 hurricanes that caused at least $1 billion in damages. The total damages by all of these hurricanes combined exceeded $176 billion.

Source: EM-DAT: The Public International Disaster Database. Available online. URL: http://www. emdat.be/advanced-search. Accessed June 11, 2010.

INTERNATIONAL CASE STUDIES

3.1 Ring of Fire

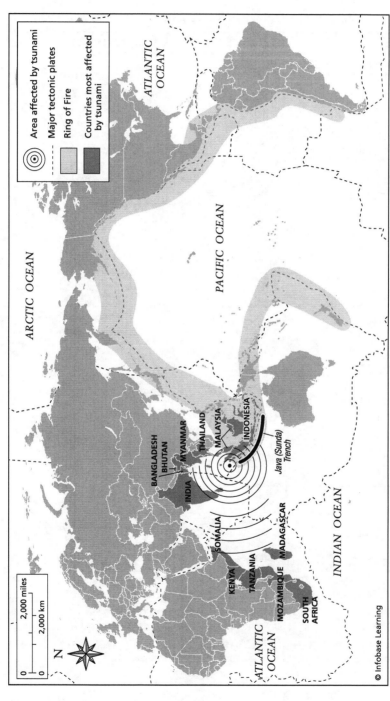

© Infobase Learning

The Ring of Fire runs along the western coasts of North America and South America and the eastern coast of Asia. More than half of the world's volcanoes exist within this region of the Pacific Ocean. A portion of the Ring of Fire, the Sunda Trench, runs along the western coast of Sumatra and is the site of the 2004 earthquake that created a devastating tsunami in the Indian Ocean.

Source: National Oceanic and Atmospheric Administration

3.2 2004 Indian Ocean Earthquake and Tsunami Casualties

Country	Confirmed Casualties
Indonesia	126,900
Sri Lanka	31,000
India	10,700
Thailand*	5,400
Somalia	300
Maldives	80
Malaysia	70
Myanmar	60
Tanzania	10
Seychelles	2
Bangladesh	2
South Africa	2
Yemen	2
Kenya	1
Total	174,500

* Estimate includes foreign tourists

The western coast of Sumatra, Indonesia, which was closest to the earthquake, experienced the greatest number of casualties from the 2004 tsunami that occurred in the Indian Ocean. Sri Lanka, India, and Thailand also reported significant deaths from the disaster. A total of 174,500 fatalities have been confirmed.

Source: Risk Management Solutions

3.3 2004 Indian Ocean Earthquake and Tsunami Economic Losses

Country	Economic Losses (millions U.S. $)	Insured Losses* (millions U.S. $)
Indonesia	4,500	500
Thailand	1,000	500
Sri Lanka	1,000	100
India	1,000	100
Maldives	500	50
Other	2,000	50
Total	10,000	1,300

* Includes property insurance only; life and health losses are estimated at $250 million total and travel losses at $50 million total

Economic losses from the 2004 tsunami in the Indian Ocean reached a disaster total of $10 billion. Insurance covered only 13 percent of these losses.

Source: Risk Management Solutions

3.4 Major Disasters in China, 1900–2010

Date	Location	Disaster Type	Fatalities
07/1931	Central China	Flood	3,700,000
1928	Shaanxi, Henan, Gansu	Drought	3,000,000
07/1959	North	Flood	2,000,000
07/1939	Henan Province	Flood	500,000
1920	North	Drought	500,000
7/27/1976	Tangshan, Beijing, Tientsin	Earthquake	242,000
5/22/1927	Nanchang (Jiangxi Province)	Earthquake	200,000
12/16/1920	Gansu Province	Earthquake	180,000
1935	Yellow River	Flood	142,000
7/27/1922	Shantou	Tropical cyclone	100,000

Many of the world's worst natural disasters have taken place in China. In the past century, the worst natural disaster occurred in 1931 when floods in central China claimed 3.7 million lives. This flood followed a severe drought, which began in 1928, that killed 3 million people.

Source: EM-DAT: The Public International Disaster Database. Available online. URL: http://www.emdat.be/advanced-search. Accessed June 18, 2010.

3.5 Three Gorges Dam

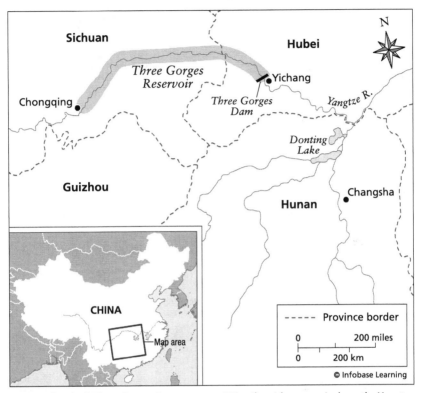

The 331-foot high Three Gorges Dam creates a 410-mile-wide reservoir along the Yangtze River and plays key roles in power production, flood control, and water management in China.

Source: Centers for Disease Control

3.6 Major Droughts in Ethiopia, 1960–2010

Date of Drought	Total Persons Affected
2003	12,600,000
May 1983	7,750,000
June 1987	7,000,000
October 1989	6,500,000
May 2008	6,400,000
September 1999	4,900,000
December 1973	3,000,000
November 2005	2,600,000
September 1969	1,700,000
July 1965	1,500,000
July 1998	986,200

In the past half-century, 11 major droughts have affected millions of people in Ethiopia. The worst drought disaster occurred in 2003 when more than 12 million people experienced food insecurity.

Source: EM-DAT: The Public International Disaster Database. Available online. URL: http://www.emdat.be/advanced-search. Accessed June 21, 2010.

3.7 Lahars from Nevado del Ruiz in 1985

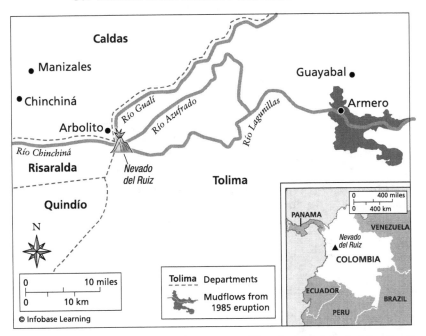

On November 13, 1985, lahars from the eruption of Nevado del Ruiz buried the city of Armero, Colombia, killing at least 20,000 of the town's 30,000 residents.

Source: Victoria Bruce. *No Apparent Danger.* New York: HarperCollins, 2001, pp. 84–85.

8

Key Players A to Z

MOHAMMED AMIN (1943–1996)   Kenyan camera operator whose film footage in a October 1984 BBC report helped attract worldwide attention to a severe famine in the Sahel zone and spark an international outpouring of aid. In a career spanning more than 30 years, Amin became a highly respected photographer and camera operator known for his coverage of major events in Africa. He died in November 1996 when hijackers took over the Ethiopian airliner he was on, crashing the aircraft in the Indian Ocean.

CLARISSA ("CLARA") HARLOWE BARTON (1821–1912)   Founder of the American Red Cross. When the American Civil War broke out in 1861, Barton was working at the U.S. Patent Office in Washington, D.C. The Sixth Massachusetts Regiment arrived there after the Baltimore Riots, and Barton organized local relief efforts for the soldiers. Afterward, she advertised for public donations of medical supplies and necessities for the soldiers. She obtained a pass allowing her to travel with army ambulances to distribute supplies and nurse the sick and wounded on battlefields and in field hospitals. After the war, she organized efforts to identify and locate missing soldiers. In 1869, Barton traveled to Europe where she learned about the International Red Cross. She founded the American Red Cross in 1881.

HUGH HAMMOND BENNETT (1881–1960)   Soil conservation pioneer who founded and headed the Soil Conservation Service (SCS) of the U.S. Department of Agriculture during America's dust bowl, serving from 1935 until his retirement in 1951. Under Bennett's leadership, the SCS developed conservation programs to retain topsoil and reduce damage to the land, including new farming methods such as strip cropping, terracing, crop rotation, contour plowing, and cover crops that are still in use today.

KATHLEEN BLANCO (1942–)   Governor of Louisiana when Hurricane Katrina struck in 2005. The hurricane struck during the second year

314

of Blanco's term, launching her into the national spotlight. Her administration suffered criticism for a slow and sometimes ineffective response to the disaster, its aftermath, and the recovery (though not nearly the severity of criticism leveled at FEMA). Following a long public record of Blanco's heated protests and criticisms of federal response to the disaster in her state, she chose not to run for reelection in 2007. She was the first woman to be elected governor of Louisiana.

MICHAEL D. BROWN (1954–) Administrator of the Federal Emergency Management Agency (FEMA) when Hurricane Katrina struck in August 2005. He resigned his position two weeks after the hurricane amid a public outcry against his management of the disaster. Brown's education and early career were in law, and in 2001 he joined FEMA as general counsel. In 2002, he became deputy director of FEMA, helping to oversee recovery efforts following the terrorist attacks of September 11, 2001. In 2003, Brown became the director of FEMA. On August 31, 2005, Brown was placed in charge of the government's response to Hurricane Katrina. However, nine days later he was relieved of those duties. He resigned from FEMA three days later. Since then, he has held jobs in the private sector, including positions at two firms that specialize in disaster recovery.

GEORGE W. BUSH (1946–) President of the United States when Hurricane Katrina struck in 2005. His administration suffered sharp criticism for slow response to the hurricane, mismanagement of resources during response and recovery, and lack of adequate preparation for such a crisis. Bush himself was also criticized for being slow to respond. (He remained on vacation in Texas for 1 1/2 days after the hurricane made landfall.) Critics also accused him of being out of touch with the reality of the situation. (On September 2, 2005, he praised FEMA head Michael D. Brown's handling of the disaster by quipping, "Brownie, you're doing a heck of a job!" The ineffectual Brown was relieved of his duties a week later.)

MARTA LUCÍA CALVACHE One of the scientists assigned by the Central Hydroelectric Company of Caldas (CHEC) in Colombia to monitor and record activities at Nevado del Ruiz, the volcano that erupted on November 13, 1985. The 26-year-old Colombian geologist had no experience with active volcanoes, but she had studied geothermal energy and, consequently, had more relevant experience than any other scientist in Colombia. After the disaster, Calvache went on to earn a Ph.D. at Arizona State University and in 2010 was assistant director of geological threats at the Colombian Institute of Geology and Mining.

MANUEL CERVERO   Colombian pilot of the DC-8 cargo transport flying 7,000 feet above Nevado del Ruiz when it erupted just after 9:00 P.M. on November 13, 1985. He later described a reddish illumination reaching to 26,000 feet and his plane's being covered with ash. With the aircraft's windows obscured by ash, he flew by instruments only and diverted his flight to the city of Cali, where he landed safely 20 minutes later.

ISAAC M. CLINE (CA. 1861–1955)   The chief meteorologist at the Galveston, Texas, office of the U.S. Weather Bureau when the hurricane of 1900 struck. Cline enlisted in the U.S. Army Signal Corps in 1882, working for the U.S. weather service, which, at that time, was part of the Signal Corps. During his distinguished career, he was chief meteorologist in five different cities, including Galveston. Cline performed groundbreaking research in hurricane warning and river and flood work. He was officially commended for his work in Galveston during the 1900 hurricane and again in the Mississippi Valley flooding of 1927. He published numerous papers on meteorology and a book, *Tropical Cyclones.*

GUSTAVE-GASPARD DE CORIOLIS (1792–1843)   French mathematician, mechanical engineer, and scientist who first described what is now called the Coriolis force. In his early career, Coriolis researched friction and hydraulics, publishing *Calculation of the Effect of Machines* in 1829. He went on to research ideas of motion in rotating systems like waterwheels. In 1835, he published "On the Equations of Relative Motion of a System of Bodies," in which he describes mathematically what happens with forces in a rotating system. Although Coriolis did not apply his ideas to the rotation of the Earth, the Coriolis force has been used to explain the rotation of winds in hurricanes, for example, and why winds follow curved paths across Earth.

HENRI DUNANT (HENRY DUNANT; JEAN-HENRI DUNANT) (1828–1910)   Swiss businessman and social activist whose firsthand account of the suffering of the wounded after the Battle of Solferino (June 24, 1859) inspired the creation of the International Committee of the Red Cross in 1863. In 1862, Dunant published *Memoir of Solferino,* describing the suffering he witnessed on the battlefield. In the book, he proposes that an agency be established to provide humanitarian aid during wartime and, in addition, a government treaty that would protect this agency's workers from harm. These proposals led to the founding of the Red Cross and to the first Geneva Convention (in 1864). The latter sets standards for international law for the humanitarian treatment of war victims. In 1901, Dunant was honored with the Nobel Peace Prize.

JAN EGELAND (1957–) The United Nations Undersecretary-General for Humanitarian Affairs from June 2003 to December 2006. He led the massive UN relief effort following the 2004 Indian Ocean earthquake and tsunami, coordinating donations and relief activities from around the world. After leaving his position at the UN he became the director of the Norwegian Institute of International Affairs; he is also a professor at the University of Stavanger in Norway.

FRANK FOURNIER (1948–) French photographer who snapped a photograph of 13-year-old Omayra Sánchez, victim of the 1985 eruption of the Nevado del Ruiz volcano in Colombia. In Fournier's photograph, Omayra sits neck-deep in water, trapped by debris, and looks at the camera. She died hours later. The image tugged at heartstrings around the world—and ignited a storm of controversy over the Colombian government's role, or lack thereof, in preparedness for the volcano's eruption. Fournier's photograph won the 1985 World Press Photo award.

TETSUYA THEODORE (TED) FUJITA (1920–1998) Japanese-American meteorologist who formulated the Fujita scale to rate the intensity of tornadoes. Fujita also discovered wind phenomena called downbursts and microbursts. Born in Kitakyushu City on the Japanese island of Kyushu, Fujita became an associate professor at Kyushu Institute of Technology. His groundbreaking research into severe storms led to an invitation from the University of Chicago to work as a visiting research associate in the Department of Meteorology. He began working there in 1953 and quickly made a name for himself in tornado research. He was appointed to the university faculty in 1955. In 1968, Fujita obtained American citizenship, and in 1971 he completed the Fujita scale.

FREDERICK FUNSTON (1865–1917) U.S. Army general in charge of the Presidio of San Francisco who, according to some historical accounts, declared martial law in San Francisco following the earthquake of April 18, 1906. (Some accounts assert that he took control of the city without technically declaring martial law.) Funston is responsible for the use of black powder to create firebreaks in an attempt to stop fires that raged through the city, ironically making the disaster worse by spreading the fires.

NÉSTOR GARCÍA One of the scientists assigned by the Central Hydroelectric Company of Caldas (CHEC) in Colombia to monitor and record activities at Nevado del Ruiz, the volcano that erupted on November 13, 1985. A chemical engineer, García was assigned to test the acidity and temperature of steam venting from the mountain. After the disaster, García went on to work as an industrial chemist at the National University in Colombia. He died on January 14, 1993, along with five other scientists when, as they were gathering data in the cone of Colombia's Galeras volcano, the volcano erupted.

ROBERT "BOB" FREDERICK ZENON GELDOF (1951–   )   An Irish singer and songwriter who earned international acclaim during 1984 and 1985 for organizing efforts that raised millions of dollars in aid for victims of famine in Ethiopia. In December 1984, his charity group Band-Aid released the hit single "Do They Know It's Christmas." In 1985, he organized the Live Aid fund-raising event, a multi-venue concert that took place on July 13 in London, England, and Philadelphia, Pennsylvania, and was broadcast worldwide. In 2010, when news broke that some funds raised through his efforts had been diverted by Ethiopian militants, he angrily denied the report.

BENO GUTENBERG (1889–1960)   A German geophysicist who collaborated with CHARLES RICHTER to invent the Richter scale, which measures the magnitude of earthquakes. Born in Germany, Gutenberg became a professor of geophysics at the University of Frankfurt-am-Main in 1926. He began publishing his research findings on the passage of earthquake tremors through the earth, becoming one of the leading seismologists in the world. In 1930, he left Germany to take a professorship at California Institute of Technology (Caltech). He also became director of the Seismological Laboratory in Pasadena, where he conducted extensive seismological research with Richter. Their joint invention, the Richter scale, was published in 1935.

PETE HALL   American volcanologist and UN representative who several times visited the scientists monitoring Nevado del Ruiz in Colombia in 1985 to make recommendations concerning the monitoring system on the mountain and evacuation procedures.

JIARANG   Engineer who in 8 B.C.E. recommended that China's Yellow River be channelized, a technique that cuts channels to straighten the run of a river and decrease silting. He also instituted programs to divert irrigation water into canals and basins to reduce flooding and projects to build higher levees. These strategies became the standard flood control methods in China for nearly 2,000 years.

WALTER C. JONES   Mayor of Galveston, Texas, when the hurricane of 1900 destroyed the city. On the day following the storm, he established the Central Relief Committee and served as chair of the committee, coordinating first-response efforts for storm survivors.

DANIEL J. MORRELL (1821–1885) The general manager of Cambria Iron Company in Johnstown, Pennsylvania, who arranged an inspection of the South Fork Dam above the town out of concern for the structure's shoddy maintenance. Despite the engineer's damaging report, the dam went unrepaired. Four years after Morrell's death in 1885, a flood destroyed Johnstown and the Gautier Works, a steel mill owned by Cambria Iron Company.

C. RAY NAGIN (1956–) Mayor of New Orleans, Louisiana, when Hurricane Katrina struck in 2005. On August 27, two days before the hurricane made landfall in Louisiana, he declared a state of emergency and issued a voluntary evacuation order for the city. The next day he issued a mandatory evacuation order. In the days following the storm's disastrous work, Nagin angrily criticized the slow response of state and federal agencies, suggesting that President Bush and Governor Blanco were delaying critical action while people in his city were dying. Nagin was elected to a second term in office in 2006, serving through May 2010.

CHRISTOPHER (CHRIS) G. NEWHALL (1948–) Co-creator of the Volcanic Explosivity Index (with STEPHEN SELF). Educated at the University of California at Davis and Dartmouth College, Newhall spent his early career working on the USGS's Volcano Hazards Team. He also was an affiliate professor at the University of Washington and a chairman of the World Organization of Volcano Observatories. In 1982, Newhall and Stephen Self published "The Volcanic Explosivity Index (VEI): An Estimate of Explosive Magnitude for Historical Volcanism" in which they proposed the VEI to measure the explosiveness of volcanic eruptions. As of early 2010, Newhall was group leader of volcanic science at the Earth Observatory of Singapore, an institute of Nanyang Technological University.

CHARLES RICHTER (1900–1985) The American seismologist and physicist who, with BENO GUTENBERG, invented the Richter scale to measure earthquake magnitude. Richter was a pioneer in the field of seismological research. After completing an A.B. degree in physics at Stanford University and a Ph.D. in theoretical physics at Caltech, Richter went to work at the newly established Seismological Laboratory in Pasadena. For decades, he collaborated on research with Beno Gutenberg, the director of the laboratory. Both men developed the Richter scale, published in 1935, although only Richter's name was associated with it.

FRANKLIN D. ROOSEVELT (1882–1945) President of the United States during the dust bowl and the economic depression of the 1930s. His relief and recovery programs, the First New Deal (1933–34) and the Second

New Deal (1935–36), included such agencies as the Civilian Conservation Core, the Agricultural Adjustment Administration, and the Works Progress Administration, designed to combat unemployment and assist desperate farmers, among other goals. Government programs, including social programs, increased rapidly during Roosevelt's administration and helped to redefine the role of government in the lives of citizens.

BENJAMIN F. RUFF (1835–1887) The purchaser in 1879 of the South Fork Dam on Lake Conemaugh, above Johnstown, Pennsylvania. Ruff and a small group of wealthy men formed the South Fork Fishing and Hunting Club and built a clubhouse, cottages, and a road atop the dam, but they neglected proper repair and upkeep of the dam. In 1889, the South Fork Dam burst, flooding Johnstown and killing 2,500 people. While many survivors blamed Ruff for the disaster, he was never held legally accountable for the event.

HERBERT SAFFIR (1917–2007) American civil engineer who, with ROBERT SIMPSON, created the Saffir-Simpson hurricane scale. Saffir began his career as assistant county engineer in Dade County, Florida, in 1947. Twelve years later he established his own consulting firm in Coral Gables, Florida. In 1969, the United Nations commissioned Saffir to do a study of windstorm damage on low-cost housing, and as part of this study Saffir created a scale to measure the force of hurricanes. Robert Simpson contributed descriptions of damages done by the storm in each category. The end result of their work was the Saffir-Simpson hurricane scale. Saffir worked at his firm, Saffir Engineering, until a few weeks before his death at age 90.

BERNARDO SALAZAR A civil engineer assigned by the Central Hydroelectric Company of Caldas (CHEC) in Colombia to monitor and record activities at Nevado del Ruiz, the volcano that erupted on November 13, 1985. When the volcano erupted, he transmitted a message to the capital of Colombia: "The volcano erupted . . . Nevado del Ruiz erupted!" and, moments later, a warning to the towns lying below the volcano: "Manizales, the mudflows are coming. Alert the state of Tolima immediately!"

OMAYRA SÁNCHEZ (CA. 1972–1985) Thirteen-year-old Colombian girl who died in the aftermath of the eruption of the Nevado del Ruiz volcano. Pinned by concrete and debris, Omayra sat neck-deep in water for three days before dying from hypothermia. Hours before her death, photographer FRANK FOURNIER snapped a photograph of Omayra. When the image was published worldwide, the young girl became the face of the disaster, a symbol of the tragic results that can occur when the signs of an imminent natural hazard are not acted upon swiftly.

STEPHEN SELF (1946–) British co-creator of the Volcanic Explosivity Index. Self has written more than 190 articles on volcanoes and volcanic rocks, based on research conducted around the world. In 1982, Self and CHRISTOPHER NEWHALL published "The Volcanic Explosivity Index (VEI): An Estimate of Explosive Magnitude for Historical Volcanism" in which they proposed the VEI to measure the explosiveness of volcanic eruptions. His early career included posts as professor in the department of geology and geophysics at the University of Hawaii and as research faculty member at the Hawaii Institute for Geophysics. Self is responsible for the term *super-eruption*, first used in a paper in 1992 and then formalized in a report and paper in 2005 and 2006, respectively. In the early 2000s, Self was chair in volcanology at the Open University, UK, and head of the volcano dynamics group in the Department of Earth and Environmental Sciences. As of early 2010, Self was a visiting professor in that department and senior volcanologist for the U.S. Nuclear Regulatory Commission.

ROBERT SIMPSON (1912–) Meteorologist who, with HERBERT SAFFIR, created the Saffir-Simpson hurricane scale. From 1967 to 1974, Simpson was director of the National Hurricane Center (NHC). During his tenure there, he and Herbert Saffir developed the Saffir-Simpson hurricane scale to rate the strength—and thus the risk of damage—of hurricanes. In 1974, Simpson left the NHC to establish Simpson Weather Associates, a consulting meteorological firm whose scientists and other specialized personnel conduct research and problem-solving in meteorology, air quality modeling and monitoring, and other environmental areas. During his long career, Simpson has written numerous tropical meteorological books and articles. In 1991, he received the Cleveland Abbe Award for distinguished Service to Atmospheric Sciences by an Individual.

YU THE GREAT Legendary founder of the Xia dynasty (China's first recorded prehistoric dynasty) who is credited with introducing dredging to control river flooding.

9

Organizations and Agencies

A vast array of organizations and agencies concern themselves with natural hazards and disasters. Research institutes and organizations investigate historical and contemporary natural disasters to gain scientific knowledge that can be used for many purposes, from education to disaster risk reduction. Relief agencies assist victims of natural disasters. Some governmental agencies not only perform disaster research for official purposes, but they also organize and explain their findings for the general public. All of these types of organizations and agencies can provide reliable and current information about natural disasters in general or specific disasters and events.

In the list below, most entries include a contact telephone number. If an organization or agency provides an e-mail address in lieu of a telephone number, the e-mail address is included. When contacting an organization, be aware that it may take several days to several weeks to receive a response.

African Centre for Disaster Studies (ACDS)
URL: http://acds.co.za
North-West University
Potchefstroom Campus
11 Hoffman Street
Potchefstroom 2531
South Africa
Phone: (011-27-18) 299-1111

Based at well respected North-West University, ACDS has the primary aim of providing world-class training, education, and research in disaster-related activities within southern Africa and the wider African context. The center publishes the journal *Jàmbá* as well as articles and research reports on disaster risk reduction.

American Radio Relay League, Inc. (ARRL)
URL: http://www.arrl.org
225 Main Street
Newington, CT 06111
Phone: (860) 594-0200

A volunteer organization of licensed radio amateurs, the ARRL-sponsored Amateur Radio Emergency Services (ARES) provides volunteer radio communications services to federal, state, county, and local governments. During and after a natural disaster, members volunteer their services and their privately owned equipment.

American Red Cross
URL: http://www.redcross.org
2025 E Street NW
Washington, DC 20006
Phone: (202) 303-5000

The American Red Cross provides food, shelter, clean-up kits, comfort kits, financial assistance, and counseling to victims of natural disasters. The organization also created the Web site http://www.prepare.org to help people prepare for disasters. In addition, the organization provides community services for the needy, educational programs, and blood donor programs.

Aon Benfield UCL Hazard Research Centre (ABUHRC)
URL: http://www.abuhrc.org
Department of Earth Sciences
University College London
136 Gower Street (Lewis Building)
London, WC1E 6BT
United Kingdom
Phone: (011-44-20) 7679-3637

Based at University College London, the Aon Benfield UCL Hazard Research Centre (ABUHRC) has three research groups: Geological Hazards, Meteorological Hazards and Seasonal Forecasting, and Disaster Studies and Management. Scientists share their findings with businesses, governments, and international agencies to promote the understanding, improvement, and mitigation of natural hazards.

Berkeley Seismological Laboratory (BSL)
URL: http://seismo.berkeley.edu
215 McCone Hall #4760

University of California at Berkeley
Berkeley, CA 94720
Phone: (510) 642-3977

The Berkeley Seismological Laboratory conducts research in earthquake processes and earth structure, provides timely and accurate earthquake information to public and private agencies, and assists in the education of students and the public in earthquake science.

Canadian Association for Earthquake Engineering (CAEE)
URL: http://www.caee.uottawa.ca
c/o Department of Civil Engineering
University of Ottawa
Ottawa, ON K1N 6N5
Canada
E-mail: caee@eng.uottawa.ca

A national nonprofit society of engineers, geoscientists, architects, researchers, educators, designers, planners, economists, social scientists, public officials, government officials, and building code officials, Canadian Association for Earthquake Engineering (CAEE)'s main goal is to foster earthquake engineering practice and research in Canada.

Canadian Centre for Emergency Preparedness (CCEP)
URL: http://www.ccep.ca
860 Harrington Court
Suite 210
Burlington, Ontario L7N 3N4
Canada
Phone: (905) 331-2552

This federal agency's purpose is to create a disaster-resilient country through the education and support of individuals, small businesses, and nonprofit organizations. Canadian Centre for Emergency Preparedness (CCEP) works with Canada's government and the natural disaster management community to form policies, standards, and practices related to disaster management and to foster research and leadership in disaster management.

Canadian Hurricane Centre (CHC)
URL: http://www.ec.gc.ca/ouragans-hurricanes/default.asp?lang=En&n
=DA74FE64-1
Environment Canada
Inquiry Centre

351 St. Joseph Boulevard
Place Vincent Massey, Eighth Floor
Gatineau, Quebec K1A 0H3
Canada
Phone: (819) 997-2800

Canadian Hurricane Centre (CHC) provides Canadians with meteorological information on hurricanes, tropical storms, and post-tropical storms to help them make informed decisions to protect their safety and secure their property.

Caribbean Disaster Emergency Response Agency (CDERA)
URL: http://www.cdera.org/index.php
Building #1, Manor Lodge
Lodge Hill, St. Michael
Barbados
Phone: (246) 425-0386

Caribbean Disaster Emergency Response Agency (CDERA)'s main function is to make an immediate and coordinated response to any disastrous event affecting any participating state, once the state requests such assistance. As of February 2010, CDERA had 16 participating states that included the Bahamas, Barbados, the British Virgin Islands, Jamaica, and others.

Centers for Disease Control and Prevention (CDC)
URL: http://www.bt.cdc.gov/disasters
1600 Clifton Road
Atlanta, GA 30333
Phone: (800) 232-4636

The Centers for Disease Control and Prevention (CDC) seek to prevent public health disasters and to prepare the public for those that do occur. In the event of a natural disaster, the CDC works with local, state, and national responders to contain resulting health threats and to improve victims' recovery. The CDC also works with other agencies to restore public health facilities and functions.

Central Pacific Hurricane Center
URL: http://www.prh.noaa.gov/cphc
2525 Correa Road
Suite 250
Honolulu, HI 96822
Phone: (808) 973-5286

The Central Pacific Hurricane Center issues tropical cyclone warnings, watches, advisories, discussions, and statements for all tropical cyclones in the Central Pacific from 140 degrees West longitude to the International Dateline.

Children's Disaster Services (CDS)
URL:http://www.brethren.org/site/PageServer?pagename=serve_
  childrens_disas ter_services
Church of the Brethren
1451 Dundee Avenue
Elgin, IL 60120
Phone: (800) 323-8039

Children's Disaster Services (CDS) sets up childcare centers in shelters and disaster-assistance centers across the nation. Trained volunteers provide a calm, safe, and reassuring presence in the midst of the chaos created by natural or human-caused disasters.

Disaster Research Center (DRC)
URL: http://www.udel.edu/DRC
University of Delaware
166 Graham Hall
Newark, DE 19716
Phone: (302) 831-6618

Disaster Research Center (DRC) conducts field interviews and research studies on group, organizational, and community preparation for, response to, and recovery from natural and human-caused disasters. Researchers have studied hurricanes, floods, earthquakes, tornadoes, hazardous chemical incidents, plane crashes, and other types of disasters. The Web site provides links to DRC research findings as well as links to other research institutions.

Earthquake Engineering Research Institute (EERI)
URL: http://www.eeri.org
499 Fourteenth Street
Suite 320
Oakland, CA 94612
Phone: (510) 451-0905

Earthquake Engineering Research Institute (EERI) is a society of engineers, geoscientists, architects, planners, public officials, and social scientists. Its goal is to reduce earthquake risk by advancing the science and practice of earthquake engineering and by improving the understanding of the impact of earthquakes on the physical, social, economic, political, and cultural environment. EERI's Web site contains news and updates on recent earthquakes worldwide.

Federal Emergency Management Agency (FEMA)
URL://www.fema.gov
500 C Street SW
Washington, DC 20472
Phone: (202) 646-2500

A part of the U.S. Department of Homeland Security (DHS), Federal Emergency Management Agency (FEMA) is responsible for the emergency response to natural and human-caused disasters. FEMA steps in when the governor of an affected state declares a state of emergency and formally requests FEMA's assistance. FEMA coordinates the response to and recovery from the disaster.

Food and Agriculture Organization (FAO)
URL: http://www.fao.org
Viale delle Terme di Caracalla
00153 Rome, Italy
Phone: (011-39-06) 57051

An agency of the United Nations, Food and Agriculture Organization (FAO)'s goal is to defeat hunger in poor and developing countries. The FAO helps countries modernize and improve agriculture, forestry, and fisheries practices and ensure good nutrition for all. Following a natural disaster, the FAO may set up programs to rehabilitate the agricultural, fishing, and forestry industries.

GeoHazards International
URL://www.geohaz.org
200 Town and Country Village
Palo Alto, CA 94301
Phone: (650) 614-9051

GeoHazards International is an organization that works to make communities safe from earthquakes and other geologic hazards through preparation, mitigation, and advocacy.

Habitat for Humanity
URL: http://www.habitat.org

Habitat for Humanity International
121 Habitat Street
Americus, GA 31709
Phone: (800) 422-4828

A nonprofit Christian ministry whose goal is to eliminate poverty housing and homelessness from the world, Habitat for Humanity relies on volunteers of all backgrounds, races, and religions to help build and renovate houses in partnership with families in need. The organization's Disaster Response provides shelter and housing for people vulnerable to or affected by natural disasters.

International Association for Earthquake Engineering (IAEE)
Ken chiku-kaikan Building, Third Floor
Minatoku Shiba 5-Chome
26-20 Tokyo 108-0014
Japan
E-mail: secretary@iaee.or.jp

An international association of scientists who represent earthquake-engineering organizations from around the world, International Association for Earthquake Engineering (IAEE)'s main goal is "to promote international cooperation among scientists and engineers in the field of earthquake engineering through interchange of knowledge, ideas, and results of research and practical experience." IAEE holds the World Conference on Earthquake Engineering every four years.

International Relief Friendship Foundation (IRFF)
URL: http://www.irff.org
International Administrative Office
880 Route 199
Red Hook, NY 12571
Phone: (917) 319-6802

International Relief Friendship Foundation (IRFF) provides support and emergency services (such as blankets, food, clothing, and relief kits) following a disaster. IRFF also assists other disaster-relief agencies by mobilizing volunteer groups. Other major projects include providing health care, nutrition, and education to those in need. IRFF also sponsors forums to educate the public about humanitarian needs and relief work, both at home and worldwide.

International Strategy for Disaster Reduction (ISDR)
URL: http://www.unisdr.org

Palais des Nations
CH-1211 Geneva 10
Switzerland
Phone: (011-41-22) 917-8908

A program of the United Nations, the International Strategy for Disaster Reduction (ISDR) promotes increased awareness of disaster reduction as an integral part of sustainable development, with the "goal of reducing human, social, economic, and environmental losses" due to disasters. ISDR provides disaster statistics, country and region information, links to organizations and Web sites, and more.

Multidisciplinary Center for Earthquake Engineering Research
(MCEER)
URL: http://mceer.buffalo.edu
State University of New York at Buffalo
Red Jacket Quadrangle
Buffalo, NY 14261
Phone: (716) 645-3391

Based at the University of Buffalo, the State University of New York, the Multidisciplinary Center for Earthquake Engineering Research (MCEER) is devoted to discovering new knowledge, tools, and technologies that can help communities become more disaster resilient.

National Drought Mitigation Center (NDMC)
URL: http://www.drought.unl.edu/
University of Nebraska at Lincoln
819 Hardin Hall
3310 Holdrege Street
Lincoln, NE 68583
Phone: (402) 472-6707

The National Drought Mitigation Center (NDMC) "helps people and institutions develop and implement measures to reduce societal vulnerability to drought" through preparedness and risk management. The center serves state, regional, federal, and tribal governments in addition to maintaining a drought portal online (http://drought.unl.edu/dm/), among other programs and projects.

National Earthquake Information Center (NEIC)
URL: http://earthquake.usgs.gov/regional/neic

1711 Illinois Street
Golden, CO 80401
Phone: (303) 273-8500

The primary goal of the National Earthquake Information Center (NEIC) is "to determine rapidly the location and size of all destructive earthquakes worldwide and to immediately disseminate this information to concerned national and international agencies, scientists, and the general public." The Web site provides links to earthquake data, maps, history, current reports, and more.

National Emergency Response Team (NERT)
URL: http://www.nert-usa.org
1058 Albion Road
Unity, ME 04988
Phone: (888) 637-8872

A volunteer-driven organization that provides shelter, food, clothing, and first aid during times of crisis and disaster, National Emergency Response Team (NERT) also provides Emergency Mobile Trailer Units (EMTUs), truck trailers outfitted as dormitories for emergency shelter.

National Hurricane Center (NHC)
URL: http://www.nhc.noaa.gov
11691 S.W. 17th Street
Miami, FL 33165
Phone: (305) 229-4470

The National Hurricane Center (NHC) issues hurricane watches, warnings, forecasts, and analyses of hazardous tropical weather.

National Institute of Standards and Technology (NST)
URL: http://www.nst.gov
100 Bureau Drive
Stop 1070
Gaithersburg, MD 20899
Phone: (301) 975-6478

Among the National Institute of Standards and Technology (NST)'s programs is the technology innovation program, whose purpose is to provide "cost-shared awards to industry, universities, and consortia for research on potentially revolutionary technologies that address critical national and societal needs" such as how to reduce the damages of natural disasters.

National Oceanic and Atmospheric Administration (NOAA)
URL: http://www.noaa.gov
1401 Constitution Avenue NW
Room 5128
Washington, DC 20230
Phone: (301) 713-1208

National Oceanic and Atmospheric Administration (NOAA)'s mission is "to understand and predict changes in Earth's environment and conserve and manage coastal and marine resources to meet our Nation's economic, social, and environmental needs." Organizations within NOAA include the National Environmental Satellite, Data, and Information Service; the National Weather Service; the Office of Oceanic and Atmospheric Research; and others. NOAA and its agencies research and provide data on the workings of the earth and its atmosphere, including natural hazards.

National Organization for Victim Assistance (NOVA)
URL: http://www.trynova.org
510 King Street
Suite 424
Alexandria, VA 22314
Phone: (703) 535-6682

National Organization for Victim Assistance (NOVA) volunteers provide crisis intervention, emotional support, and spiritual care to help victims, survivors, and disaster-relief workers to cope with the emotional and psychological pain inflicted by disasters and crime.

National Science Foundation (NSF)
URL: http://www.nsf.gov
4201 Wilson Boulevard
Arlington, VA 22230
Phone: (703) 292-5111

The National Science Foundation (NSF) exists "to promote the progress of science; to advance the national health, prosperity, and welfare; [and] to secure the national defense. . . ." Among other duties, the NSF strives to keep the United States at the leading edge of research in geology and to ensure that scientists and engineers receive training in revolutionary methods and technologies in understanding and dealing with natural disasters.

331

National Snow and Ice Data Center (NSIDC)
URL: http://nsidc.org
449 UCB
University of Colorado at Boulder
Boulder, CO 80309-0449
Phone: (303) 492-6199

National Snow and Ice Data Center (NSIDC) supports "research into our world's frozen realms: the snow, ice, glacier, frozen ground, and climate interactions that make up Earth's cryosphere. Scientific data, whether taken in the field or relayed from satellites orbiting Earth, form the foundation for the scientific research that informs the world about our planet and our climate systems."

National Weather Service (NWS)
URL: http://www.nws.noaa.gov
1325 East West Highway
Silver Spring, MD 20910
E-mail: w-nws.webmaster@noaa.gov

The National Weather Service provides weather forecasts and warnings for the United States and its territories. Visitors can find information on current weather hazard assessments, tsunami preparedness, maps, charts, long-range forecasts, and more.

Natural Hazards Center
URL: http://www.colorado.edu/hazards
University of Colorado at Boulder
482 UCB
Boulder, CO 80309-0482
Phone: (303) 492-6818

A national and international clearinghouse of knowledge about the social science and policy aspects of disasters, the Natural Hazards Center collects and shares research related to preparedness for, response to, and recovery from disasters. Among other publications, the center publishes the *Natural Hazards Observer*, a free bimonthly periodical on disaster issues, available online.

NOAA Center for Tsunami Research
URL: http://nctr.pmel.noaa.gov/index.html
7600 Sand Point Way NE
Seattle, WA 98115
Email: Jana.Goldman@noaa.gov

The NOAA Center for Tsunami Research's researchers develop methods and tools for predicting, monitoring, and forecasting tsunamis.

Northeast States Emergency Consortium (NSEC)
URL: http://www.nesec.org
1 West Water Street
Suite 205
Wakefield, MA 01880
Phone: (781) 224-9876

An emergency management organization that responds to hazards in Connecticut, Maine, Massachusetts, New Hampshire, New Jersey, New York, Rhode Island and Vermont. Northeast States Emergency Consortium (NESEC) is active in all phases of emergency management, including preparedness, response, recovery, and mitigation. NESEC works in partnership with federal, state and local governments, and private organizations and is supported and funded by FEMA.

Oxfam America
URL://www.oxfamamerica.org
226 Causeway Street
Fifth Floor
Boston, MA 02114
Phone: (800) 776-9326

Oxfam America is a branch of Oxfam International, a confederation of 14 member organizations whose mission is to help communities find solutions to problems resulting from natural disasters, poverty, armed conflict, and other problems.

Pacific Tsunami Warning Center (PTWC)
URL: http://www.prh.noaa.gov/pr/ptwc
91-270 Fort Weaver Road
Ewa Beach, HI 96706
Phone: (808) 689-8207

The Pacific Tsunami Warning Center (PTWC) "provides warnings for Pacific basin teletsunamis (tsunamis that can cause damage far away from their source) to almost every country around the Pacific rim and to most of the Pacific island states."

ProVention Consortium
URL: http://www.proventionconsortium.org

333

IFRC
17, chemin des Crêts
CH-1211 Geneva 19
Switzerland
E-mail: provention@ifrc.org

ProVention Consortium is an organization established by the World Bank in 2000 to reduce the risk and social, economic, and environmental impacts of natural hazards in developing countries. Since 2003, the organization has been hosted by the International Federation of the Red Cross and Red Crescent Societies.

Public Entity Risk Institute (PERI)
URL: https://www.riskinstitute.org/peri
11350 Random Hills Road
Suite 210
Fairfax, VA 22030
Phone: (703) 352-1846

Public Entity Risk Institute (PERI) provides risk management information, training, data, and data analysis to public entities, small business, and non-profit organizations. PERI's resources include videos, articles, training guides, and other publications on disaster preparedness, response, and recovery.

ReliefWeb
URL: http://www.reliefweb.int/rw/dbc.nsf/doc100?OpenForm
ReliefWeb New York
Office for the Coordination of Humanitarian Affairs
United Nations
New York, NY 10017
Phone: (212) 963-1234

ReliefWeb provides documents and maps pertaining to humanitarian emergencies. Reaching approximately 70,000 subscribers, ReliefWeb offers around-the-clock updates, posting approximately 150 maps and documents each day and maintaining a database of close to 300,000 maps and documents going back to 1981.

Salvation Army
URL: http://www.salvationarmyusa.org/usn/www_usn_2.nsf
615 Slaters Lane
Alexandria, VA 22313

The Salvation Army provides emergency assistance including food, shelter, counseling, missing person services, medical assistance, and distribution of donated goods. The organization also provides referrals to government and private agencies for special services. To contact the Salvation Army, type a zip code into the Locations search box on the Web site, and it will provide contact information for an office nearby.

UNICEF
URL: http://www.unicef.org
New York Headquarters
URL: http://unicefusa.org
UNICEF House
3 United Nations Plaza
New York, NY 10017
(212) 326-7000

During and after natural and human-caused disasters, UNICEF provides children with short-term and long-term emergency aid, including food, water, shelter, protection, and healthcare. UNICEF also provides children in need with healthcare (such as immunizations), clean water, nutrition, and education when they would not otherwise receive these necessities.

United States Agency for International Development (USAID)
URL: http://www.usaid.gov
Ronald Reagan Building
Washington, DC 20523
(202) 712-4810

United States Agency for International Development (USAID) is the principal U.S. agency to extend assistance to countries recovering from disaster, trying to escape poverty, and engaging in democratic reforms. The organization provides humanitarian assistance, education, healthcare, natural resource management, and more. USAID works in close partnership with more than 3,000 private and governmental organizations and agencies in the United States to accomplish its goals.

United States Geological Survey (USGS)
URL: http://www.usgs.gov
12201 Sunrise Valley Drive
Reston, VA 20192
Phone: (703) 648-4000

An agency of the U.S. government that collects and analyzes information and data to describe and understand the Earth, the United States Geological Survey (USGS)'s research, reports, and maps are used to understand and respond to natural disasters, manage natural resources, and compare U.S. data to data from around the world.

University of Delaware
URL: http://www.udel.edu
Newark, DE 19716
Phone: (302) 831-2792

This university offers undergraduate research opportunities in the study of disasters, funded by the Disaster Research Center and the National Science Foundation. In addition, degree programs include the following: BA in Sociology with a Concentration in Emergency and Environmental Management; Ph.D. in Sociology with emphases on Collective Behavior, Social Movements, and Disaster; Interdisciplinary MS in Disaster Science and Management; and Interdisciplinary Ph.D. in Disaster Science and Management.

U.S. Army Corps of Engineers
URL: http://www.usace.army.mil
441 G. Street, NW
Washington, DC 20314
Email: hq-publicaffairs@usace.army.mil

U.S. Army Corps of Engineers helps to mitigate risks of and damage from natural disasters by developing hurricane and storm damage reduction systems, such as designing and building levees along waterways.

West Coast and Alaska Tsunami Warning Center
URL: http://wcatwc.arh.noaa.gov
910 S. Felton Street
Palmer, AK 99645
Phone: (907) 745-4212

The West Coast and Alaska Tsunami Warning Center announces tsunami watches, warnings, and advisories for the U.S. mainland, Canada, Puerto Rico, and the Virgin Islands. The center conducts tsunami response exercises to help train emergency-response personnel to respond effectively when there is a tsunami emergency.

World Bank
URL: http://www.worldbank.org
1818 H Street NW
Washington, DC 20433
Phone: (202) 473-1000

The World Bank is made up of two institutions, the International Bank for Reconstruction and Development (IBRD) and the International Development Association (IDA). The World Bank provides low-interest loans, interest-free credits, and grants to developing countries for purposes including investments in education, health, public administration, infrastructure, agriculture, and environmental and natural resource management. As of February 2010, the World Bank had 186 member countries.

World Food Programme (WFP)
URL: http://www.wfp.org
Via C.G.Viola 68
00148 Rome
Italy
Phone: (011-39-06) 65131

Part of the United Nations system, the World Food Programme (WFP) acts in response to natural disasters, wars, civil conflicts, and other disasters. The organization gets food to those who need it and, following the emergency, uses food to help communities rebuild.

World Health Organization (WHO)
URL: http://www.who.int/en
Avenue Appia 20
1211 Geneva 27
Switzerland
Phone: (011-41-22) 791-21-11

World Health Organization (WHO) is the directing and coordinating authority for health within the United Nations system. It provides leadership on global health matters, shaping the health research agenda, setting norms and standards, articulating evidence-based policy options, providing technical support to countries, and monitoring and assessing health trends.

World Meteorological Organization (WMO)
URL: http://www.wmo.int/pages/index_en.html
7bis Avenue de la Paix

Case Postal 2300
CH-1211 Geneva 2
Switzerland
Phone: (011-41-22) 730-81-11

An agency of the United Nations, the World Meteorological Organization (WMO) has 189 member states and territories (as of December 2009). The WMO coordinates the international development of meteorology and operational hydrology to protect life and property against natural disasters, to safeguard the environment, and to enhance the economic and social well-being of all sectors of society in areas such as food security, water resources, and transport.

10

Annotated Bibliography

The following annotated bibliography focuses on natural disasters, including the science of specific hazards; the history of natural disasters worldwide; case studies of specific disasters; and the research and development of technologies for studying, predicting, and mitigating disasters. Entries are grouped into the following categories:

General Works

Droughts and Famines

Earthquakes and Tsunamis

Storms (Fires, Floods, Hurricanes, Typhoons, Cyclones)

Tornadoes

Volcanoes

Disaster Preparedness and Response

U.S. Case Studies

   Johnstown Flood, 1889

   Galveston Hurricane of 1900

   1906 San Francisco Earthquake

   Dust Bowl of 1934–1941

   Eruption of Mount St. Helens, 1980

   Hurricane Katrina, 2005

International Case Studies

   Sumatra-Andaman Earthquake and Tsunami, 2004–2005

   Water Resource Management in China

   The Famine in Ethiopia in 1984–1985

   Eruption of Nevado del Ruiz, Colombia, 1985

Each category, except for specific case studies, is subdivided into the following sections: Books, Articles and Papers, Web sites, and Other Media.

Your school or local library may or may not have specific books from the following lists. Remember that you can request a book through interlibrary loan at your local public library. Just ask any librarian for help or check for an online form on the library's Web site. You will need a current library card to use this service.

Keep in mind that even the most reliable Web sites reorganize their pages, shut down, or merge with another Web site. The URLs provided in the bibliography below were "live" and functioning at the time this book was published. However, if you find that the URL for an article or Web site no longer works, try searching for the article's title or the Web page's name in a search engine. You will probably find the information in a new location.

GENERAL WORKS

Books

Abbott, Patrick. *Natural Disasters.* Seventh edition. Boston: McGraw-Hill Higher Education, 2008. Written by a professor at San Diego State University, this thick book explores and explains how natural Earth processes can damage, destroy, and kill humans and their surroundings. Key topics include energy sources underlying disasters; plate tectonics and climate change; Earth processes operating in earth, water, and atmosphere; and detailed case studies, among other topics.

Collier, Michael, and Robert H. Webb. *Floods, Droughts, and Climate Change.* Tucson: University of Arizona Press, 2002. An introduction to Earth's climate and its changes and effects. Written at the college freshman reading level.

Gallant, Roy A. *Plates: Restless Earth.* New York: Benchmark Books, 2003. Information about plate tectonics, continental drift, volcano plumes and hot spots, seafloor spreading, and more. Includes full-color photographs, maps, and diagrams.

Grace, Catherine O'Neill. *Forces of Nature: The Awesome Power of Volcanoes, Earthquakes, and Tornadoes.* Washington, D.C.: National Geographic, 2004. Dominated by full-color photographs, this book examines the science of volcanoes, earthquakes, and tornadoes and gives case studies of specific events.

Kiesbye, Stefan, editor. *Are Natural Disasters Increasing?* Detroit: Greenhaven Press, 2009. A collection of articles by different authors on topics including the increase of natural disasters, global warming, human-caused climate change, and social politics and natural disasters.

———. *Disasters.* Detroit: Greenhaven Press, 2009. A collection of short articles by different authors on specific natural disasters, disaster response, and ways to improve relief efforts. Coverage includes the 1960 tsunami in Hawaii, Hurricane Katrina, Cyclone Nargis, the 1998 winter storm in the northeastern United States, the Turkish earthquake of May 2003, and more.

Kovach, Robert, and Bill McGuire. *Firefly Guide to Global Hazards.* Buffalo, N.Y.: Firefly Books, 2004. A guide to global hazards, including volcanoes, earthquakes, tsunamis, flooding, hurricanes, tornadoes, desertification, and many more. Includes case studies of specific events. Packed with full-color photographs and maps.

Median, Phil. *Earth Science.* Homework Helpers series. Franklin Lakes, N.J.: Career Press, 2005. See chapter 4, "Earthquakes and Volcanoes." The chapter includes a series of lessons that include tables, charts, diagrams, and other graphical organizers, review tests, a chapter exam, and an answer key.

Reed, Jim. *Storm Chaser: A Photographer's Journey.* Revised edition. New York: Abrams, 2009. A collection of extreme weather photographs, including images of hurricanes, tornadoes, and floods.

Stille, Darlene R. *Plate Tectonics.* Minneapolis, Minn.: Compass Point Books, 2007. This short book is a study of the earth's tectonic plates and how they move, including the connection to earthquakes and volcanoes. Includes color photographs and diagrams.

Thompson, Graham R., and Jonathan Turk. *Modern Physical Geology.* Second Edition. Fort Worth, Tex.: Saunders College Publishing, 1997. This textbook provides an overview of geology, including information on how geologic processes change and shape Earth. Includes chapters on plate tectonics, volcanoes, and earthquakes and sections on floods, dams, landslides, and the dust bowl. Written for college freshmen students.

Viegas, Jennifer L., ed. *Critical Perspectives on Natural Disasters.* New York: Rosen Publishing Group, 2007. A collection of articles by different authors on earthquakes, tsunamis, volcanic eruptions, hurricanes, tornadoes, and other disaster-related topics.

Articles and Papers

FEMA. "A Citizen's Guide to Disaster Assistance." Available online. URL: http://training.fema.gov/EMIWeb/Is/is7.asp. Accessed February 22, 2010. An online study course that provides a basic understanding of the roles and responsibilities of the local community, the state, and the federal government in providing disaster assistance. Download course materials in PDF format. Unit topics include "Introduction to Disaster Assistance," "Overview of Federal Disaster Assistance," and others.

———. "Are You Ready? An In-depth Guide to Citizen Preparedness." Available online. URL: http://www.fema.gov/areyouready/. Accessed February 22, 2010. An online booklet prepared by FEMA to educate the general public about natural hazards and how to prepare for them and recover from them. Scroll down the page to click on links to individual sections of the booklet (which are in PDF format).

———. "Voluntary Agencies Active in the United States." Available online. URL: http://training.fema.gov/EMIWeb/Is/is7.asp. Accessed February 22, 2010. In the course materials, click on "Appendices, Glossary, and Resource List." See Appendix B, which gives an extensive annotated list of agencies that provide disaster relief.

Kious, Jacqulyne, and Robert I. Tilling. "This Dynamic Earth: The Story of Plate Tectonics." Available online. URL: http://pubs.usgs.gov/gip/dynamic/dynamic.html. Accessed February 18, 2010. An online version of a book published through the USGS. Click on a link in the Contents to go to "Understanding Plate Motions," "'Hotspots': Mantle Thermal Plumes," and other sections. Each section includes definitions and explanations along with plentiful maps and diagrams.

Williams, Jack. "Understanding the Coriolis Force." Available online. URL: http://www.usatoday.com/weather/resources/basics/coriolis-understanding.htm. Accessed March 10, 2010. The author, former *USA Today* weather editor, explains Coriolis force and its role in wind patterns (e.g., tornadoes and hurricanes).

Web sites

Disaster Center. Available online. URL: http://www.disastercenter.com. Accessed February 22, 2010. This Web site provides links to news, graphical data, reports, and other information pertaining to disasters in the United States. Click through to a current U.S. drought map, recent earthquake reports, worldwide volcanic activity reports, and more.

International Strategy for Disaster Reduction. "Stop Disasters! A Disaster Simulation Game from the UN/ISDR." Available online. URL: http://www.stopdisastersgame.org/en/home.html. Accessed February 18, 2010. In this game, plan and build a safer environment for your population, assess disaster risk, and respond when natural hazards strike. Along the way, receive advice that may or may not be useful. Disasters include floods, tsunamis, hurricanes, wildfires, and earthquakes.

NASA Earth Observatory. "Natural Disasters." Available online. URL: http://earthobservatory.nasa.gov/NaturalHazards/. Accessed January 29, 2010. Provides news, articles, and satellite imagery of natural hazards.

Other Media

National Archives. *Natural Disasters.* 538 minutes. Topics Entertainment. 2008. DVD. This set of six DVDs uses materials from the National Archives to chronicle some of the world's most devastating earthquakes, tornadoes, hurricanes, forest fires, and volcanoes, as well as the history of natural disasters. Also recounted are the heroic efforts of relief workers to return cities to normal and the attempts of local governments to better prepare for future events.

National Geographic. *Forces of Nature.* Directed by George Casey. 40 minutes. National Geographic Video, 2004. DVD. Originally released at IMAX theaters, this documentary examines volcanoes, earthquakes, and tornadoes by following scientists on expeditions to discover how natural disasters are triggered.

———. *Nature's Fury.* 55 minutes. National Geographic Video, 2000. DVD. This documentary delves into the destruction wrought by earthquakes, tornadoes, hurricanes, and floods, including the San Francisco earthquakes of 1906 and 1989, and Hurricane Andrew in 1992. Footage includes scenes of disasters as they unfold and interviews with survivors.

DROUGHTS AND FAMINES
Books

Becker, Jasper. *Hungry Ghosts: Mao's Secret Famine.* New York: Henry Holt, 1996. A well-researched account of the famine in China that resulted from Mao Zedong's Great Leap Forward that killed millions of people in 1959–61.

Burgan, Michael. *Not a Drop to Drink: Water for a Thirsty World.* Washington, D.C.: National Geographic, 2008. Investigation into global warming and potential consequences of desertification and drought around the globe.

Clarke, Robin, and Jannet King. *The Water Atlas.* New York: The New Press, 2004. Gives statistics about water availability and usage around the world, including information on floods and droughts. Full-color maps, charts, and diagrams bring the facts to life.

Edwards, Ruth Dudley, and Bridget Hourican. *An Atlas of Irish History.* Third edition. London: Routledge, 2005. Use the index to locate the numerous places in the book where the authors explain the development and effects of the Great Famine of 1845–50.

Fradin, Judy, and Dennis Fradin. *Droughts: Witness to Disaster.* Washington, D.C.: National Geographic, 2008. An explanation of droughts, case studies of some deadly droughts, and information about drought mitigation.

La Bella, Laura. *Not Enough to Drink: Pollution, Drought, and Tainted Water Supplies.* New York: Rosen Publishing Group, 2009. An examination of water supplies on Earth, the renewable quality of freshwater, and the companies and organizations that are fighting to preserve sources of freshwater on the planet.

Litton, Helen. *Irish Famine: An Illustrated History.* Dublin, Ireland: Wolfhound Press, 2003. A chronicle of Ireland's Great Famine, from the coming of the blight in 1845 to the suffering and relief efforts during the famine to the aftermath of the disaster. Statistical tables, quotations, and reproductions of contemporary reports and other materials support the author's account.

Wilhite, Donald A., ed. *Drought and Water Crises: Science, Technology, and Management Issues.* Boca Raton, Fla.: CRC Press, 2005. A collection of articles by different authors on understanding drought, predicting and monitoring droughts, and managing droughts and water demand. The role of science and technology runs throughout the book, and case studies of droughts are given.

Articles and Papers

Avery, Dennis T. "Drought: The Real and Unstoppable Danger." Available online. URL: http://www.cgfi.org/2009/10/05/drought-the-real-and-unstoppable-danger-by-dennis-t-avery. Accessed February 2, 2010. In this short article, the author discusses the link between global warming, drought, and food shortage.

Fominyen, George. "Denial Will Not Solve West Africa's Hunger Problems." Available online. URL: http://www.alertnet.org/db/blogs/58388/2010/00/29-185031-1.htm. Accessed February 1, 2010. Written by AlertNet's humanitarian affairs correspondent for West and Central Africa, this blog examines the hunger crisis in West Africa's Sahel belt, including Nigeria and other nations.

Grogg, Patricia. "World Responds to Drought in Cuba." Available online. URL: http://www.fairelectionsinternational.org/countries/americas/cuba/3231.html. Accessed February 1, 2010. Published in 2005, this article describes international relief efforts for a severe drought in Cuba. Includes information and data regarding water supply in Cuba.

Lovgren, Stefan. "Climate Change Killed Off Maya Civilization, Study Says." Available online. URL: http://news.nationalgeographic.com/news/2003/03/0313_030313_mayadrought.html. Accessed March 10, 2010. The author examines research suggesting that climate change and drought led to the downfall of the Maya civilization ("some 15 million people") in Mesoamerica sometime during the period 800–1000 C.E.

Overland, Martha Ann. "Vietnam Feels the Heat of a 100-Year Drought." Available online. URL: http://www.time.com/time/world/article/0,8599,1969630,00.html. Accessed March 10, 2010. In this article published on March 4, 2010, the author reports on Vietnam's worst drought in more than 100 years.

Spence, Jonathan D. "Mao Zedong." Available online. URL: http://www.time.com/time/magazine/article/0,9171,988161,00.html. Accessed March 10, 2010. A mini-biography of Mao Zedong, the Communist leader responsible for the Great Leap Famine during 1959–61.

Web sites

National Drought Mitigation Center. "U.S. Drought Monitor." Available online. URL: http://www.drought.unl.edu/dm/monitor.html. Accessed February 1, 2010. A clickable map of the United States showing current drought intensity, if present, across the country. Also includes a National Drought Summary.

U.S. Drought Portal. Available online. URL: http://www.drought.gov. Accessed February 2, 2010. Provides maps and data on current droughts. Also provides drought forecasting and early warnings, links to news articles, and other resources. In early 2010, the Web site's research section was under development, promising information and tools in the areas of climatology, hydrology, natural resources, and social sciences.

Other Media

Understanding Drought and Desertification. 29 minutes. Educational Video Network, Inc., 2004. DVD. Examines areas of the Earth that receive low amounts of annual precipitation and how human exploitation in such areas can trigger desertification.

EARTHQUAKES AND TSUNAMIS
Books

Altwater, Brian F., et al. *The Orphan Tsunami of 1700: Japanese Clues to a Parent Earthquake in North America.* Seattle: University of Washington Press, 2005. Tells about a tsunami that struck Japan in 1700 without the normal warning sign of a nearby earthquake. The authors present scientific evidence linking the tsunami to an earthquake that occurred in North America.

Clague, John, Chris Yorath, Richard Franklin, and Bob Turner. *At Risk: Earthquakes and Tsunamis on the West Coast.* Vancouver, B.C., Canada: Tricouni Press, 2006. Information about earthquakes and tsunamis that have occurred or could occur in British Columbia (in Canada) and Washington and Oregon (in the United States). Discusses earthquake prediction and preparedness. Includes numerous full-color photographs, diagrams, and maps.

Clague, John J., Adam Munroe, and Tad Murty. "Tsunami Hazard and Risk in Canada." *Natural Hazards,* 28 (2003): 433–461. A history of tsunamis in Canada and an assessment of the risk of future tsunamis.

de Boer, Jelle Zeilinga, and Donald Theodore Sanders. *Earthquakes in Human History: The Far-Reaching Effects of Seismic Disruptions.* Princeton, N.J.: Princeton University Press, 2005. A history of earthquakes around the world and how they have shaped societies.

Fradkin, Philip L. *Magnitude 8: Earthquakes and Life along the San Andreas Fault.* Berkeley: University of California Press, 1998. A guide to the San Andreas Fault and the powerful earthquakes that have occurred in California since the mid-19th-century. Includes accounts of legendary earthquakes in New York, New England, the central Mississippi River valley, Europe, and the Far East.

Harris, Nancy, ed. *Earthquakes.* San Diego, Calif.: Greenhaven Press, 2003. A collection of articles by different authors on the science and study of earthquakes, specific earthquakes around the world, and prediction and preparedness.

Kusky, Timothy. *Tsunamis: Giant Waves from the Sea.* New York: Facts On File, 2008. A detailed study of the causes and physics of tsunamis, historical tsunami disasters, the 2004 Indian Ocean tsunami, and tsunami preparedness.

Vogel, Carole Garbuny. *Shock Waves through Los Angeles: The Northridge Earthquake.* Boston: Little, Brown and Company, 1996. Account of the January 17, 1994, earthquake in southern California approximately 20 minutes northwest of downtown Los Angeles. Includes photographs, maps, and diagrams.

Articles and Papers

Achenbach, Joel. "The Next Big One." *National Geographic* (April 2006): 120–147. Gives details about the location and current effects of the Hayward Fault, the San Andreas Fault, and other faults in California. Examines the scientific debate over the accuracy of earthquake prediction.

Adler, Jerry, and Mary Carmichael. "The Tsunami Threat." Available online. URL: http://www.newsweek.com/id/48160/page/1. Accessed February 4, 2010. Discusses tsunami risk around the world.

Broad, William. "Deadly and Yet Necessary, Quakes Renew the Planet." Available online. URL: http://www.nytimes.com/2005/01/11/science/11plat.html?_r=1. Accessed February 19, 2010. The author explains why some scientists "argue that in the very long view, the global process behind great earthquakes is quite advantageous for life on earth—especially human life."

Davidson, Sarah. "Mystery of Deadly 1946 Tsunami Deepens." Available online. URL: http://www.livescience.com/environment/041206_tsunami_new.html. Accessed February 4, 2010. Explains scientists' speculations about what caused the April 1, 1946, tsunami in Hilo, Hawaii, including their surprise at not finding the evidence they sought during a seafloor-mapping project.

Geist, Eric L., et al. "Tsunami: Wave of Change." *Scientific American* (January 2006): 56–63. Discusses what scientists have learned about tsunami prediction and preparedness following the 2004 Indian Ocean earthquake and tsunami.

Harmon, Katherine. "Haiti Earthquake Disaster No Surprise to Some Seismologists." Available online. URL: http://www.scientificamerican.com/article.cfm?id=haiti-earthquake-prediction. Accessed February 18, 2010. Explains why some seismologists were concerned about the likelihood of a devastating earthquake in Haiti before the January 2010 quake occurred.

National Geographic News. "The Deadliest Tsunami in History?" Available online. URL: http://news.nationalgeographic.com/news/2004/12/1227_041226_tsunami.html. Accessed February 4, 2010. Gives information about the Indian Ocean tsunami of 2004.

New York Times. "Sichuan Earthquake." Available online. URL: http://topics.nytimes.com/topics/news/science/topics/earthquakes/sichuan_province_china/index.html. Accessed March 10, 2010. This article describes the magnitude 7.9 earthquake that struck Sichuan Province in China on May 12, 2008, killing at least 70,000. The article examines responses to China's one-child policy in the aftermath of the disaster that killed the only child in many families.

NOVA Online. "Wave That Shook the World." Available online. URL: http://www.pbs.org/wgbh/nova/tsunami. Accessed February 16, 2010. This Web page contains links to information about the 2004 Indian Ocean tsunami. Tsunami expert Lori Dengler answers questions about tsunamis. Read about tsunami detection and hazard mitigation. Use interactive models and graphics that document the 2004 Indian Ocean tsunami and other tsunamis.

O'Hanlon, Larry. "Tsunami! How to Catch a Wave before It Kills." Available online. URL: http://dsc.discovery.com/news/2008/03/13/tsunami-buoy.html. Accessed February 4, 2010. The author reports on the completion of the 39-buoy tsunami warning system for the Indian Ocean.

PBS Online. "Savage Earth: The Restless Planet: Earthquakes." Available online. URL: http://www.pbs.org/wnet/savageearth/earthquakes/index.html. Accessed February 25, 2010. A scientific explanation of why earthquakes occur and the devastation they can cause. Includes an animation with short explanations to help viewers visualize an earthquake as it happens, beginning deep within Earth.

———. "Savage Earth: Waves of Destruction: Tsunamis." Available online. URL: http://www.pbs.org/wnet/savageearth/tsunami. Accessed February 25, 2010. A scientific explanation of why tsunamis occur and the devastation they can cause. Includes an animation with short explanations that illustrates a tsunami in action.

Annotated Bibliography

USGS. "Surviving a Tsunami—Lessons from Chile, Hawaii, and Japan." Available online. URL: http://pubs.usgs.gov/circ/c1187. Accessed February 4, 2010. Based on interviews with tsunami survivors, this document explains "how to survive—and how not to survive—a tsunami." Includes numerous photographs, maps, diagrams, charts, and other graphical data.

Web sites

International Tsunami Information Center. Available online. URL: http://ioc3.unesco.org/itic. Accessed February 25, 2010. This Web site gives a scientific overview of tsunamis, explains tsunami warnings, provides official tsunami bulletins, gives a clickable list of tsunami events from 1929 to the present (clicking on a year will take you to paragraph-length summaries of tsunamis that occurred that year), and more.

National Earthquake Hazards Reduction Program. Available online. URL: http://www.nehrp.gov. Accessed February 24, 2010. The NEHRP was established by Congress "to reduce the risks of life and property from future earthquakes in the United States." The program is administered by representatives from the United States Geological Survey, the National Institute of Standards and Technology, the National Science Foundation, and the Federal Emergency Management Agency.

Natural Resources Canada. "Earthquakes." Available online. URL: http://earthquakescanada.nrcan.gc.ca/index-eng.php. Accessed February 1, 2010. Provides recent news and reports on earthquakes in Canada. Also includes information on historic earthquakes, a hazard calculator, a seismogram viewer, links to the Web sites of seismic monitoring stations, and more.

NOAA Center for Tsunami Research. "DART (Deep-Ocean Assessment and Reporting of Tsunamis." Available online. URL: http://nctr.pmel.noaa.gov/Dart/index.html. Accessed February 2, 2010. Explains the DART real-time tsunami warning system. Includes a map showing the locations of the DART buoys.

Other Media

America's Tsunami: Are We Next? Discovery Channel, 2008. DVD. A team of scientists investigate whether a devastating tsunami could strike America. They journey below the ocean's surface to obtain new data from an ocean ridge taller than the Alps, seeking to learn whether a tsunami like the 2004 Indian Ocean tsunami could happen again—and where.

National Geographic: Tsunami—Killer Wave. 52 minutes. National Geographic Television, 2005. DVD. "The documentary explores the causes of tsunami waves. . . . Survivors and scientists tell gripping tales of past tsunami disasters in Hawaii, Japan, and the Pacific Northwest."

Tsunami 2004: Waves of Death. 50 minutes. A&E Home Video, 2005. DVD. Chronicles the 2004 Indian Ocean tsunami as it moves from coast to coast, killing people in 14 countries. Draws on amateur video and interviews with scientists.

STORMS (FIRES, FLOODS, HURRICANES, TYPHOONS, CYCLONES)

Books

Amundson, Mavis. *The Great Forks Fire*. Port Angeles, Wash.: Western Gull Publishing, 2003. Factual narrative about a wildfire that threatened Forks, Washington, in September 1951.

Fox, William Price. *Lunatic Wind: Surviving the Storm of the Century*. Chapel Hill, N.C.: Algonquin Books of Chapel Hill, 1992. For this book, the author interviewed survivors of Hurricane Hugo and wrote stories to illustrate what it was like to experience the most damaging hurricane the United States had suffered to that date (1989). The fact-based stories are interspersed with nonfiction essays that document the storm.

Lauber, Patricia. *Flood: Wrestling with the Mississippi*. Washington, D.C.: National Geographic Society, 1996. A photographic essay giving information about changes in the course of the Mississippi River, levees and floodwalls, and the river's flooding. The floods of the 1880s, 1932, and 1993 receive special attention. Includes full-color photographs, maps, and diagrams.

Toomey, David. *Stormchasers: The Hurricane Hunters and Their Fateful Flight into Hurricane Janet*. New York: W. W. Norton, 2002. The author uses navy documents and interviews to reconstruct the September 1955 mission in which U.S. Navy lieutenant commander Grover B. Windham and a crew of eight flew into the eye of Hurricane Janet for research purposes. They never returned.

Articles and Papers

Carter, Ann. "Q&A with Herbert Saffir." Available online. URL: http://www.novalynx.com/saffir-interview.html. Accessed March 10, 2010. Ann Carter interviews Herbert Saffir, co-creator of the Saffir-Simpson Hurricane Scale. Saffir gives information about his background as a structural engineer and how he got involved in the investigation of windstorms. He explains how he and Robert Simpson developed the Saffir-Simpson Hurricane Scale.

Lacovelli, Debi. "The Saffir/Simpson Hurricane Scale: An Interview with Dr. Robert Simpson." Available online. URL: http://www.novalynx.com/simpson-interview.html. Accessed on March 10, 2010. Most of this article consists of the author's interview with Robert Simpson, co-creator of the Saffir-Simpson Hurricane Scale. Simpson explains how he and Herbert Saffir developed the scale.

Nash, J. Madeleine. "What's Behind California's Wild Weather?" Available online. URL: http://www.time.com/time/magazine/article/0,9171,1018082,00.html. Accessed January 27, 2010. An explanation of the Pineapple Express, a weather phenomenon that brings "powerful cyclonic systems over warm waters near Hawaii, where they tank up with moisture before slamming into the West Coast."

National Weather Service Eastern Region Headquarters. "Historical Floods in the Northeast." Available online. URL: http://www.erh.noaa.gov/er/nerfc/historical.

Accessed February 2, 2010. A list of dates of floods in the northeastern United States from 1927–96. Click on a date to go to the Web page devoted to information, data, and images relating to that flood.

NEMO. "NEMO Remembers the Great Hurricane of 1780." Available online. URL: http://www.cdera.org/cunews/news/saint_lucia/article_1314.php. Accessed January 27, 2010. A gathering of quotations going back to 1780 that describe the destruction by a hurricane that hit the Caribbean during October 10–16, 1780.

O'Connor, Jim E., and John E. Costa. "Large Floods in the United States: Where They Happen and Why." Circular 1245. Washington, D.C.: U.S. Geological Survey, 2003. Explains why floods are a natural hazard and explains features of landscapes and climate that contribute to large floods in the United States. Includes maps, charts, diagrams, and photographs.

———. "The World's Largest Floods, Past and Present: Their Causes and Magnitudes." Circular 1254. Washington, D.C.: U.S. Geological Survey, 2004. The authors describe "the causes and magnitudes of the world's largest floods, including those measured and described by modern methods in historic times, as well as floods of prehistoric times, for which the only records are those left by the floods themselves." Includes maps, charts, diagrams, and photographs.

Perry, Charles A. "Significant Floods in the United States during the 20th Century: USGS Measures a Century of Floods." Available online. URL: http://ks.water. usgs.gov/pubs/fact-sheets/fs.024-00.html. Accessed February 2, 2010. The author defines different kinds of floods, from storm-surge floods to regional floods. Includes a U.S. map showing locations and years of significant floods. Also includes a chart listing significant floods by type, along with data such as deaths reported and economic cost.

Reuters AlertNet. "Eyewitness: Hurricane Mitch Five Years On." Available online. URL: http://www.alertnet.org/thefacts/reliefresources/106759809664.htm. Accessed January 28, 2010. A firsthand account of one Nicaraguan woman's survival of Hurricane Mitch, a powerful hurricane that struck Central America in 1998.

Sharp, Deborah. "Floridians Recall Andrew's Rage." Available online. URL: http://www. usatoday.com/news/nation/2002-08-22-andrew-anniversary_x.htm. Accessed January 27, 2010. In 2002, 10 years after Hurricane Andrew, southern Floridians describe how their community and lives were permanently changed by the costliest natural disaster in history at the time.

Tremlett, Giles, and Larry Elliott. "Madeira Floods: Death Toll Rises to 40." Available online. URL: http://www.guardian.co.uk/world/2010/feb/21/madeira-floods-death-toll-rises. Accessed March 10, 2010. This article reports on a flash flood that occurred on February 20, 2010, on the Portuguese island of Madeira. The disaster killed at least 40 people and injured 120.

UNESCO Courier. "Dancing with the Rivers." Available online. URL: http://www. unesco.org/courier/1999_08/uk/dici/txt1.htm. Accessed January 28, 2010. An article and photographs documenting the lives of Bangladeshis who live on chars, which are "islands that are periodically submerged by the nation's mighty rivers."

Web sites

Dartmouth Flood Observatory. Available online. URL: http://www.dartmouth. edu/~floods/index.html. Accessed February 2, 2010. The observatory provides space-based measurements of surface water on Earth by using orbital remote sensing to detect, measure, and map river discharge and river flooding. The information is used for flood detection, flood response, future risk assessment, and water resources research.

Hipke, Deana C. "The Great Peshtigo Fire of 1871." Available online. URL: http://www. peshtigofire.info. Accessed February 25, 2010. This Web site gathers and organizes extensive information about the forest fires of 1871 in northeastern Wisconsin and Upper Michigan. Sections include the Cause, the Victims, Bibliography, Picture Gallery, Museum, Peshtigo Today, and FAQs.

Long Island Express: The Great Hurricane of 1938. Available online. URL: http:// www2.sunysuffolk.edu/mandias/38hurricane. Accessed January 28, 2010. Created and maintained by Professor Scott A. Mandia, this Web site gives a factual account of the category 3 hurricane that struck New England in 1938. Includes information on the hurricane's weather, damages, and geological impact. Also includes human interest stories and more.

NOVA Online. "Fire Wars." Available online. URL: http://www.pbs.org/wgbh/nova/ fire. Accessed February 16, 2010. This Web page contains links to an assortment of wildfire information. You can take a look behind the scenes as film crews document wildfires, review the 2000 fire season on six continents, investigate the wildfire-wildlife bond, use a wildfire simulator, and explore the science behind wildfires.

———. "Flood!" Available online. URL: http://www.pbs.org/wgbh/nova/flood. Accessed February 16, 2010. This Web page contains links to an assortment of flood information. Read about floods on the Nile River, the Yellow River, and the Mississippi River. Read about the 1997 Red River Flood and the Great Flood of 1993 (both in the United States). Listen to audio files of floodwaters in action.

Other Media

The Hurricane of '38. Written and produced by Thomas Lennon and Michael Epstein. 53 minutes. American Experience, 1993. Reissued by WGBH Boston Video, 2007. DVD. This documentary examines the destruction wrought by a hurricane that the National Weather Bureau predicted would blow out in Cape Hatteras, North Carolina. Instead, it began an unexpected sprint north along the coast. Over 600 people were killed. Another 100 were never found.

TORNADOES

Books

Bluestein, Howard B. Tornado Alley: Monster Storms of the Great Plains. New York: Oxford University Press, 1999. Historical account of the study of tornadoes and the great thunderstorms that spawn them. The author spent 20 years as a storm chaser and photographer. Includes color and black-and-white photographs and diagrams.

Mathis, Nancy. *Storm Warning: The Story of a Killer Tornado.* New York: Touchstone, 2007. Written by a journalist who grew up in Tornado Alley, this well-researched book chronicles a series of 71 tornadoes that tore through Oklahoma on May 3, 1999. One of these tornadoes spanned a mile, making it the largest tornado in recorded history.

Articles and Papers

Vesilind, Priit J. "Chasing Tornadoes." Available online. URL: http://environment. nationalgeographic.com/environment/natural-disasters/chasing-tornadoes. Accessed February 28, 2010. The author describes how and why professional storm-chasers and a photographer chase tornadoes. Details about specific tornadoes, descriptions of specialized equipment, and quotations from the participants make this long article both informative and entertaining.

Web sites

Tornado Project Online. Available online. URL: http://www.tornadoproject.com. Accessed February 28, 2010. Researched and written by a small company named the Tornado Project, this Web site gives extensive information on tornadoes, from case studies to storm shelters. As of February 2010, the "Recent Tornadoes" list included data from storms in 2008, while the "Past Tornadoes" list covered storms from 1896 and 1995–2007.

Other Media

Tornado Intercept. Written and produced by Lawrence Cumbo. 52 minutes. National Geographic, 2006. DVD. A scientist and a filmmaker explore the intricate and potentially deadly forces of a tornado by driving into the heart of an actual storm in a specially designed vehicle made of 8,000 pounds of steel plating and bullet-proof glass.

VOLCANOES

Books

Bourseiller, Philippe, photographer. *Volcanoes: Journey to the Crater's Edge.* Adapted by Robert Burleigh. Text by Hélène Montardre. Drawings by David Giraudon. New York: Harry N. Abrams, 2003. Large, full-color photographs of volcanoes, ash fallout, underwater lava tunnels, mud bowls, craters, volcanic eruptions, and more, with short explanatory texts.

Decker, Robert, and Barbara Decker. *Volcanoes in America's National Parks.* Sheung Wan, Hong Kong: Airphoto International, 2001. A travel guide to active and inactive volcanoes in the United States. Includes maps and full-color photographs.

Duffield, Wendell A. *Chasing Lava: A Geologist's Adventures at the Hawaiian Volcano Observatory.* Missoula, Mont.: Mountain Press Publishing, 2003. An entertaining account of living and working at Kīlauea, one of the world's most active volcanoes.

Harris, Stephen L. *Fire Mountains of the West: The Cascade and Mono Lake Volcanoes.* Third edition. Missoula, Mont.: Mountain Press Publishing, 2005. Detailed introduction to the geology, history, and hazards of volcanic activity from California to southwestern British Columbia.

Hill, Richard L. *Volcanoes of the Cascades: Their Rise and Their Risks.* Guilford, Conn.: Globe Pequot Press, 2004. A study of the Cascades' 13 major volcanoes, including Mount Baker, Mount Rainier, and Mount St. Helens. Color photographs.

Lopes, Rosaly. *The Volcano Adventure Guide.* Cambridge: Cambridge University Press, 2005. Informative guide to visiting 42 different volcanoes around the world. Briefly discusses eruption styles of different types of volcanoes, how to prepare for a volcano trip, and how to avoid volcanic dangers.

Rosi, Mauro, et al. *Volcanoes.* Buffalo, N.Y.: Firefly Books, 2003. Describes and illustrates 100 volcanoes from around the world, including their structure and history. Packed full of full-color photographs, maps, and diagrams.

Scarth, Alwyn. *Vesuvius: A Biography.* Princeton, N.J.: Princeton University Press, 2009. An extensively researched historical account of the volcano Vesuvius and nearby Naples and the province of Campania. Examines the science of the volcano's behavior and "the changing social, religious, and intellectual impact that the volcano has always had upon the population." Includes photographs, diagrams, and maps.

Winchester, Simon. *Krakatoa: The Day the World Exploded: August 27, 1883.* New York: HarperCollins, 2003. A well-researched book that examines the world-changing effects of the eruption of Krakatoa off the coast of Java. Includes maps, diagrams, and black-and-white photographs.

Articles and Papers

Bindeman, Ilya N. "The Secrets of Supervolcanoes." *Scientific American* (June 2006): 36–43. Explains what supervolcanoes are and what microscopic crystals of volcanic ash reveal about past eruptions of the world's supervolcanoes.

Camp, Vic. "How Volcanoes Work: Mt. Pelee Eruption (1902)." Available online. URL: http://www.geology.sdsu.edu/how_volcanoes_work/Pelee.html. Accessed March 10, 2010. The author, a geology professor at San Diego State University, chronicles events surrounding the eruption of Mt. Pelee, including descriptions of the eruption and accounts of survivors.

Krystak, Lee. "Is the Super Volcano beneath Yellowstone Ready to Blow?" Available online. URL: http://www.unmuseum.org/supervol.htm. Accessed February 18, 2010. Explains what a super volcano is and gives information about the historical eruption of the super volcano beneath Yellowstone National Park. Speculates about the likelihood of a similar eruption in the future.

Monterey Bay Aquarium Research Institute. "Submarine Volcanism." Available online. URL: http://www.mbari.org/volcanism. Accessed February 18, 2010. Short articles on numerous topics relating to submarine volcano activity. Topics include hot spots, volcanic processes, explosive eruptions, mid-ocean ridges, and others. Articles include photographs as well as summaries and bibliographical citations of the institute's published research.

Newhall, Christopher G., and Stephen Self. "The Volcanic Explosivity Index (VEI): An Estimate of Explosive Magnitude for Historical Volcanism." *Journal of Geophysical Research* 87 (C2) (1982): 1,231–1,238. The authors introduce the Volcanic Explosivity Index, a scale for ranking the strength of a volcanic eruption according to its magnitude, and propose the use of the VEI in the study of volcanoes.

NOVA Online. "Volcano under the City." Available online. URL: http://www.pbs. org/wgbh/nova/volcanocity. Accessed February 16, 2010. Read a documentary filmmaker's account of filming inside the crater of the volcano Nyiragongo, just north of the city of Goma in the Democratic Republic of the Congo. Explore an interactive diagram of Nyiragongo. Read about the debate over whether we can accurately forecast volcanic eruptions, and read about some of the worst volcanic disasters of the past 400 years.

Web sites

Alaska Volcano Observatory. Available online. URL: http://www.avo.alaska.edu. Accessed February 25, 2010. An information center for current and past activity of Alaska's volcanoes. Current activity reports are posted on the home page, and libraries of photographs are available. An interactive map lets you locate and view webcams and webicorders; you can also plot the locations of seismic stations, recent earthquakes, monitored volcanoes, and other volcanoes.

NOVA Online. "Volcano's Deadly Warning." Available online. URL: http://www.pbs. org/wgbh/nova/volcano. Accessed February 16, 2010. This Web page contains links to an assortment of volcano information. Read an interview with Bernard Chouet, a volcano seismologist, in which he describes his theory of predicting volcanic eruptions. Read an interview with Dan Miller, chief of the Volcano Disaster Assistance Program, in which he explains what the team does. View photographs illustrating volcanic terminology, from ash to magma. View a seismograph's recordings of volcano-related earthquakes.

Smithsonian Institution. "Global Volcanism Program: Worldwide Halocene Volcano and Eruption Information." Available online. URL: http://www.volcano.si.edu. Accessed February 18, 2010. Provides volcano news, activity reports, maps, interactive maps, and photographs.

University of Rhode Island, Graduate School of Oceanography. "Exploring Earth's Volcanic Environments: Vesuvius Volcano 79 A.D. Eruption." Available online. URL: http://www.gso.uri.edu/vesuvius/Home/index.html. Accessed February 18, 2010. An online course in the study of volcanoes, based on the eruption of Vesuvius. By working through the lessons, you can reconstruct the eruption of Vesuvius by examining evidence in the volcanic material, carry out observations and measurements during a virtual field trip, and form your own hypotheses about what happened.

Other Media

Exploring the Fire Below Us: Mount St. Helens and the Cascade Volcanoes. Global Net Productions, 1996. CD-ROM. Includes 45 minutes of video footage, more than 50 photographs, an interactive eruption time line, survivor reenactments, and a review of volcanoes by Dr. Stephen Harris.

Killer Volcano: Eruption of Mount Pinatubo. Produced and directed by Rob Whittlesey and Noel Buckner. 56 minutes. NOVA, 1993. Reissued 2006. DVD. This documentary follows a group of geologists that remains behind when the Philippines's Mount Pinatubo, a stratovolcano, erupts in June 1991. Includes footage of the world's largest volcanic eruption in 80 years.

Siebert, Lee, et al. *Volcanoes of Central America.* Smithsonian Institution, Global Volcanism Program, 2006. CD-ROM. "Extensive compilation of data and images for the volcanoes of Central America. A map-driven interface allows selection of each of the 80 Holocene and 46 Pleistocene volcanoes of Central America. Users can view data about individual volcanoes, chronologies of known eruptions in the past 10,000 years, more than 950 images, over three decades of activity reports, and GVP's petrologic, bibliographic, and map databases."

DISASTER PREPAREDNESS AND RESPONSE
Books

Cherry, Katie E., ed. *Lifespan Perspectives on Natural Disasters: Coping with Katrina, Rita, and Other Storms.* New York: Springer, 2009. A collection of articles by different authors—many written by teams of authors—that examine the psychological effects of disasters on different age groups, from children and adolescents through senior citizens. The emphasis is on disaster recovery and coping as well as on how trained personnel can use this information for age-appropriate disaster preparedness and recovery.

Coppola, Damon. *Introduction to International Disaster Management.* Boston: Butterworth Heinemann, 2007. Includes sections on the history of disaster management, hazard analysis, risk and vulnerability, and disaster mitigation.

Drabek, Thomas E. *The Human Side of Disaster.* Boca Raton, Fla.: CRC Press, 2010. An evaluation of human responses to natural and human-caused disasters, particularly the overwhelming desire for survival and protection of loved ones. The author highlights the role of human beings in disaster situations and how emergency response systems can use that role in disaster preparedness and response.

Goldstein, Natalie. *Drought and Heat Waves: A Practical Survival Guide.* New York: Rosen Publishing, 2006. Includes information on how to prepare and store drinking water and what to do in case of heat-related illness, as well as how to start good habits of water and energy conservation.

Housely, Jennifer, and Larry E. Beutler. *Treating Victims of Mass Disaster and Terrorism.* Cambridge, Mass.: Hogrefe and Huber Publishers, 2007. This short volume is a scholarly explanation of disorders, such as post-traumatic stress disorder, and how to diagnose and treat them following mass casualty events.

Pampel, Fred C. *Disaster Response.* New York: Facts On File, 2007. Explores natural and human-made disasters and the nation's response to them, beginning with an overview of the history of disaster response and opinions about the topic. Contemporary case studies range from the terrorist attacks of 9/11 to natural disasters such as Hurricane Katrina. Also includes biographies, an annotated bibliography, a chronology, organization and agency listings, and a glossary.

Smith, Keith, and David N. Petley. *Environmental Hazards: Assessing Risk and Reducing Disaster.* Fifth edition. New York: Routledge, 2009. In Part 1, this respected textbook examines the nature of hazard, including the chapters "Risk Assessment and Management" and "Reducing the Impacts of Disasters." In Part 2, the authors closely examine nine categories of rapid-onset hazards, both natural and human-caused.

Yeats, Robert S. *Living with Earthquakes in California: A Survivor's Guide.* Second edition. Corvallis: Oregon State University Press, 2004. A how-to manual for earthquake preparedness and survival. Reviews California's history of earthquake hazards and examines major faults that threaten California and Nevada. Discusses earthquake forecasting, catastrophe insurance, and tsunamis.

Articles and Papers

Austin, Diane E. "Coastal Exploitation, Land Loss, and Hurricanes: A Recipe for Disaster." *American Anthropologist* 108, no. 4 (2006): 671–691. The author examines southern Louisiana as a "dynamic landscape, marked by coastal wetlands interrupted by both natural and human-made levees, and vulnerable to both the Mississippi and Atchafalaya rivers and major storms coming off the Gulf of Mexico." She explains how human activities along the Mississippi River, such as levee construction and canal dredging, contribute to coastal land loss. She discusses hurricanes "in light of the relationship between Louisiana and the rest of the United States and the environmental and community degradation that has occurred along the coast."

Comfort, Louise K. "Risk, Security, and Disaster Management." *Annual Review of Political Science* 8, no. 1 (2005): 335–356. This scholarly article "examines the policies and practices that address the evolving conditions of risk, security, and disaster management in U.S. society."

Cyranoski, David. "Earthquake Prediction: A Seismic Shift in Thinking." *Nature* (October 2004): 431. The author examines the debate about the scientific possibility and feasibility of earthquake prediction, a hot topic within the U.S. scientific community.

Gaffney, Donna A. "The Aftermath of Disaster: Children in Crisis." *Journal of Clinical Psychology* 62, no. 8 (2006): 1,001–1,016. The author uses the disasters of Hurricane Katrina, Hurricane Rita, and the terrorist attacks of September 11, 2001, as a means of illustrating the impact of natural and human-caused disasters in the lives of children and adolescents.

Hayden, Thomas. "Preventing Disaster: An Earthquake Expert Who Saw the Warning Signs." *U.S. News and World Report* (January 2005). Available online. URL: http://www.usnews.com/usnews/news/articles/050124/24tsunami.htm. Accessed February 28, 2010. In 2003, at a UN meeting, an American seismologist Phil Cummins and other scientists began discussing a tsunami warning system for the Indian Ocean. This article explains their ideas, developed before the 2004 tsunami, and how plans for a warning system began to take shape in early 2005.

Oneworld.net. "Cyclone Evacuation Saves Thousands in Bangladesh." Available online. URL: http://us.oneworld.net/article/363636-evacuation-measures-save-lives-bangladesh. Accessed January 28, 2010. This article tells how a cyclone early warning system saved thousands of lives when Cyclone Aila struck Bangladesh in 2009. Includes photographs, a map, and links to further information.

Web sites

Annenberg Media. "Volcanoes: Can We Predict Volcanic Eruptions?" Available online. URL: http://www.learner.org/interactives/volcanoes. Accessed February 28, 2010. This Web site explains how volcanoes form, why and how they erupt, hazards of volcanoes, forecasting eruptions, and reducing risk. It also includes a page of related resources for further online study.

CDC. "Emergency Preparedness and Response: Natural Disasters and Severe Weather." Available online. URL: http://www.bt.cdc.gov/disasters. Accessed February 25, 2010. From the perspective of hygiene and sanitation, the CDC explains health and safety concerns related to earthquakes, floods, hurricanes, landslides and mudslides, tornadoes, tsunamis, volcanoes, and wildfires.

National Weather Service. "TsunamiReady." Available online. URL: http://www.tsunami ready.noaa.gov. Accessed February 2, 2010. This Web page provides information and links to articles "designed to help cities, towns, counties, universities, and other large sites in coastal areas reduce the potential for disastrous tsunami-related consequences."

United States Federal Government. DisasterAssistance.gov. Available online. URL: http://disasterhelp.gov. Accessed February 24, 2010. This is a Web portal that consolidates disaster assistance information from multiple federal programs. Following a presidentially declared disaster, citizens can go to this Web site to find out if they are eligible for assistance and to apply for assistance. Citizens affected by a natural disaster while out of the country can use this Web site to find assistance.

U.S. CASE STUDIES

Johnstown Flood, 1889

Dreyfuss, Richard, narrator. *Johnstown Flood.* Directed by Richard Burkert. 84 minutes. Johnstown Area Heritage Association, 2003, DVD. Based on firsthand accounts of survivors, this account of the Johnstown Flood examines the disaster, its physical destruction, and its aftermath.

Annotated Bibliography

Johnstown Flood Museum. "Facts about the Johnstown Flood." Available online. URL: http://www.jaha.org/FloodMuseum/facts.html. Accessed May 18, 2010. A compilation of statistics about the deaths and damages caused by the flood.

———. "Survivor Stories of the Johnstown Flood." Available online. URL: http://www.jaha.org/FloodMuseum/survivors.html. Accessed May 18, 2010. A collection of short biographies and photographs of individuals and families who survived the flood. Included are Anna Fenn Maxwell, Victor Heiser, the Waters family, the Reverend H. L. Chapman, and others.

———. "The South Fork Fishing and Hunting Club and the South Fork Dam." Available online. URL: http://www.jaha.org/FloodMuseum/clubanddam.html. Accessed May 18, 2010. Information about the pleasure club that occupied the dam above Johnstown, including a complete list of club members, details of club activities, and an account of how some club members responded to the disaster by donating blankets and money and volunteering.

JohnstownPA.com. "1889 Johnstown Flood by the New York Times." Available online. URL: http://www.johnstownpa.com/History/hist30.html. Accessed May 18, 2010. A collection of *New York Times* articles on the Johnstown Flood, published in the days following the disaster. The articles report on the "appalling catastrophe," a temporary martial government, starving survivors, the inadequate supply of disinfectants, and other topics.

McCullough, David. *The Johnstown Flood: The Incredible Story behind One of the Most Devastating 'Natural' Disasters America Has Ever Known.* New York: Simon and Schuster, 1987. This well-researched, chronological account of the flood draws upon records, diaries, letters, interviews with survivors, and the transcript of a private investigation into the disaster by the Pennsylvania Railroad. The account includes details of the wealthy men who patronized the club atop the dam, the failure of the dam, and the disaster that followed. Back matter includes a list of victims.

Galveston Hurricane of 1900

Bixel, Patricia Bellis, and Elizabeth Hayes Turner. *Galveston and the 1900 Storm: Catastrophe and Catalyst.* Austin: University of Texas Press, 2000. An account of the hurricane, its destruction, relief efforts, and the rebuilding of the island. Written by two historians, this account highlights the changes in Galveston's civic culture that were prompted by the natural disaster, including discrimination against African Americans. Includes numerous illustrations.

Galveston County Daily News. "The 1900 Storm: Galveston Island, Texas." Available online. URL: http://www.1900storm.com/index.lasso. Accessed January 27, 2010. A Web site containing extensive information on the 1900 Galveston hurricane. Includes newspaper accounts, survivors' stories, photographs, and film clips.

Greene, Casey Edward, and Shelly Henley Kelly, eds. *Through a Night of Horrors: Voices from the 1900 Galveston Storm.* College Station: Texas A&M University Press, 2002. In this volume the editors have gathered a collection of survivors' letters and memoirs about the Galveston hurricane, some written in the days following the hurricane and others written years later.

Larson, Erik. *Isaac's Storm: A Man, a Time, and the Deadliest Hurricane in History.* New York: Vintage, 2000. Based on Dr. Isaac Cline's personal letters, telegrams, reports, and diaries, as well as the testimonies of survivors, this account of the Galveston hurricane of 1900 is meticulously researched and well written. Told mainly from Dr. Cline's point of view, the account takes the reader through the events of the hurricane, from events leading up to the storm to the aftermath. Includes information on Dr. Cline's role in the U.S. Weather Bureau and the bureau's disregard for hurricane warnings from Cuba.

1906 San Francisco Earthquake

Bronson, William. *The Earth Shook, the Sky Burned: A Photographic Record of the 1906 San Francisco Earthquake and Fire, 100th Anniversary Edition.* San Francisco: Chronicle Books, 2006. Illustrated with more than 400 photographs of the aftermath of the earthquake and fire, this essay chronicles the events of the disaster that brought one of America's greatest cities to its knees.

Fradkin, Philip L. *The Great Earthquake and Firestorms of 1906: How San Francisco Nearly Destroyed Itself.* Berkeley: University of California Press, 2005. This examination of the earthquake and fires in San Francisco is divided into three main sections: Before, During, and After. The first section examines the history, science, politics, and culture of the city. The second section examines the disaster as it unfolded day by day. The final section examines relief efforts and the rebuilding of the city, including issues of race and politics. Throughout the volume the author illustrates how human error and negligence in building standards and construction contributed to the catastrophe.

Jeffers, H. Paul. *Disaster by the Bay: The Great San Francisco Earthquake and Fire of 1906.* Guilford, Conn.: Lyons Press, 2003. Written by a journalist, this book tells the story of the 1906 San Francisco earthquake and fire. Illustrated with photographs.

Winchester, Simon. *A Crack in the Edge of the World: America and the Great California Earthquake of 1906.* New York: HarperCollins, 2005. A chronicle of the events of the April 1906 earthquake in California. The author places this event in the context of other natural disasters that occurred that year, including violent earthquakes in the Pacific Ocean and in Chile and the eruption of Vesuvius.

Dust Bowl of 1934–1941

Cooper, Michael L. *Dust to Eat: Drought and Depression in the 1930s.* New York: Clarion Books, 2004. Chronicles the struggle for survival during the Great Depression and the dust bowl in America. The author describes the desperate exodus of Okies and others from the Great Plains westward and how, in response to the two crises, the federal government established public works programs and the Social Security program.

Annotated Bibliography

Library of Congress. "American Memory." Available online. URL: http://memory.loc.
gov/ammem/index.html. Accessed February 3, 2010. In the search box, type dust
bowl to get a list of links to photographs of scenes and victims of the dust bowl,
many taken by Dorothea Lange. Many photos are accompanied by a quotation
from the person shown or a description of the scene.

———. "Voices from the Dust Bowl: The Charles L. Todd and Robert Sonkin Migrant
Worker Collection, 1940–1941." Available online. URL: http://memory.loc.gov/
ammem/afctshtml/tshome.html. Accessed February 3, 2010. A collection of
audio recordings, images, and print materials documenting the everyday life of
residents of migrant work camps in California.

Surviving the Dust Bowl. Written and directed by Chana Gazit. Co-produced by David
Steward. 55 minutes. *American Experience,* 1998. Reissued by WGBH Boston
Video, 2007. A documentary that features interviews with witnesses to and victims of the dust bowl, archival film footage, and photographs. Shows how the
drought of 1931 brought financial and emotional ruin to thousands of families in
the southern plains of the United States.

The Woody Guthrie Foundation. "Woody Guthrie: The Official Woody Guthrie Web
site!" Available online. URL: http://www.woodyguthrie.org/index.htm. Accessed
March 10, 2010. Read the Biography section for an account of how Woody Guthrie, folk singer, became the voice of the Dust Bowl migrants.

Eruption of Mount St. Helens, 1980

Carson, Rob. *Mount St. Helens: The Eruption and Recovery of a Volcano.* 20th edition.
Seattle, Wash.: Sasquatch Books, 2002. A photographic essay documenting the
1980 eruption of Mount St. Helens. Includes information about the 540 million-
ton ashfall, human and animal deaths, blast and scorch zones, the avalanche, the
mudflows, flooding, property damage, and the environment's path to recovery
over the 20 years following the eruption.

Dale, Virginia H., Frederick J. Swanson, and Charles M. Crisafulli, eds. *Ecological
Responses to the 1980 Eruption of Mount St. Helens.* New York: Springer Science
Business Media, 2005. "The eruption of Mount St. Helens on May 18, 1980,
had a momentous impact on the fungal, plant, animal, and human life from
the mountain to the far reaches of the explosion's ash cloud and mudflows. . . .
Based on one of the most studied areas of volcanic activity, this book synthesizes
the ecological research that has been conducted for twenty-five years since the
eruption."

Goodrich, Charles, Kathleen Dean Moore, and Frederick J. Swanson, eds. *In the Blast
Zone: Catastrophe and Renewal on Mount St. Helens.* Corvallis: Oregon State
University Press, 2009. In prose and poetry, leading literary and scientific think-
ers tell the story of Mount St. Helens' destruction and renewal. Includes map and
line drawings.

Hickson, Catherine. *Mt. St. Helens: Surviving the Stone Wind.* Vancouver, B.C.,
Canada: Tricouni Press, 2005. The author's personal account of witnessing the

eruption of Mount St. Helens and observing the regrowth and renewal of the region over time.

Hurricane Katrina, 2005

Allen, Barbara. "Environmental Justice and Expert Knowledge in the Wake of a Disaster." *Social Studies of Science* 37, no. 1 (2007): 103–110. An examination of the social and environmental damages caused by Hurricane Katrina.

Banipal, Kulwinder. "Strategic Approach to Disaster Management: Lessons Learned from Hurricane Katrina." *Disaster Prevention and Management* 15, no. 3 (2006): 484–494. This article examines the performance of communication networks and information systems during Hurricane Katrina, lists causes of failure, and proposes designs for reliable and scalable networks.

Bates, Kristin, and Richelle Swan, eds. *Through the Eye of Katrina: Social Justice in the United States.* Durham, N.C.: Carolina Academic Press, 2007. A collection of scholarly articles that examine the links among classism, racism, disaster response, disaster recovery, and Hurricane Katrina.

Brasch, Walter. *"Unacceptable": The Federal Response to Hurricane Katrina.* Charleston, S.C.: Book-Surge, 2006. Written by an award-winning journalist, this book examines not just the facts of the disaster of Hurricane Katrina, but also why the federal response was inefficient. Chapters include "Nauseous, Sweaty, Tired, Dehydrated, and Hungry: Trapped in Shelters of Last Resort" and "Diversion of Financial Assistance," among others.

City Pages. "New Orleans: Survivor Stories." Available online. URL: http://www.citypages.com/2005-09-21/news/new-orleans-survivor-stories. Accessed January 27, 2010. A collection of firsthand accounts of survivors of Hurricane Katrina, documented just weeks after the disaster.

Inside Hurricane Katrina. Produced by Sarah Huisenga and Jennifer Maidtti. 120 minutes. National Geographic. 2006. DVD. Using analysis of events, hours of government audio tapes, and personal interviews, National Geographic takes viewers into the eye of Katrina to uncover the decisions and circumstances that determined the fate of the Gulf residents.

Rose, Chris. *1 Dead in Attic.* Photographs by Charlie Varley. New Orleans: Chris Rose Books, 2005. A collection of newspaper columns the author wrote for the *Times-Picayune* between August 29, 2005, and January 1, 2006. The columns chronicle the first four months in New Orleans after Hurricane Katrina.

Shrum, Wesley. "Hurricane Stories, from Within." *Social Studies of Science* 37, no. 1 (2007): 97–102. Written by a video ethnographer (one who describes a culture using information gleaned through participation or field research), this article explains why and how the author and his team filmed interviews with survivors of Hurricane Katrina immediately following the storm.

TIME editors. *Hurricane Katrina: The Storm That Changed America.* New York: TIME Incorporated Home Entertainment, 2005. A written account of Hurricane Katrina interspersed with portfolios of photographs taken by *Time* photographers. Also includes maps and a detailed time line of the storm.

INTERNATIONAL CASE STUDIES
Sumatra-Andaman Earthquake and Tsunami, 2004–2005

Bhattacharjee, Yudhijit. "Indian Ocean Tsunami: In Wake of Disaster, Scientists Seek out Clues to Prevention." *Science* (January 2005): 22–23. Written soon after the 2004 Indian Ocean earthquake and tsunami, this article explains why and how scientists and national leaders immediately began to discuss the creation of a tsunami early warning system for the Indian Ocean.

Gezari, Vanessa. "A Reporter's Account of Tsunami Crisis." Available online. URL: http://www.poynter.org/column.asp?id=60&aid=77249. Accessed May 25, 2010. In this first-person account, the author explains why the Indian Ocean tsunami was "beyond any natural disaster that I or any reporter had ever seen." She describes some of the most notable, memorable, and tragic experiences she had while covering locations in Indonesia and Sri Lanka in the three weeks following the tsunami.

Hyndman, Jennifer. *Dual Disasters: Humanitarian Aid after the 2004 Tsunami.* Sterling, Va.: Kumarian Press, 2010. Written by an associate professor in geography, this volume "describes what happens when 'man-made' and 'natural' disasters meet. Focusing specifically on Indonesia and Sri Lanka, countries that had complex emergencies long before the tsunami arrived, Hyndman shows how the storm's arrival shifted the goals of international aid, altered relations between and within states and accelerated or slowed peacebuilding efforts. With updated comments on the 2010 Haiti earthquake, the book guides readers deftly through the multifaceted forces at work in modern humanitarian disasters."

Krauss, Erich. *Wave of Destruction: The Stories of Four Families and History's Deadliest Tsunami.* Emmaus, Pa.: Rodale, 2006. The author takes readers deep into events of the 2004 Indian Ocean earthquake and tsunami by chronicling the experiences of four families who survived the tsunami that destroyed their community and went on to rebuild their lives in the overwhelming chaos that followed.

RMS. Managing Tsunami Risk in the Aftermath of the 2004 Indian Ocean Earthquake and Tsunami. Available online. URL: http://www.rms.com/publications/indianoceantsunamireport.pdf. Accessed May 19, 2010. This publication examines the earthquake and tsunami in detail, including the events themselves, the physical and human impacts, and specifics of economic and insured losses. Also includes sections on assessing the global tsunami hazard and on managing tsunami risk. Includes maps, satellite imagery, charts, and other graphical information.

Save the Children. "Five Years Later: Rebuilding Lives after the Tsunami: The Children's Road to Recovery." Available online. URL: http://www.ioc-tsunami.org/components/com_pdffarm/files/tsunami-report-2009.pdf. Accessed May 25, 2010. In this report published five years after the Indian Ocean tsunami, Save the Children explains the work it has accomplished in more than 1,000 villages in Indonesia, Sri Lanka, India, Thailand, and Somalia. Key topics include child protection, education, health, livelihoods, and disaster risk reduction.

Strand, Carl, and John Masek, eds. *Sumatra-Andaman Island Earthquake and Tsunami of December 26, 2004: Lifeline Performance.* Reston, Va.: American Society of Civil Engineers, 2008. This volume examines lifeline infrastructures—the systems that provide utilities, transportation, communication, and social services within communities—in regions damaged by the 2004 earthquake and tsunami. Topics include roadway transportation systems, electrical power systems, water systems, wastewater systems, railway systems, airports, seaports and harbors, telecommunications systems, social services, and tsunami warnings, de-alerts, and warning systems.

Tan, Ngoh Tiong, Allison Rowlands, and Francis K. O. Yuen, eds. *Asian Tsunami and Social Work Practice: Recovery and Rebuilding.* Binghamton, N.Y.: Haworth Press, 2006. This volume collects eight essays examining the role of social work in the aftermath of the 2004 earthquake and tsunami. The volume "presents an inside look at the complicated nature of disaster preparedness and how it relates to poverty, trauma, community development, and service delivery systems." The contributors, who represent the fields of health, human services, and mental health, "reflect on the challenges facing survivors, the effects of the disaster, and interventions by the community and social work professionals."

UNESCO. "Indian Ocean Tsunami Warning System Up and Running." Available online. URL: http://portal.unesco.org/en/ev.php-URL_ID=33442&URL_DO=DO_TOPIC&URL_SECTION=201.html. Accessed May 25, 2010. In this press release issued June 28, 2006, UNESCO announces that the Indian Ocean Tsunami Warning System is up and running, gives details of how the system works and will work in the future, and stresses the need for international cooperation in the gathering and dissemination of information.

United States Senate Committee on Foreign Relations. "Tsunami Response: Lessons Learned." S. Hrg. 109–153. Available online. URL: http://www.access.gpo.gov/congress/senate/pdf/109hrg/23942.pdf. Accessed May 19, 2010. A collection of reports submitted to the U.S. Senate Committee on Foreign Relation regarding the Indian Ocean tsunami. Contributors include several U.S. senators; Daniel Toole, director of the Office of Emergency Programs at UNICEF; and Veena Siddhartha, Washington director of Human Rights Watch/Asia; among others. The reports, statements, and testimonies include announcements of an estimated $700 million in aid from American citizens, businesses, and organizations; President George W. Bush's request for $950 million in tsunami disaster relief; reports of official visits to Indian Ocean countries to observe relief and reconstruction efforts and the needs of special groups such as the elderly; and progress reports of relief organizations seeking to provide clean water, food, shelter, and medical aid to survivors. Includes maps, charts, photographs, and other graphical information.

Water Resource Management in China

China Three Gorges Corporation. "China Three Gorges Project." Available online. URL: http://www.ctgpc.com.cn/en. Accessed May 20, 2010. This Web site provides information about the Three Gorges Dam from the perspective of the dam's owners and operators. A time line chronicles the building of the dam from its inception to when the 12th generator was brought into operation in 2005. It is worth noting that the time line stops short of recording the completion of the dam, attained in October 2008 when the 26th generator was brought into operation. The Web site also explains benefits of the dam, including flood control, power generation, and environmental protection. Effects of the dam that may be perceived as negative are either not mentioned or are glossed over, such as the forced relocation of hundreds of thousands of people.

Cowen, Richard. "Huanghe, the Yellow River." Available online. URL: http://mygeology page.ucdavis.edu/cowen/~GEL115/115CHXXYellow.html. Accessed May 20, 2010. Written by a professor of geology at the University of California at Davis, this essay summarizes the history of the Yellow River, including information about its major changes in course over the past 2,000 years and the efforts of specific dynasties to control the river and its flooding. The latter part of the essay summarizes the role of the Yellow River in China today, including its importance to the nation's agricultural production and current flood control measures.

Gleick, Peter H. "China and Water." *The World's Water 2008–2009: The Biennial Report on Freshwater Resources*. Peter H. Gleick, et al. Washington, D.C.: Island Press, 2009, pp. 79–100. A report on the state of China's freshwater resources including an analysis of specific problems, regional conflicts over water, and proposed solutions.

Jun, Ma. *China's Water Crisis*. Translated by Nancy Yang Liu and Lawrence R. Sullivan. Norwalk, Conn.: EastBridge, 2004. Written by a respected Chinese environmentalist, this book explains the water resource shortage that faces China and proposes solutions for water management. The author discusses China's major drainage basins, focusing on the Yellow River and its "massive reductions in water flow caused by a variety of man-made programs" and the Yangzi River basin, where "chronic soil erosion resulting from deforestation together with dam construction has led to a cycle of flood and drought." The author also "documents the persistent drought conditions in the southeast, the impact of pollutants on the Tibetan plateau, the defects in China's large-scale reservoirs, steadily diminishing underground water tables, and the growing abuse of aquifers for urbanization and industrialization."

Lohmar, Bryan, et al. "China's Agricultural Water Policy Reforms: Increasing Investment, Resolving Conflicts, and Revising Incentives." Agriculture Information Bulletin Number 782. U.S. Department of Agriculture, 2003. Available online. URL: http://www.ers.usda.gov/publications/aib782/aib782.pdf. Accessed May 20, 2010. This report examines recent changes in China's water management policies in response to water shortages in important grain-producing regions, falling groundwater tables, and disruption of surface-water deliveries to important industrial and agricultural regions. The authors explain how water conservation policies could help to mitigate or avert a significant water crisis in China.

Sun, Xuetao, Robert Speed, and Shen, Dajun, eds. *Water Resources Management in the People's Republic of China.* New York: Routledge, 2010. Written by three experts in water resources management, this volume "describes the development of a water rights system in the People's Republic of China. It covers different aspects of water resources management in China—including water planning, the provision of environmental flows, urban water management, and irrigation district management—and examines how these are being addressed through a rights-based approach."

Xie, Jian. *Addressing China's Water Scarcity: A Synthesis of Recommendations for Selected Water Resource Management Issues.* Washington, D.C.: World Bank, 2009. This 150-page report explains the current status of "China's water scarcity situation, assesses the policy and institutional requirements for addressing it, and recommends key areas for strengthening and reform. The issues covered in the report are water governance, water rights, water pricing and affordability, watershed ecological compensation, water pollution control, and emergency prevention." Includes numerous charts, diagrams, and other graphical information.

The Famine in Ethiopia in 1984–1985

Barrett, Christopher, and Daniel G. Maxwell. *Food Aid after Fifty Years: Recasting Its Role.* New York: Routledge, 2005. This volume extensively examines the worldwide system of food aid including donors in the United States and beyond, international regulatory mechanisms, food insecurity, consequences of poor food aid management, and other topics. Use the index to find specific information on Africa, the Sahel, and famine.

de Waal, Alex. *Famine Crimes: Politics and the Disaster Relief Industry in Africa.* Bloomington: Indiana University Press, 1997. Beginning with the claim, "For almost a century there has been no excuse for famine," the author examines "the persistence of famine in Africa" and "lays the blame for Africa's problems with starvation on the political failings of African governments, western donors, and the misguided policies of international relief agencies." The author is co-director of African Rights, London, an activist group.

Mortimore, Michael. *Adapting to Drought: Farmers, Famines, and Desertification in West Africa.* Cambridge: Cambridge University Press, 1989. Drawing on 13 years of research in drought-prone rural areas in northern Nigeria, the author examines community responses to drought and the processes of desertification. He also critiques theories and proposed solutions for the area.

Omilola, Babatunde. "Patterns and Trends of Child and Maternal Nutrition Inequalities in Nigeria." IFPRI Discussion Paper 00968. Washington, DC: International Food Policy Research Institute, 2010. Available online. URL: http://www.ifpri.org/sites/default/files/publications/ifpridp00968.pdf. Accessed May 20, 2010. The author examines patterns in inequalities in child and maternal nutrition in Nigeria to determine the most vulnerable groups, reporting that "child and maternal malnutrition are concentrated among the least educated households, the rural population, the north (in particular its Hausa ethnic group), and those who drink water from public wells." Particularly useful are the data presented in the numerous charts and tables in sections 5, "Results and Discussion."

Runge, C. Ford, et al. *Ending Hunger in Our Lifetime: Food Security and Globalization.* Washington, D.C.: International Food Policy Research Institute, 2003. Written by experts on food security and hunger, this volume examines agricultural problems confronting hungry populations around the world. Use the index to find extensive information on Africa and sub-Saharan Africa.

United Nations International Strategy for Disaster Reduction Secretariat-Africa. *Disaster Reduction in Africa: ISDR Informs 2009 Issue.* Available online. URL: http://www.unisdr.org/publications/v.php?id=13709. Accessed May 24, 2010. Published in 2010, this 45-page publication examines disaster-risk reduction activities in sub-Saharan Africa (also known as the Sahel) at regional, subregional, and national levels.

von Braun, Joachim, Tesfaye Teklu, and Patrick Webb. *Famine in Africa: Causes, Responses, and Prevention.* Washington, D.C.: International Food Policy Research Institute, 2000. "The authors present the results of field work and other research from numerous parts of Africa, with a particular focus on Botswana, Ethiopia, Niger, Rwanda, Sudan, and Zimbabwe. With these data, the authors explain the factors that cause famines and assess efforts to mitigate and prevent them."

Eruption of Nevado del Ruiz, Colombia, 1985

Bruce, Victoria. *No Apparent Danger: The True Story of Volcanic Disaster at Galeras and Nevado del Ruiz.* New York: HarperCollins, 2001. In Part 1 of this volume, the author examines the geography and geology of Nevado del Ruiz, the eruption in 1985, and the catastrophic lahars that destroyed Armero. In Part 2 of the volume, the author performs a similar study of Galeras, another Colombian volcano, which erupted in January 1993.

Global Volcanism Program. "Nevado del Ruiz: Index of Monthly Reports." Available online. URL: http://www.volcano.si.edu/world/volcano.cfm?vnum=1501-02=&volpage=var#sean_1010. Accessed May 20, 2010. A chronological list of monthly reports on Nevado del Ruiz's activity, beginning in mid-1985 before the eruption of November 13. In the list of reports, click on a report to read its contents. The entry "10/1985 (SEAN 10:10)" and those that follow include information on the volcano's 1985 eruption. (The abbreviation SEAN refers to *Scientific Event Alert Network Bulletin,* a publication of the Smithsonian Institution, from which information for the report was drawn.)

Mileti, Dennis S. *The Eruption of Nevado del Ruiz Volcano, Colombia, South America, November 13, 1985.* Natural Disaster Studies vol. 4. Washington, DC: National Academy Press, 1991. Available online. URL: http://www.nap.edu/openbook.php?isbn=0309044774. Accessed May 20, 2010. Prepared by a team of researchers, this 110-page volume examines the physical setting and geological history of the Nevado del Ruiz volcano; the eruption and lahars of November 13, 1985;

the performance of structures in the lahar paths; the warning period before the disaster; the disaster's impact; and the recovery program in the hours and months following the disaster.

Russell, George. "Colombia's Mortal Agony." Available online. URL: http://www.time.com/time/magazine/article/0,9171,1050626,00.html. Accessed May 20, 2010. In this article published on November 25, 1985 (12 days after the volcano's eruption), the author gives a detailed chronological account of the eruption, situating the natural disaster in the context of Colombian politics including the president's November 6 decision to send military troops against guerrillas who had taken over the Bogotá Palace of Justice. He also reports on contemporary speculation about whether the tragedy could have been averted or mitigated.

Serrill, Michael S. "Colombia: Aftermath of a Disaster." Available online. URL: http://www.time.com/time/magazine/article/0,9171,1074794,00.html. Accessed May 20, 2010. In this article published on December 2, 1985, the writer examines relief efforts, including monetary and medical aid, in Armero following the eruption of Nevado del Ruiz. He also reports on criticisms of the Colombian government's response to the disaster.

Chronology

- During the Permian–Triassic (P–Tr) extinction event, 90 percent of all life on Earth dies. During this Great Dying, an estimated 70 percent of the planet's reptile, amphibian, and plant species become extinct, and up to 96 percent of marine life dies out. Scientists speculate that factors such as climate change, sea-level change, ocean stagnation, carbon dioxide buildup, or the impact of an asteroid may have contributed to the Great Dying—but no one knows for certain.

65 MILLION YEARS AGO

- The Cretaceous–Tertiary (K–T) extinction event kills 50 percent of all species. This catastrophic event is best known for its destruction of the dinosaurs. In geological terms, quite suddenly all dinosaurs, plesiosaurs, mosasaurs, and pterosaurs become extinct. Like the dinosaurs, most large animals on land, in the air, and in the sea become extinct. In the oceans, most plankton and tropical invertebrates become extinct. Most other groups of organisms, however, survive and go on to diversify. No one knows for sure what caused the disaster, but scientists have suggested a giant volcanic eruption or a large asteroid impact as the cause.

CA. 16TH CENTURY B.C.E.

- In the Aegean Sea, a massive eruption of Thera (also called Santorini) devastates the volcanic island. Scientists are unable to pinpoint the exact year of the eruption, but most agree that it occurred around 3,500 years ago, possibly between 1650 B.C.E. and 1500 B.C.E. Damage from the eruption, earthquakes, and tsunamis to the island of Crete possibly led to the fall of the Minoan civilization, which at the time was the dominant civilization in the Mediterranean. The legend of the lost city of Atlantis and several biblical stories of plagues have been linked to this catastrophic event, which geologists suggest was the most explosive event the planet has ever experienced.

367

340 B.C.E.

- Aristotle writes *Meteorologica*, which contains his theories concerning weather patterns and earthquakes.

79 C.E.

- The eruption of Mount Vesuvius destroys the Roman cities of Pompeii and Herculaneum, killing an estimated 10,000–25,000 people. Pliny the Younger witnesses the eruption with his uncle, Pliny the Elder, who dies during the catastrophe. Pliny the Younger later describes the eruption in a letter to the Roman historian Tacitus.

132

- The Chinese scientist Zhang Heng creates the first seismometer.

526

- *May:* An earthquake strikes Antioch (modern-day Antakya, Turkey), killing approximately 250,000 people. Situated near the Mediterranean Sea, Antioch was filled with Roman temples and aqueducts, palaces decorated with mosaics and statues, and Christian churches. The earthquake destroys many of these structures, and the fire that follows the quake destroys the remaining structures.

CA. 800–1000

- A series of intense droughts cause the collapse of the Mayan civilization in Mesoamerica, in the area of modern-day Honduras and Mexico's Yucatán Peninsula. The Mayan people numbered an estimated 15 million, and they thrived in huge cities linked by trade routes. Their towering pyramids still stand. The civilization reached its peak around 800–900 C.E. and then suddenly collapsed, leaving deserted cities and abandoned trade routes. Scientists speculate that the civilization died in famines, probably caused by a combination of natural drought and deforestation.

856

- *December 22:* The Damghan earthquake in present-day Iran, estimated at 8.0 on the Richter scale, causes approximately 200,000 deaths along a 200-mile stretch of fertile land. The city of Damghan suffers an estimated 45,000 casualties and is half destroyed, and one-third of the city of Bustam collapses. Many villages along the affected region are completely destroyed.

893

- *March 23:* An earthquake in the city of Ardabil, in modern-day Iran, kills 150,000.

Chronology

1064–1071

- Water shortages along the Nile River in Egypt cause a seven-year famine that kills 40,000 people. Farmers along the Nile relied on the river's overflow to irrigate their crops. When low water levels prevented this overflow, crops failed. The disaster occurred year after year for seven years straight (some sources say eight years). The resulting famine caused the starvation of tens of thousands of people. In their desperation, some people resorted to cannibalism.

1138

- *August 9:* An earthquake in the city of Aleppo, located in modern-day Syria, kills 230,000 people.

1275–1289

- Severe famine devastates the Anasazi, a Native American people, contributing to the collapse of their civilization. The Anasazi thrived in the area that is now the American Southwest, building distinctive multilevel apartments in the sides of cliffs. In a region already dry, the Anasazi suffered through several periods of drought. In 1274, a severe drought struck, driving the Anasazi to abandon their cities. Archaeologists speculate that drought, combined with other factors such as religious conflict, political conflict, or warfare, caused the downfall of the civilization.

1315–1317

- The Great Famine of 1315–1317 causes millions of deaths in Europe. The trouble begins in the spring of 1315, when rains made fields too wet to plow and rotted much of the seed planted. Consequently, harvests are small and people begin to suffer malnutrition. The spring and summer of 1316 are also wet, with similar results. By now, people are perishing from starvation and dying of diseases (e.g., pneumonia, bronchitis, tuberculosis) that, in their malnourished state, they are unable to recover from. In 1317, the weather returns to normal patterns, but people struggle to recover with less seed, fewer work animals, and weaker bodies.

1441

- Jang Yeong-sil creates a standardized rain gauge used to collect yearly precipitation amounts in Korea.

1556

- *January 23:* One of the deadliest earthquakes on record kills approximately 830,000 people in China's Shaanxi Province and the neighboring Shanxi Province. The quake causes mountains to crumble, changes the course of rivers, topples stone buildings, and causes fires that burn for days. An estimated 60 percent of the residents of these provinces are killed.

1667

• *November:* Eighty thousand people lose their lives in an earthquake that strikes Shemakha, a city in the Caucasus region of Eurasia.

1693

• *January 11:* A magnitude 7.5 earthquake kills 60,000 people in Sicily, a large island in the Mediterranean Sea.

1727

• *November 18:* An earthquake in Tabriz in Persia (Iran) kills 77,000 people.

1755

• *November 1:* The Great Lisbon Earthquake (magnitude 8.7) strikes Lisbon, the capital of Portugal, on All Saint's Day. A quarter of the city's inhabitants perish, many of them crushed in collapsing churches and others drowned by a tsunami on the riverfront. Still others died in the fire that broke out following the earthquake.

1769–1773

• The Bengal famine in the lower Gangetic plain of India kills an estimated 10 million people—one-third of the area's population at the time. Precursors to the disaster include low crop yields in the years 1768 and 1769. In the latter part of 1769, a severe drought made things worse. During 1770, people perish of starvation in great numbers (thus, the disaster is often called the Bengal Famine of 1770). Late in 1770, a good rainfall provides a brief respite, but 1771 again provides low crop yields. Survivors of the famine migrate en masse in search of food.

1780

• *October:* A hurricane strikes the eastern Caribbean and kills an estimated 22,000 people.

1783

• *February 4:* Fifty thousand people lose their lives in an earthquake that strikes Calabria, a region in the "toe" of the Italian peninsula.

1789

• *December:* In Coringa, India, 20,000 people lose their lives when a cyclone and three tidal waves strike the city. The city is completely destroyed, and most ships in the harbor are destroyed.

Chronology

1792

- In western Kyushu, a Japanese island, Mount Unzen erupts, leading to a landslide and a tsunami that kill 15,000 people. Mount Unzen is made up of a group of composite volcanoes. After the initial eruption, an earthquake triggered a landslide from the Mayuyama peak in this group. The landslide rushed down upon the city of Shimabara and on to the Ariake Sea. Here, it set off a tsunami that devastated nearby areas. Most of the 15,000 deaths are attributed to the landslide and tsunami.

1811–1812

- Three (some say four) large earthquakes strike the area near New Madrid in the Louisiana Territory (present-day Missouri) during December 1811–February 1812. The shaking tears down trees, topples chimneys, and sends people running from their houses. The ground ripples like water and the earth splits with long fissures. Entire buildings topple into the holes, and sulfurous fumes arise. The earthquakes and more than 1,800 aftershocks affect 50,000 square miles in the United States and parts of Canada.

1815

- *April:* Mount Tambora in Indonesia produces the largest volcanic eruption ever recorded. An estimated 90,000 people lose their lives, and the ash cloud circles the globe and blocks sunlight, causing the Year without Summer.

1839

- *November:* In Coringa, India, 300,000 people lose their lives when a cyclone and a 40-foot tidal wave strike the city. In the harbor, 20,000 vessels are destroyed.

1840

- *May 7:* In Natchez, Mississippi, the Great Natchez Tornado kills 317 people and injures 109. It sucks boats up out of the Mississippi River, sinks others, and uproots trees along the shore. When the massive tornado—now carrying timbers and other debris—enters Natchez, it destroys banks, stores, and homes. Many buildings burst open. The twister leaves behind a 20-mile path of destruction, including the ruined city of Natchez. The tornado is ranked as the second deadliest in U.S. recorded history.

1845–1850

- Over 1 million people starve as potato blight causes widespread food shortages in Ireland. During the disaster, now known as the Irish Potato Famine, Ireland loses one-fourth of its entire population. About half this loss—approximately

1 million people—results from starvation, typhoid, typhus, scurvy, and other health-related issues. In addition, an estimated 1 to 1.5 million people emigrate to escape the disaster.

1850

- This year marks the beginning of reliable instrumental recording of global temperatures.

1863

- The International Committee of the Red Cross is formed.

1871

- *October 8:* The Great Peshtigo Fire burns over a million acres in northeastern Wisconsin and upper Michigan and kills an estimated 1,200 to 2,400 people. Though the deadliest fire in recorded U.S. history, it is often overlooked because the Great Chicago Fire started on the same day.

1875–1878

- A famine in northern China kills 9 million people.

1876–1878

- A famine in southern India kills 5.2 million people.

1880

- Sir James Alfred Ewing, Thomas Gray, and John Milne found the Seismological Society of Japan, which funds the creation of the seismograph.

1881

- Clara Barton founds the American Red Cross.

1883

- *August:* In Indonesia, the eruption of Krakatoa leaves a 960-foot crater and creates tsunamis that cause massive damages along the coasts of Sumatra and Java. Approximately 36,000 people die.

1884

- *February 19:* A series of 37 tornadoes sweep across the southeastern United States, killing at least 167 people (some reports say 182) and injuring more than 1,000 others. Deaths are reported in Georgia, Alabama, North Carolina, South Carolina, Kentucky, and Mississippi. The tornadoes cause $4 million in damages. The event has been called the Enigma Tornado Outbreak.

Chronology

1887

- **September:** The Yellow River (Huang He) in China breaks dikes near Zhengzhou and floods 50,000 square miles. The flood leaves 2 million people homeless and kills at least 900,000. An epidemic that follows the disaster kills upwards of another million people. Some sources place the flood's death toll at between 1.5 and 7 million people. With the latter death toll, it surpasses the Yellow River flood of 1931 as the deadliest recorded natural disaster worldwide.

1889

- **May 31:** Heavy rainfall causes the South Fork Dam near Johnstown, Pennsylvania, to fail. Johnstown and smaller villages are flooded, and 2,200 people lose their lives.

1896

- **May 27:** A tornado kills 256 people in the towns of St. Louis and East St. Louis, Missouri. Nearly 200 people are reported missing and presumed dead. Another 1,000 people are injured, and more than 8,800 buildings are damaged or destroyed. The tornado causes damages estimated at $10 million. The disaster has been called the St. Louis Cyclone.

1899

- **June 12:** A tornado sweeps through the center of New Richmond, Wisconsin, damaging an area measuring 1,000 feet wide by 3,000 feet long. It kills 117 people, injures 150 people, and destroys significant parts of the city. The disaster has been called the New Richmond Cyclone.

1900

- In India, a drought results in the deaths of between 250,000–3.25 million people.
- **September 8:** A category 4 hurricane causes storm surges in Galveston, Texas, that kill an estimated 6,000 people. An additional 6,000 are injured, and tens of thousands of people are made homeless. The storm causes $20 million in damages.

1902

- **May 8:** Mount Pelée erupts on the French-owned island of Martinique in the Caribbean Sea. The city of St. Pierre is wiped out by pyroclastic flows, with only two known survivors from a population of approximately 28,000.
- **May 18:** A tornado outbreak kills 114 people in Goliad, Texas.

NATURAL DISASTERS

1906

- **April 18:** A magnitude 7.8 earthquake occurs off the northern California coast, shaking the ground from Coos Bay, Oregon, in the north to Los Angeles, California, in the south. The city of San Francisco suffers catastrophic damage, with buildings crumbling and collapsing and a fire sweeping through the central business district. In this city, an estimated 700 people die, 500 city blocks are destroyed, and nearly $500 million in damages occur.

1907

- During a drought in China, approximately 24 million people perish from starvation.

1908

- **April 23–26:** The 1908 Dixie Tornado Outbreak produces 34 tornadoes in 13 states. The storms kill at least 320 people and injure more than 1,000 more.
- **June 30:** The Tunguska Event levels 850 square miles of forest in Russia. The cause is commonly attributed to a large meteorite exploding at an altitude of three to six miles.
- **December 28:** A magnitude 7.2 earthquake strikes Messina, Italy, killing 72,000 people in the area. The earthquake triggers a tsunami, whose waves are 20–39 feet high on the coast of Sicily south of Messina and 20–33 feet high along the coast of Calabria.

1920

- **December 16:** A magnitude 7.8 earthquake strikes the Chinese province of Gansu, killing a total of 200,000 people. Damage occurs in seven provinces and regions, including the major cities of Lanzhou, Taiyuan, Xi'an, Xining, and Yinchuan. Huge death tolls accrue in Haiyuan County (73,000 dead) and Guyuan County (more than 30,000 dead). The quake causes a landslide that buries a village, and it changes the course of rivers.

1921–1922

- Famine in the former USSR kills 2 million people.

1923

- **September 1:** The Great Kanto (also *Kwanto*) Earthquake in Japan, measuring 7.9 on the Richter scale, causes widespread damage in Tokyo and kills an estimated 142,800 people. The disaster destroys 694,000 houses, including 381,000 houses that burn in quake-caused firestorms in the Tokyo-Yokohama region. This disaster is also known as the Great Tokyo Earthquake and the Great Tokyo Fire.

Chronology

1925

- *March 18:* The Tri-State Tornado, with a diameter of more than a mile, devastates 164 square miles in eastern Missouri, southern Illinois, and southern Indiana. It travels 219 miles and spends more than three hours on the ground. The disaster kills 695 people, making it the deadliest tornado in U.S. history. It injures an additional 13,000 people and causes $17 million in property damage.

1928

- *September 16:* A hurricane—now known as the Lake Okeechobee Hurricane—strikes the Caribbean and moves upward to Florida's Atlantic coast. After striking Palm Beach County and the Treasure Coast (which includes the counties of Indian River, St. Lucie, and Martin), it takes an unlikely route inland across the Florida Everglades and Lake Okeechobee. The lake floods, swamping the surrounding farmland and washing people away to their deaths. Many victims are never found; many of the dead are never identified. Fearful of disease, survivors bury hundreds of bodies in mass graves and burn piles of other bodies. The disaster's death toll has been placed at 2,500 people, making it the second deadliest disaster in U.S. history (behind the Galveston, Texas, hurricane of 1900).

1931

- Between 1 and 4 million people die in China when the Yellow River (Huang He) floods. During 1928–30, China suffered a drought, followed by heavy snowfall in the winter of 1930–31. As the snow began to melt, heavy rains set in. River levels rose, and still the rain fell. The Yangtze and Huai Rivers flooded as well. By November 1931, millions of people had drowned in the floodwaters, with additional victims dying of starvation and diseases such as cholera and typhus. An estimated 51 million people were affected by the floods. This disaster is considered the worst recorded natural disaster worldwide.

1934–1941

- Years of drought and famine known as the dust bowl bring disaster to Kansas, Oklahoma, Texas, Arkansas, and Missouri. Beginning in 1918, thousands of acres of farmland, stripped of grass, trees, and shrubs, were turned over to pastures for cattle. With cattle trampling it, vegetation was slow to take hold. When gales of wind gusted in from the northwest, the rich topsoil began to blow away. Then drought struck from 1934–37. Winds raked across the land drying up every drop of moisture, emptying ponds and stock tanks, and creating enormous dust clouds called black blizzards. People and cattle died from breathing in dirt. Unable to grow crops, and with their cattle dying, farmers lost their incomes and then their farms. An estimated 60 percent of the population of the region were driven away by the disaster.

1935

- Charles Francis Richter creates the Richter scale, which is used to measure the severity of an earthquake.
- **September:** The Labor Day Hurricane, a category 5, strikes the Florida Keys and leaves 400 dead. Damages are estimated to be $6 million.

1936

- **April 5–6:** The Tupelo-Gainesville tornado outbreak includes 17 tornadoes that race across northern Mississippi and northern Georgia, killing 203 people and injuring another 1,600.

1938

- **September 21:** A category 3 hurricane strikes the East Coast along New York, Connecticut, Rhode Island, and Massachusetts. Storm surges are 12 to 15 feet high, and 600 people lose their lives. The storm causes $306 million in damages.

1943

- **February 20:** In central Mexico, a fissure opens in a cornfield owned by Dionisio Pulido. Over the next year, eruptions of lava and ash create a volcano over 1,000 feet tall.

1946

- **April 1:** An earthquake of magnitude 7.4 occurs approximately 90 miles south of Unimak Island in the Aleutian Island chain. It produces a Pacific-wide tsunami. A 100-foot tsunami strikes the Scotch Cap area on Unimak Island, killing the five-member lighthouse crew there. Tsunami waves varying in height from 55 feet to 33 feet strike the Hawaiian Islands, doing extensive damage and killing 159 people. Smaller waves strike the west coast of the United States, drowning one person in Santa Cruz, California. Waves estimated to be 10 to 14 feet high cause $20,000 in damages in the Half Moon Bay area south of San Francisco.
- **December 11:** The United Nations establishes the United Nations Children's Fund (UNICEF).

1947

- **April 9:** The Glazier-Higgins-Woodward tornadoes sweep through Texas, Oklahoma, and Kansas. One hundred and eighty-one people lose their lives, and 970 others are injured.

1948

- *March 25:* Captain Robert C. Miller and Major Ernest Fawbush of the United States Air Force issue the first official tornado forecast.
- *April 7:* The United Nations creates the World Health Organization (WHO).
- *October 5:* A magnitude 7.3 earthquake does severe damage in Ashgabat, capital of the Turkmen Soviet Socialist Republic (Turkmen SSR) in the Soviet Union (USSR). In Ashgabat and nearby villages, brick and concrete structures fall apart and freight trains derail. Additional deaths and damage occur in the Darreh Gaz area of Iran. An official report places the earthquake's death toll at 110,000.

1949

- The Pacific Tsunami Warning Center is formed in response to the 1946 Aleutian Islands earthquake.

1953

- *January 31–February 1:* On the east coast of Britain and the coast of the Netherlands, a storm surge that is 8 to 10 feet above normal kills more than 1,800 people.
- *May 11:* A violent tornado strikes downtown Waco, Texas. The storm kills 144 people and destroys 200 business buildings.
- *June 8–9:* The Flint–Worcester Tornadoes kill 210 people. The first of the two tornadoes strikes Flint, Michigan, on June 8, killing 116 people. The next day, the same storm system spawns another tornado, this time in the area of Worcester, Massachusetts. The tornado is on the ground for 1.5 hours and travels for 46 miles. It kills a total of 94 people, 60 of them in Worcester. The two tornadoes are part of a larger outbreak of 46 tornadoes, all spawned by the same storm system between June 7–9.

1954

- Heavy rains in Iran result in the Great Iran Flood, which causes the deaths of 10,000 people.

1958

- *July 9:* An earthquake loosens about 40 million cubic yards of rock above the northeastern shore of Lituya Bay in Alaska, producing a massive landslide that plunges into Gilbert Inlet. The impact creates a mega-tsunami that is 1,720 feet tall, the largest tsunami in the recorded history of the world. The wave crashes against the opposite shoreline of Gilbert Inlet and speeds through Lituya Bay and into the Gulf of Alaska. As it does so, it rakes the shoreline up to a height of 1,720 feet, sweeping millions of trees away.

NATURAL DISASTERS

1959

- **September 26–27:** Typhoon Vera strikes Honshu, Japan, killing 4,580 people and causing over $261 million in damages. Nearly 1.5 million people are rendered homeless. Winds, floods, and landslides damage and destroy roads, bridges, and communications. This category 5 storm is considered to be one of Japan's most destructive natural disasters and is the strongest typhoon to hit the country in its recorded history.

1959–1961

- In China, between 16 and 40 million people starve to death during the Great Leap Famine. The famine's name originates from an economic program called the Great Leap Forward begun by Mao Zedong in 1957. Under this program, mismanagement of agricultural resources resulted in the deadly famine.

1960

- **April 1:** *TIROS-1*, the first weather satellite, is launched from Cape Canaveral, Florida.
- **May 22:** The most powerful earthquake ever recorded strikes Chile. Measuring 9.5 on the moment magnitude scale, the earthquake creates tsunamis that affect the entire Pacific basin. In southern Chile, approximately 1,655 people are killed, 3,000 are injured, and 2 million are left homeless. Damages there are estimated to be $550 million. Tsunami waves cause 61 deaths and $75 million in damages in Hawaii, 138 deaths and $50 million in damages in Japan, up to 32 deaths in the Philippines, and $500,000 in damages to the west coast of the United States.

1965

- **August–September:** Hurricane Betsy—more than 600 miles in diameter at one time—causes 75 death in the Bahamas, Florida, and Louisiana and causes $1 billion in damages.

1968–1984

- During these years, two extended periods of drought and famine occur in Ethiopia, Sudan, and other parts of the Sahel region of Africa. The exact dates of the droughts and related famines differ among sources, but sources consistently report starvation conditions resulting in the deaths of hundreds of thousands (some say more than a million) of people. The first drought occurs during the general period of 1968–73, and the second drought occurs during the period of 1979–84.

Chronology

1969

- *August:* Hurricane Camille, a category 5 storm, kills approximately 250 people in Louisiana, Mississippi, and Virginia. The storm causes $1.42 billion in damages.

1970

- *May 31:* A magnitude 7.9 earthquake occurs in Chimbote, Peru, killing an estimated 70,000 and injuring an additional 150,000. A million people are left homeless. The earthquake causes a massive landslide that travels 11 miles at up to 100 miles per hour.

- *November 12:* The Bhola cyclone strikes East Pakistan (now Bangladesh) and West Bengal, India. This disaster, also called the Ganges-Brahmaputra delta cyclone, kills an estimated 500,000 people in the Ganges-Brahmaputra delta region. It is the deadliest tropical cyclone on record, and it ranks among the top five deadliest natural disasters on record.

1971

- The Fujita scale, which rates the intensity of tornadoes, is introduced by Theodore Fujita.

- In North Vietnam, heavy rains cause flooding that kills 100,000 people.

1972

- *February:* In Iran, a blizzard ends a four-year drought but causes the deaths of 4,000 people.

1974

- *April 3–4:* During a 16-hour period, 148 tornadoes strike areas in Illinois, Indiana, Michigan, Ohio, Tennessee, Mississippi, and Alabama. Called the "Super Outbreak of April 3–4, 1974," the disaster includes 30 tornadoes causing F4 damage or worse. The outbreak kills 315 people and injures 6,142.

1976

- *February 4:* A magnitude 7.5 earthquake kills 23,000 people in Guatemala, injures an additional 70,000, and makes 1 million people homeless. The mighty quake twists railroad tracks, destroys adobe dwellings, topples hotels, triggers deadly landslides, and causes extensive flooding of rivers.

- *May 6:* An earthquake causes more than 900 deaths in Friuli, Italy.

- *June 25:* A magnitude 7.1 earthquake in Iran kills 6,000 people.

- *July 28:* The Great Tangshan Earthquake in China measures 7.5 on the Richter scale and kills at least 255,000 people. Another 164,000 are injured. (Some sources estimate the number of dead to be as high as 655,000.) The epicenter of the earthquake is near Tangshan, an industrial city of 1 million residents located 95 miles east of Beijing (Peking). The city virtually crumbles to the ground, with mud and brick buildings collapsing. The massive quake's tremors are felt nearly 100 miles away.

- *August 17:* An earthquake and tsunami in the Philippines kill more than 4,000 people.

- *November 24:* An earthquake near Muradiye, Turkey, kills more than 4,000 people.

1979

- *April 10:* A tornado rated F4 causes $400 million in damages to Wichita Falls, Texas. Forty-two people lose their lives, and 1,740 people are injured.

1980

- *May 18:* Mount St. Helens in the U.S. Pacific Northwest erupts, killing 57 people and countless wildlife, including 7,000 big game animals and 12 million salmon fingerlings in hatcheries. Twenty-seven bridges and nearly 200 homes are destroyed. In forests, 4 billion board feet of timber are blasted down The blast reduces the elevation of the volcano's summit by more than 1,000 feet. The 520-million-ton cloud of ash spreads across the United States in three days and circles the earth in 15 days.

1982

- Christopher G. Newhall and Stephen Self introduce the Volcanic Explosivity Index (VEI).

1982–1983

- An El Niño event causes an estimated $8 billion in damages to the world economy. In South America, the fishing industries of Ecuador and Peru suffer severely. Hawaii and Tahiti experience tropical storms that are unusual for the area. Droughts and forest fires occur in Indonesia and Australia. Widespread flooding occurs across the southern United States.

1985

- *September 19:* An earthquake measuring 8.1 on the Richter scale strikes 250 miles west of Mexico City, sending devastating tremors through the unstable dirt and sand beneath the city. Nearly three minutes of shaking result in the deaths of 10,000 people. An additional 30,000 are injured, and thousands more are left homeless. Buildings crumble and topple, gas mains break, and fires ignite. Thirty thousand buildings are completely destroyed, and 100,000 more are badly damaged.

- *November 13:* The Nevado del Ruiz volcano in Colombia erupts, creating lahars that kill an estimated 25,000 people. The town of Armero is completely destroyed.

1988

- *Summer–Fall:* In the "Great Fires of 1988," wildfires scorch 1.2 million acres in the Yellowstone area. Nearly 800,000 of these acres are within Yellowstone National Park, amounting to just over a third of the park's acreage. Property damages are estimated at $3 million. Of the more than 25,000 firefighters who battled the blazes, only two were killed.

- *December 7:* In the Soviet Union, a magnitude 6.9 earthquake strikes northwestern Armenia. Four minutes later a magnitude 5.8 aftershock strikes. More than 20 towns and 342 villages are affected, and 58 of them suffer heavy damage. Spitak, a major population center, is almost completely destroyed. More than 60,000 people die, and an additional 15,000 are injured. Half a million Armenians are left homeless. Direct economic losses are estimated at $14 billion.

1989

- *April 26:* The Daulatpur-Salturia Tornado kills an estimated 1,300 in the Manikganj District in central Bangladesh. An additional 12,000 people were injured, and 80,000 were left homeless. The cities of Daulatpur and Salturia are hardest hit, with trees uprooted and homes completely destroyed.

1990

- *June 20:* A magnitude 7.4 earthquake strikes western Iran, killing 40,000–50,000 people. An additional 60,000 are injured, and at least 400,000 are left homeless.

1991

- *April 29–31:* Tropical Cyclone O2B strikes the Chittagong, a coastal city in the Bay of Bengal in Bangladesh. More than 138,000 people lose their lives, and the storm causes more than $1.5 billion in damages.

- *October:* In Oakland and Berkeley, California, wildfires kill 25 people, injure 150 others, destroy 3,000 homes, and burn 1,500 acres. The fires cause $1.5 billion in damages.

- *November 5–8:* In the western North Pacific, Typhoon Thelma strikes the Visayan Islands in the central Philippines. Six thousand people lose their lives in this category 5 storm, which causes flooding, dam failures, and landslides.

1992

- *August 22–28:* Hurricane Andrew, a category 5 storm, strikes the Bahamas and the southern United States. Sixty-five people lose their lives. The storm causes $26 billion in damages, with some estimates placing the figure closer to $34 billion. Until Hurricane Katrina in 2005, Andrew was the costliest hurricane in U.S. history.

1994

- *January 17:* The Northridge earthquake near Los Angeles measures 6.7 on the Richter scale and causes an estimated $20 billion in damage. More than 9,000 people are injured, and 72 people lose their lives.

1995

- *January 17:* The Great Hanshin earthquake, or Kobe earthquake, measures 6.9 on the Richter scale, kills over 6,000 people, injures more than 26,000, and causes an estimated $100 billion in damage. It is one of the most devastating earthquakes in Japan's recorded history.
- *July:* A heat wave in Chicago kill over 700, hospitalizes thousands more, and causes widespread power outages.

1998

- *October 26–November 4:* Hurricane Mitch, a category 5 storm, causes flooding and mudslides in the mountainous regions of Central America. Approximately 11,000 people lose their lives and 3 million are impacted. The storm causes over $5 billion in damages.

1999

- *May 3:* During the 1999 Oklahoma/Kansas Tornado Outbreak, 74 tornadoes touch down in the two states in less than 21 hours. The disaster kills 46 people and injures another 800. Estimated property damages are near $1.5 billion.

2000

- The Big Dry is Australia's worst drought in a century.

2002

- *January 17:* Mount Nyiragongo in the Democratic Republic of the Congo erupts. The destruction leaves 120,000 homeless.

2003

- *May 3–11:* In the United States, 401 tornadoes cause damage in 19 states and one Canadian province during the May 2003 Tornado Outbreak. The disaster kills 41 people.
- *Summer:* A heat wave in Europe causes the deaths of approximately 35,000 people. France suffers 14,802 of these deaths.

Chronology

- *October 20–November 3:* Fourteen wildfires sweep across southern California. The fires kill 22 people, destroy 3,600 homes, scorch nearly 740,000 acres of land, and cause more than $2 billion in property damages. One of these fires, the Cedar Fire, is responsible for deaths and damages in the San Diego metropolitan area, the hardest-hit area.

2004

- *December 26:* A magnitude 9.1 earthquake off Sumatra causes massive tsunamis in the Indian Ocean. Entire cities in Indonesia are destroyed, and Thailand, Sri Lanka, India, and the Republic of the Maldives suffer significant damages. The earthquake and waves kill an estimated 227,898 people. Approximately 1.7 million people in 14 countries in South Asia and East Africa are displaced by the disaster. The USGS reports this to be the third largest earthquake in the world since 1900.

2005

- *August:* Hurricane Katrina makes landfall in southern Florida and Louisiana and causes deadly storm surges along the Gulf Coast of the United States. About 80 percent of New Orleans is flooded when the city's levees fail. At least 1,836 people lose their lives in the storm. With property damages at an estimated $100 billion, Katrina is the costliest hurricane in the country's history.
- *October 8:* A magnitude 7.6 earthquake strikes northern Pakistan and kills at least 86,000 people. An additional 69,000 are injured. Kashmir suffers the heaviest damages, with entire villages completely destroyed and many more nearly destroyed.

2005–2006

- Lack of rain and swarms of locusts cause the Niger food crisis in Africa and threaten the food security of approximately 3 million people.

2007–2008

- Dramatic increases in the price of food threatens food security around the world. During the crisis the price of rice rises 217 percent.

2008

- *May:* Cyclone Nargis makes landfall in Myanmar (the official name of Burma), killing more than 100,000 people and leaving more than a million homeless. Some reports place the death toll at over 138,000. It is the worst disaster in the recorded history of the nation.
- *May 12:* A magnitude 7.9 earthquake strikes the Sichuan Province of China, killing an estimated 87,587 people and injuring more than 370,000 others. The

disaster affects more than 45.5 million people in 10 provinces and regions. Schools, factories, hospitals, and homes collapse—an estimated 5.36 million buildings destroyed. An additional 21 buildings are damaged. A landslide in Qingchuan County buries at least 700 people. Other landslides block the flow of the Jian River, forming more than 30 "quake lakes" that are flood hazards to at least 700,000 people living downstream.

- *June:* Typhoon Fengshen strikes the Philippines and China. More than 1,350 deaths are reported, with at least 40 additional people declared missing. The typhoon capsizes the *Princess of the Stars,* a Filipino ferry, killing 846 of the 922 people on board.

2010

- *January 12:* A magnitude-7.0 earthquake strikes the Caribbean island nation of Haiti, its epicenter approximately 15 miles southwest of the capital, Port-au-Prince. The poor nation is ill-prepared for an earthquake. A fifth of the structures in Port-au-Prince collapse outright, with 80 percent of those still standing suffering serious damage. According to official reports, at least 222,570 people are killed, 300,000 are injured, 1.3 million are displaced, 97,294 houses are destroyed and 188,383 damaged in the Port-au-Prince area and in much of southern Haiti.

- *February 27:* A magnitude 8.8 earthquake strikes central Chile, tearing down houses, buildings, roads, and bridges. An estimated 279 people are confirmed dead, with additional people missing and 2 million people made homeless. The quake is felt 1,800 miles away in São Paulo, Brazil. The quake causes tsunami waves, 8.5 feet at their highest, that race across the Pacific. The tsunami destroys San Juan Bautista village on Robinson Crusoe Island off Chile's coast, where 4 die and 11 are reported missing. The tsunami sets off alarms of the Pacific tsunami warning system but does little damage across the Pacific.

- *Spring:* A series of severe droughts in southwestern China affected 10 provinces, in what is called the worst drought in a century in that region. The drought also affects Vietnam, Thailand, and other parts of Southeast Asia. Dust storms in March and April affect much of East Asia.

- *March–May:* In Iceland, eruptions of the volcano Eyjafjallajökull, while rated low on the VEI, produce ash that disrupts air travel across western and northern Europe during April and May. The first eruption, on March 20, was rated 1 on the VEI. The second eruption, on April 14, was rated 4 on the VEI.

- *June–August:* In parts of Russia, temperatures rise higher than they have since records began more than 100 years ago, and the head of Russia's weather service says the heat wave is likely Russia's worst in 1,000 years. Extreme temperatures contribute to hundreds of wildfires that produce poisonous smog, drought that destroys approximately 25 million acres (10 million hectares) of crops, and the deaths of thousands of people. In Moscow, the death rate nearly doubles to an average of 700 deaths per day during the height of the heat wave.

- *July–August:* Unusually heavy monsoon rains cause severe flooding in Pakistan, submerging more than one-fifth of the total land area of the country and severely affecting more than 20 million people (about an eighth of the population). Nearly 2,000 people die, more than 1 million homes and hundreds of thousands of acres of vital crops are destroyed. The floods, deemed by Pakistani officials as the worst in the country's history, cause an estimated $43 billion in damages. By August 27, more than $680 million in relief monies are received, with expectations for relief and rebuilding efforts in the country to continue for months or years.

- *March:* In Australia, monsoon rains cause heavy flooding in western and central Queensland, destroying roads, railways, and crops in what is called the worst flood in 120 years.

- *August 8:* In China, in Zhugqu County, Gannan Tibetan Autonomous Prefecture, heavy rainfall and flooding trigger a deadly mudslide at midnight. The death toll includes 1,471 confirmed dead and another 294 who are missing and presumed dead. More than 1,700 people evacuate their homes. The province receives US$17.7 million in relief funds. A rare national day of mourning is observed in China on August 15.

- *September:* A series of floods across Victoria, Australia, causes hundreds of people to evacuate their homes and results in millions of dollars in damage. The state had been experiencing severe drought since the late 1990s.

- *October–November:* Mount Merapi in Central Java, Indonesia, experiences a violent series of eruptions that shoots lava and ash into the air and sends pyroclastic flows down its populated slopes. Seismic activity around the volcano had increased beginning in September, and more than 350,000 people were evacuated from the area. Some people returned to their homes before the eruptions had stopped, resulting in 353 deaths.

- *December 16–21:* Record-breaking rainfall triggers severe floods along the Gascoyne River in western Australia. Damage is estimated at A$100 million.

- *December 2010–January 2011:* In Australia, heavy precipitation caused by Tropical Cyclone Tasha combines with the 2010 La Nina pattern to trigger catastrophic flooding in central and southern Queensland. Thousands of people evacuate their towns and cities, with floodwaters affecting more than 200,000 people in at least 70 towns. Three-fourths of the state is declared a disaster zone. As of January 2011, 31 people have perished, with another 40 people still missing, and the damage is estimated to be well over A$1 billion. By mid-January traveling floodwaters combine with heavy rainfall to cause major flooding across western and central Victoria in one of that state's worst recorded floods. Thousands of people evacuate their homes, with more than 51 communities affected. Damages are estimated to be in the hundreds of millions of dollars.

2011

- *January:* A series of floods and mudslides in the mountainous region of Rio de Janeiro, Brazil, causes more than 900 deaths and an estimated US$1.2 billion in damages. The disaster is thought to be the second most deadly in Brazil's history.

- *March:* On March 11, a 9.0 magnitude earthquake strikes 250 miles (400 km) northeast of Tokyo, triggering a tsunami that flattens towns all along the northeast coast of Japan. Tens of thousands are feared dead, and more than 530,000 people are evacuated nationwide. Emergencies are declared at two nuclear power plants. As Japan struggles to avert nuclear disaster, Prime Minister Naoto Kan calls the crisis Japan's worst since World War II.

Glossary

active volcano a VOLCANO that has recently erupted, is erupting, or is likely to erupt in the future.

acute malnutrition a condition in which energy and nutrients from food intake are insufficient, and the body begins to break down fat and muscle for bodily functions.

aftershock a smaller EARTHQUAKE that comes after the main shock of an earthquake. Aftershocks may continue for months.

alluvial plain a flat area of land formed by the deposition of sediment from a river during periodic flooding over a long period of time.

anemometer a device that measures wind velocity.

ash also known as volcanic ash, ash is very fine rock and mineral fragments created during a volcanic eruption.

atmosphere the blanket of gases (mainly oxygen, hydrogen, and nitrogen) that surrounds Earth.

blizzard a heavy snowstorm with winds over 36 miles per hour.

building codes construction rules and standards.

caldera a large basin-shaped CRATER formed by the collapse of a VOLCANO's cone.

Cascadia subduction zone a region about 620 yards long off the coasts of northern California, Oregon, Washington, and Vancouver Island, where the JUAN DE FUCA PLATE is sliding under the NORTH AMERICA PLATE.

casualty a person injured or killed by a DISASTER.

continent one of Earth's seven large landmasses.

core the central part of Earth, below the MANTLE.

Coriolis force also known as Coriolis effect, Coriolis force is a term that describes how the velocity of Earth's rotation changes with latitude. It helps to explain why countries within 10 degrees north or south of the equator do not experience CYCLONES or STORM SURGES.

crater a bowl-shaped depression, or hollow, at the mouth of a VOLCANO.

crust the outermost layer of Earth, composed of rocks.

cyclone the name for a HURRICANE that forms near Australia or in the Indian Ocean.

database a large set of records organized in a methodical manner.

debris broken pieces of buildings, trees, rocks, etc.

debris flow a mass of water, dirt, rocks, and plants that rushes down a slope; often caused by heavy rain.

desert a region, either hot or cold, with less than 10 inches of precipitation per year.

desertification a process by which SEMIARID LAND is converted to DESERT, often by improper farming or by climate change.

developing country a poor country that is gradually developing better conditions for its people.

diameter the width across a circle or circular shape, such as a HURRICANE.

dike a wall, usually of earth reinforced with stone or concrete, built to contain or hold back water.

disaster a natural or human-caused event, usually unexpected, that causes human, economic, or environmental losses that exceed the area's ability to cope without outside help.

disaster risk the potential losses in lives, health, and economy that could occur at or during a future time, such as during an EARTHQUAKE.

disaster risk management the use of administrative policies, procedures, and organizations to lessen the damage of HAZARDS and the possibility of DISASTER.

disaster risk reduction decreasing DISASTER RISKS through the study and management of causes of disasters. Examples include reducing EXPOSURE to HAZARDS, managing land wisely, and improving disaster PREPAREDNESS.

Doppler radar a device that shows and measures wind directions and speeds by using reflected radio waves. Scientists use the readings to learn about the formation and movement of storms.

dormant volcano a VOLCANO that has not erupted in a long time but could erupt again.

drought prolonged shortage of water, caused when less rain than normal falls.

dust storm a windstorm that blows clouds of dust across an area.

dyke a British spelling of DIKE.

early warning system a set of equipment and people whose purpose is to generate and distribute timely warning information to help an area threatened by a HAZARD prepare and respond in time to reduce harm or loss.

earthquake vibrations in the earth's CRUST caused by movements of a FAULT in or between TECTONIC PLATES.

emergency management   the administration of resources and responsibilities for emergency PREPAREDNESS, response, and recovery.

Enhanced Fujita Scale (EF-scale)   a means of classifying the intensity of tornadoes; the scale assigns a class of EF0 to EF5 to denote the destructive force of the storm.

epicenter   the point on Earth's surface directly above the focus of an EARTHQUAKE.

erosion   the wearing away of soil by water, wind, or glacial ice.

eruption   an outpouring of lava, hot gases, and other material from a VOLCANO.

estuary   an arm of the sea where the tide meets the current of a river.

Eurasia plate   the TECTONIC PLATE that includes Europe, Asia, and adjoining seafloor areas.

evacuation   an organized departure of people from an area threatened by a DISASTER.

exposure   people, property, or other things in HAZARD zones that could be lost or damaged in a DISASTER.

extinct volcano   a VOLCANO that has not erupted for thousands of years and which scientists do not think will erupt again.

eye   a HURRICANE's core. It is very calm in the middle of the eye.

eye wall   a bank of swirling clouds that surrounds the EYE of a HURRICANE.

famine   a severe shortage of food that causes many people to die from hunger; often caused by a DROUGHT.

famine belt   a group of adjoining regions that historically have suffered famines.

fatality   a death.

fault   a fracture in Earth's CRUST along which rocks have moved past one another.

fertile   excellent for growing crops.

fissure   a long, thin crack in the earth through which LAVA and steam escape.

flash flood   a sudden overflow of water, usually caused by heavy rainfall in a short period.

flood   an overflow of water onto ground that is normally dry.

flood plain   the area of flat land around a river where the river naturally floods.

food-demand model   a method of predicting famine that compares the market availability and price of food with people's buying power. Sudden increases in prices or decreases in income signal the beginning of famine.

food-supply model   a method of predicting famine that uses harvest predictions, measures of food storage, and expected food consumption rates to determine if adequate food is available.

forecast   a statement or estimate of the likely occurrence of a future event or conditions for a specific area.

forecaster   a person who predicts the weather.

frostbite   damage to skin caused by prolonged exposure to freezing temperatures.

fumarole   a hole or VENT near a VOLCANO from which steam and gases escape.

funnel   the spiraling center part of a TORNADO.

geological hazard   internal earth processes, such as EARTHQUAKES and volcanic activity, that can generate a DISASTER.

geologist   a scientist who studies the earth by examining its layers of rocks.

hazard   a dangerous geophysical event, substance, human activity, or condition that could cause a DISASTER.

hazard reduction planning   actions and procedures designed to reduce or eliminate the danger of damage from a HAZARD.

hot spot   a location on Earth's surface that has experienced volcanic activity for a long time.

hurricane   a TROPICAL STORM with swirling winds at sustained speeds of 74 miles per hour or greater, bringing torrential rain. Hurricanes are called CYCLONES in the Indian Ocean and TYPHOONS in the Pacific Ocean.

hypothermia   dangerously low body temperature caused by exposure to cold and wind.

irrigation   the supply of water to fields by means of pipes, streams, or channels.

Juan de Fuca Plate   a TECTONIC PLATE located between the PACIFIC PLATE and the NORTH AMERICA PLATE, off the west coast of the United States.

Juan de Fuca Ridge   the part of a ridge on the ocean floor that forms the boundary between the PACIFIC PLATE and the JUAN DE FUCA PLATE.

lahar   mudflow or mud river caused by rapidly melting snow and ice, following a volcanic eruption.

landfall   to come onto shore.

landslide   the rapid collapse of rocks and soil down a slope.

lava   MOLTEN rock that erupts from a VOLCANO. When this material is underground, it is called MAGMA.

levee   a natural or artificial embankment that prevents a river from overflowing.

limnic eruption   a rare type of NATURAL DISASTER in which gases such as carbon dioxide from volcanoes or other sources, trapped at the bottom of a deep lake, suddenly rush to the surface and erupt, suffocating people and animals.

linear erosion   a type of EROSION that occur when rainfall on already saturated ground forms small streams that cut into the ground.

liquefaction   the process in which soil behaves like a dense fluid rather than a solid mass during an EARTHQUAKE.

magma   MOLTEN rock beneath the earth's surface. When it erupts from a VOLCANO, it is called LAVA.

magnitude a measurement of the amount of energy released in an EARTHQUAKE.

malnutrition a state of poor nourishment due to inadequate intake of nutrients from food.

mantle the thick part of Earth that lies between the CORE and the CRUST.

map a drawing that shows the layout of a place as seen from overhead.

mass extinction a period during which life on land and sea suffered large reductions of population or certain types of life completely disappeared.

megathrust earthquake a type of EARTHQUAKE that occurs when the PLATE that rests on top of a SUBDUCTION ZONE suddenly moves forcefully upward.

meteorologist a scientist who studies the weather.

mitigation the lessening or limitation of the negative results of HAZARDS and DISASTERS. The verb form is *mitigate.*

molten melted to the form of a hot liquid.

mudslide a collapse of mud down a slope, often carrying rocks, trees, and other DEBRIS.

National Guard volunteer soldiers trained by each state of the United States. They serve during emergencies and times of war.

natural disaster a DISASTER with its cause or origin in nature, as opposed to being human-caused.

natural hazard a HAZARD with its cause or origin in nature, as opposed to being human-caused.

North America Plate the TECTONIC PLATE that includes North America, part of Siberia, Greenland, and the western half of the Atlantic Ocean.

observatory a room or building set up for the purpose of scientific study.

one-hundred-year flood a severe FLOOD that happens, on average, once every 100 years.

Pacific Plate the TECTONIC PLATE that includes much of the seafloor of the Pacific Ocean.

Pacific Ring of Fire see RING OF FIRE.

plate see TECTONIC PLATE.

plate tectonics the theory that TECTONIC PLATES move about on Earth's upper MANTLE, grinding against one another and slipping under the edges of adjacent plates.

plume also known as a volcanic plime, a plume is the column of ash and gas that shoots upward from a VOLCANO during an ERUPTION.

prediction the forecasting in time, place, and intensity of an event such as a HAZARD.

preparedness the knowledge, planning, and ability to anticipate, respond to, and recover from HAZARDS.

prescribed fire   a fire that firefighters set to manage fire threats in a controlled manner.

prevention   the avoidance of negative impacts of HAZARDS and DISASTERS.

pumice   lightweight rock thrown from a VOLCANO during an ERUPTION.

pyroclastic flow   a fast-flowing and destructive outpouring of hot ash, molten rock, and gases, ejected from a VOLCANO during an explosive ERUPTION.

rapid onset   occurring suddenly.

rations   also known as general rations, rations are the main component of food relief efforts, composed of grains, oils, protein sources, and a mix of vitamins. The energy content of a ration ranges from 2,250 to 2,500 calories.

recovery   the rebuilding and improvement of facilities, livelihoods, and living conditions in a community following a DISASTER.

reservoir   a body of water collected in a human-made or natural lake.

reservoir-induced seismicity (RIS)   an EARTHQUAKE triggered by the weight of a RESERVOIR or water seeping into a fault under the reservoir. RIS is still a debated phenomenon.

response   the provision of emergency services and public assistance during or immediately after a DISASTER in order to save lives, reduce health impacts, ensure public safety, and meet people's basic needs.

Richter scale   a scale for measuring the intensity of EARTHQUAKES, with 9+ indicating the strongest TREMORS.

Ring of Fire   a ring-shaped zone of ACTIVE VOLCANOES and frequent earthquakes that roughly follows the boundary where the PACIFIC PLATE meets several other TECTONIC PLATES (including the NORTH AMERICA PLATE).

risk   the losses (deaths, damages, etc.) that would be caused by a DISASTER if it occurred.

risk assessment   the examination of HAZARDS and existing conditions of VULNERABILITY that together could damage or destroy people, property, or the environment.

risk management   establishing procedures and practices to reduce potential harm and loss.

runoff   water from a storm, FLOOD, or other source that is not absorbed by soil or vegetation.

run-up   the height of water on shore above sea level. This height is measured during a TSUNAMI, for example.

rupture   the action of TECTONIC PLATES shifting suddenly and violently.

Saffir-Simpson Hurricane Scale   a scale for measuring the strength of HURRICANES. The scale assigns a category to a HURRICANE, with Category 1 being the weakest and Category 5 being the strongest.

sediment   deposits of sand, rocks, and other material that were carried by water.

seismic caused by an EARTHQUAKE.

seismic sea wave another term for TSUNAMI.

seismic wave a wave that runs through the Earth, usually caused by an EARTHQUAKE.

seismogram a record of SEISMIC activity on paper or computer, produced by a SEISMOGRAPH.

seismograph also known as a seismometer, a seismograph is an instrument for detecting, recording, and measuring the strength and direction of SEISMIC WAVES.

seismologist a scientist who studies EARTHQUAKES.

semiarid land an area of land that receives between 10 and 20 inches of precipitation per year.

sheet erosion occurs as rain impacts the ground and dislodges particles of earth that the runoff carries away as SILT.

silt a type of SEDIMENT. A mud-like material made from tiny pieces of rock, found in rivers and TIDAL FLATS, among other places.

spout a cone-shaped tube. A TORNADO can also be called a spout.

storm surge an overflow of water onto coastal areas, caused by HURRICANE winds.

strata layers of hardened LAVA and volcanic ASH that form a STRATOVOLCANO.

stratovolcano also known as a composite volcano, a stratovolcano is a steep-sided VOLCANO formed by an alternating series of LAVA flows and PYRO-CLASTIC flows.

subduction the process of one TECTONIC PLATE moving beneath an adjacent plate.

subduction zone a long, narrow area of the Earth's CRUST where the SUBDUC-TION of a TECTONIC PLATE occurs. Most subduction zones lie along deep ocean trenches.

super-eruption a volcanic ERUPTION that produces [GT]300 km3 of magma. The ERUPTION of a SUPERVOLCANO.

super volcano a VOLCANO whose ERUPTION is exceptionally powerful, comparable to the force of a small asteroid colliding with Earth.

tectonic plate one of about 20 pieces that make up Earth's CRUST and upper MANTLE. The parts of the plates that carry the continents are denser and thicker than the parts beneath the oceans.

tidal flat a broad, mostly flat area of mud and silt that is exposed at low tide but submerged at high tide.

tidal wave *see* TSUNAMI.

tornado a twisting wind that appears as a dark funnel-shaped cloud reaching down to the ground at its tip.

Tornado Alley the southern part of the middle of the United States, where TORNADOES occur most frequently.

tornado season the time of year when TORNADOES are most likely to occur.

tornado warning a weather alert given when a TORNADO has been spotted in the area or has shown up on local radar.

tornado watch a weather alert given when conditions are right for producing TORNADOES. A TORNADO has not yet been spotted.

tremor a shaking or vibrating movement, as in earth tremors.

trench a long narrow ditch in the ocean floor.

tropical depression a low-pressure area surrounded by winds that spin in circles. When the wind goes above 38 miles per hour, the depression becomes a TROPICAL STORM.

tropical storm a storm characterized by winds whirling at sustained speeds of 39–73 miles per hour. When wind speeds reach a sustained speed of 74 miles per hour, the storm becomes a HURRICANE, CYCLONE, or TYPHOON.

tsunami A fast-moving ocean wave or series of waves, usually generated by submarine EARTHQUAKES, volcanic ERUPTIONS, or landslides. Tsunamis have been incorrectly referred to as *tidal waves;* however, tsunamis are not tidal.

tsunami warning system an EARLY WARNING SYSTEM that monitors a specific area for tsunamis and issues warnings to countries that may be affected by the tsunami.

typhoon the name for a HURRICANE that forms near the Philippines or in the Pacific Ocean.

vector-borne disease an illness, such as malaria, that is transmitted through a secondary factor, such as mosquitoes.

vent an opening in a VOLCANO through which LAVA or volcanic gases escape.

Volcanic Explosivity Index (VEI) a scale that scientists use to rank the strength of a volcanic ERUPTION according to its magnitude, from zero to nine.

volcanic material ash, lava, gas, and rock that erupt from a VOLCANO.

volcano an opening in the earth's CRUST through which ASH, LAVA, gases, and hot rock erupt.

volcanologist a scientist who studies VOLCANOES.

vulnerability the characteristics and circumstances of a community or place that expose it to the damaging effects of a HAZARD.

vulnerability mapping an assessment of regions or places with a predisposition to a HAZARD.

wildfire an uncontrolled fire in a forest, plain, or other wilderness area. The source of the fire may be natural or human-caused.

wind chill also known as wind chill factor, wind chill is a calculation that takes into account the actual temperature and the skin's heat loss due to wind. Wind chill is always colder than the actual temperature.

Index

Note: Page numbers in **boldface** indicate major treatment of a subject. Page numbers followed by *c* indicate chronology entries. Page numbers followed by *f* indicate figures. Page numbers followed by *g* indicate glossary entries. Page numbers followed by *m* indicate maps.

Index

Index

Index

Index

405

Index

Index

Index

Index